MANAGEMENT PRACTICES FOR ENGAGING A DIVERSE WORKFORCE

Tools to Enhance Workplace Culture

MANAGEMENT PRACTICES FOR ENGAGING A DIVERSE WORKFORCE

Tools to Enhance Workplace Culture

Edited by
Manish Gupta, PhD

Apple Academic Press Inc.
4164 Lakeshore Road
Burlington ON L7L 1A4
Canada

Apple Academic Press, Inc.
1265 Goldenrod Circle NE
Palm Bay, Florida 32905
USA

© 2021 by Apple Academic Press, Inc.

Exclusive co-publishing with CRC Press, a Taylor & Francis Group

No claim to original U.S. Government works

International Standard Book Number-13: 978-1-77188-863-9 (Hardcover)
International Standard Book Number-13: 978-0-36780-841-9 (eBook)

All rights reserved. No part of this work may be reprinted or reproduced or utilized in any form or by any electric, mechanical or other means, now known or hereafter invented, including photocopying and recording, or in any information storage or retrieval system, without permission in writing from the publisher or its distributor, except in the case of brief excerpts or quotations for use in reviews or critical articles.

This book contains information obtained from authentic and highly regarded sources. Reprinted material is quoted with permission and sources are indicated. Copyright for individual articles remains with the authors as indicated. A wide variety of references are listed. Reasonable efforts have been made to publish reliable data and information, but the authors, editors, and the publisher cannot assume responsibility for the validity of all materials or the consequences of their use. The authors, editors, and the publisher have attempted to trace the copyright holders of all material reproduced in this publication and apologize to copyright holders if permission to publish in this form has not been obtained. If any copyright material has not been acknowledged, please write and let us know so we may rectify in any future reprint.

Trademark Notice: Registered trademark of products or corporate names are used only for explanation and identification without intent to infringe.

Library and Archives Canada Cataloguing in Publication

Title: Management practices for engaging a diverse workforce : tools to enhance workplace culture / edited by Manish Gupta, PhD.
Names: Gupta, Manish, 1987- editor.
Description: Includes bibliographical references and index.
Identifiers: Canadiana (print) 20200270567 | Canadiana (ebook) 20200270699 | ISBN 9781771888639 (hardcover) | ISBN 9780367808419 (ebook)
Subjects: LCSH: Personnel management. | LCSH: Diversity in the workplace.
Classification: LCC HF5549 .M313427 2020 | DDC 658.3—dc23

CIP data on file with US Library of Congress

Apple Academic Press also publishes its books in a variety of electronic formats. Some content that appears in print may not be available in electronic format. For information about Apple Academic Press products, visit our website at **www.appleacademicpress.com** and the CRC Press website at **www.crcpress.com**

Dedication

I would like to dedicate this book to my parents and my wife.

—Manish Gupta, PhD

About the Editor

Manish Gupta, PhD
*Assistant Professor, Department of HR, IBS Hyderabad,
a Constituent of IFHE, Deemed to be University, Telangana, India*

Manish Gupta, PhD, is an Assistant Professor in the Department of HR, IBS Hyderabad, a Constituent of IFHE, Deemed to be University, Telangana, India. He has both teaching and industry experience. He is an editorial team member of the *Australasian Journal of Information Systems, International Journal of Knowledge Management*, and *Journal of Electronic Commerce in Organizations*. Recently, he has completed special issue editorial assignments for *Advances in Developing Human Resources, Journal of Global Operations and Strategic Sourcing, Industrial and Commercial Training, Australasian Journal of Information Systems*, and *International Journal of Knowledge Management*. He is also editing special issues for *Human Resource Management Review, International Journal of Nonprofit and Voluntary Sector Marketing, International Journal of Sustainability in Higher Education, Management Decision, International Journal of Productivity and Performance Management*, and *Journal of Electronic Commerce in Organizations*. He is also a consulting editor of *The HRM Review*. He has co-authored a book, *Practical Applications of HR Analytics: A Step-by-Step Guide*, and has co-edited a book, *Qualitative Techniques for Workplace Data Analysis*. He has conducted a workshop on academic writing and has presented papers at various conferences. His work has received media coverage in international magazines, newspapers, blogs, and news channels. He has published book chapters and over 30 research papers in the journal of international repute. He has recently won the AIMS International Outstanding Young Management Researcher Award 2019 and the *Advances in Developing Human Resources* best issue of 2018 award. He is a recipient of a Junior Research Fellowship and National Eligibility Test certificate awarded by the University Grants Commission, Government of India. His PhD thesis was in the area of work engagement.

Contents

Contributors .... *xi*
Abbreviations .... *xiii*
Preface .... *xv*
Introduction .... *xvii*

1. **Engaging an LGBT Workforce: Inclusion Through Workplace Culture** 1
   Ruby Sengar, Narendra Singh Chaudhary, Bharat Bhushan, and Santosh Rangnekar

2. **Active Aging: Engaging the Aged** 33
   Ritu Gupta, Shruthi J. Mayur, and Swati Hans

3. **Engaging Diverse Races at Work** 73
   Arti Sharma and Rajesh Mokale

4. **Education Divide at Work** 107
   Neha Gangwar

5. **Gen-Z, the Future Workforce: Confrontation of Expectations, Efforts, and Engagement** 131
   Sheema Tarab

6. **Cross-Generational Engagement Strategies** 157
   Musarrat Shaheen and Farrah Zeba

7. **Managing Married Employees** 183
   Anugamini P. Srivastava

8. **Cross-Border Mergers: The Use of Employment Engagement Tools in Overcoming Challenges of Workforce Cultural Diversity** 213
   Shweta Lalwani

9. **Engaging People with Physical Disability at Work** 237
   Anitha Acharya

10. **Engaging Different Income Classes at Work** 267
    Anitha Acharya

11. **Is Work Engagement Gender Oriented? A Man/Woman Perspective** 295
    Sindhu Ravindranath

12. **Volunteering for Community: Learning and
    Challenges in Diversity** 319
    Trilok Kumar Jain

Index *343*

Contributors

Anitha Acharya
Assistant Professor, Marketing and Strategy Department, ICFAI Business School (IBS), Hyderabad, The ICFAI Foundation for Higher Education (IFHE), Hyderabad, India, Tel.: (+91) 8712290557, E-mail: anitha.acharya@ibsindia.org

Bharat Bhushan
Symbiosis Centre for Management Studies, Noida–[A Constituent of Symbiosis International (Deemed University), Pune, Maharashtra, India] Plot–47 & 48, Block A, Industrial Area, Sector-62, Noida, Uttar Pradesh–201301, India, Tel.: (+91) 9810182974, E-mail: bb.bhushan@gmail.com

Narendra Singh Chaudhary
Symbiosis Centre for Management Studies, Noida–[A Constituent of Symbiosis International (Deemed University), Pune, Maharashtra, India] Plot–47 & 48, Block A, Industrial Area, Sector-62, Noida, Uttar Pradesh–201301, India, Tel.: (+91) 9997771938, E-mail: narendraiet15@gmail.com

Neha Gangwar
Doctoral Fellow, National Institute of Industrial Engineering (NITIE), Mumbai; Room No 411, Gilbreth Hall, NITIE, Powai, Mumbai, Tel: 9871623987/9415230212, E-mail: gangwar.n12@gmail.com

Manish Gupta
Assistant Professor, Department of Human Resources, IBS Hyderabad, Telangana, India, E-mail: manish.gupta.research@gmail.com

Ritu Gupta
Associate Professor, T A Pai Management Institute, Badagabettu Road, Manipal, Karnataka–576104, India, Tel.: (+91) 9652081180, E-mail: rgritu@gmail.com

Swati Hans
ICFAI Business School (IBS), Hyderabad, The ICFAI Foundation for Higher Education (IFHE), Hyderabad, Telangana–501203, India, E-mail: swati.hans@ibsindia.org

Trilok Kumar Jain
Professor, School of Business and Commerce, Manipal University Jaipur, Rajasthan, India, Tel.: (+91) 9414430763, E-mail: jain.tk@gmail.com

Shweta Lalwani
Assistant Professor, School of Management, Sir Padampat Singhania University, Bhatewar, Udaipur (Rajasthan), PIN–313601, India, Tel.: (+91) 9950396133, E-mail: shweta.lalwani@spsu.ac.in

Shruthi J. Mayur
Associate Professor, T A Pai Management Institute, Badagabettu Road, Manipal–576104, Karnataka, India, Tel.: (+91) 9447703833, E-mail: shruthi.mayur@tapmi.edu.in

Rajesh Mokale
FPM; FPM (OB & HRM), Indian Institute of Management, Indore, Madhya Pradesh–452003, India, Tel.: (+91) 8108441844, E-mail: f17rajeshm@iimidr.ac.in, rajeshmokale.mils@gmail.com

Santosh Rangnekar
Professor, Department of Management Studies, IIT Roorkee, Roorkee, Uttarakhand–247667, India,
Tel.: (+91) 9410543454, E-mail: srangnekar1@gmail.com

Sindhu Ravindranath
Assistant Professor, ICFAI Business School (IBS), Hyderabad,
The ICFAI Foundation for Higher Education (IFHE), Hyderabad, Telangana, India,
Tel.: (+91) 9032865740, E-mail: sindhur.inc@gmail.com

Ruby Sengar
Symbiosis Centre for Management Studies, Noida–[A Constituent of Symbiosis International (Deemed University), Pune, Maharashtra, India] Plot–47 & 48, Block A, Industrial Area, Sector-62, Noida, Uttar Pradesh–201301, India, Tel.: (+91) 8447815224,
E-mail: rubysengar@gmail.com

Musarrat Shaheen
Assistant Professor, Department of Human Resource, ICFAI Business School (IBS), Hyderabad, a constituent of IFHE (Deemed to-be-university), India.
Tel.: (+91) 8978219231, E-mail: shaheen.musarrat@gmail.com (M. Shaheen)

Arti Sharma
FPM, Indian Institute of Management, Indore, Madhya Pradesh–452003, India,
Tel.: (+91) 9001270997, E-mail: f15artis@iimidr.ac.in

Anugamini P. Srivastava
Symbiosis Institute of Business Management Pune, Symbiosis International Deemed University, Maharashtra, India, E-mail: srivastavaanu0@gmail.com

Sheema Tarab
Assistant Professor, Department of Commerce, Aligarh Muslim University, Aligarh,
Uttar Pradesh–202002, India, Tel.: (+91) 8791537797, E-mail: sheematarab@gmail.com

Farrah Zeba
Farrah Zeba, Assistant Professor, Department of Marketing and Strategy,
ICFAI Business School (IBS), Hyderabad, a constituent of IFHE (Deemed to-be-university), India.
Tel.: (+91) 8367574051, E-mail: drfarrahzeba@gmail.com (F. Zeba)

Abbreviations

AAI	active aging index
ADA	Americans with Disabilities Act
ADEA	age discrimination in employment act
ASD	autism spectrum disorder
CEC	Commission of the European Communities
CEM	categorization-elaboration model
DOL	department of labor
EEO	equal employment opportunity
EEOC	Equal Employment Opportunity Commission
EPGDP	Executive Post-Graduate Diploma Program
ERG	employee resource group
HR	human resource
HRM	human resource management
ICCPR	International Covenant on Civil and Political Rights
ICESCR	International Covenant on Economic, Social, and Cultural Rights
ILO	International Labor Organization
IPC	Indian Penal Code
LGBT	lesbians, gays, bisexual, and transgender
M&A	mergers and acquisition
MNCs	multi-national companies
NSS	national service scheme
ODT	optimal distinctive theory
OECD	Organization for Economic Co-operation and Development
RRI	retirement resources inventory
TCL	the creative league
WHO	World Health Organization
WIC	Workplace Intergenerational Climate

Preface

This book, *Management Practices for Engaging a Diverse Workforce: Tools to Enhance Workplace Culture,* is in the field of human resource management (HRM), commonly a mandatory course in any management program. Unlike typical HRM textbooks, this book gives greater emphasis to short cases and lesser to describing the theory. It is specific to the challenges and mechanisms adopted by organizations for engaging a diverse workforce.

Teachers also need to constantly upgrade their knowledge and devise new ways and means of keeping students engaged in the class. One of the most effective techniques of engaging students is case-based learning, which shifts the focus of the learning from teachers to students. However, the length of the case leads to boredom. Moreover, teachers often face difficulty in directing the case in the direction of achieving the learning objectives of the chapter. The students and teachers tend to lose interest in the case. Therefore, this book is a step in the direction of resolving this issue by offering two remedies. To reduce boredom and keep the user engaged, the cases included in this book are short in length. Similarly, to enable teachers to direct the case in a particular direction, each case in this book is coupled with a case discussion that consists of the timeline, teaching objectives, theory involved, and board plan among others.

The present book includes a brief description of the concept on a particular theme of workforce diversity. But it essentially focuses on the repercussions of engaging diverse workforce on the different stages of HR process, including recruitment, selection, performance appraisal, demand forecasting, supply forecasting, job description, job specification, job analysis, job evaluation, training, and development, career planning and development, succession planning, etc.

I thank the authors for their efforts and for contributing chapters for this book. Readers are likely to benefit from authors' experience and expertise in the area of employee engagement and workforce diversity. I hope that readers will gain insights into the different ways of engaging a workforce from combination of diversity that this book illustrates.

—**Manish Gupta, PhD**

Introduction

Engaging the workforce becomes difficult when it is diverse. It is primarily because of the different barriers that diversity creates in fully investing one's energy into the work. Indeed, an engaged workforce is beneficial to both employees and employers, but at the same time, the diversity may damage the company's bottom line if the workforce is disengaged. This book is unique because it addresses an identified need of engaging a diverse employee base at work. Moreover, it essentially focuses on the repercussions of engaging a diverse workforce at the different stages of the HR process, including recruitment, selection, performance appraisal, demand forecasting, supply forecasting, job description, job specification, job analysis, job evaluation, training, and development, career planning and development, succession planning, etc.

This book will enable readers to comprehend what exactly it means to have a diverse workforce and how to engage such a workforce for the betterment of the employees as well as the employer.

The specific features of the book are as follows:
- Based on cases using real or hypothetical incidents.
- Brief descriptions of a particular type of workforce diversity.
- Case discussions to help teachers direct a case toward concept and make the student know the possible case solution.
- Critical questions to brainstorm each concept.
- Short cases that require only a 15-minutes reading.

This book covers 12 different themes on workforce diversity. The first chapter has an emphasis on lesbians, gays, bisexual, and transgender (LGBT). The second chapter focuses on old age individuals. The third chapter is about engaging diverse races at work. The fourth chapter describes the workforce with diverse educational backgrounds. The fifth chapter specifically elaborates on the Z-generation employees, whereas the sixth chapter compares different generations. The seventh chapter is on marital status. The eighth chapter is on cultural diversity. The ninth chapter focuses on engaging physically disabled children. The tenth chapter is on engaging different income class employees at work. The eleventh chapter takes the evergreen man vs.

woman perspective at work. The last chapter discusses workforce diversity in general and summarizes the entire book.

The book espouses a standard structure. In that, each chapter introduces a particular theme of workforce diversity followed by the significance of the theme, review of the theme, the implications of the theme, and conclusion.

CHAPTER 1

Engaging an LGBT Workforce: Inclusion Through Workplace Culture

RUBY SENGAR,[1] NARENDRA SINGH CHAUDHARY,[1]
BHARAT BHUSHAN,[1] and SANTOSH RANGNEKAR[2]

[1]*Symbiosis Centre for Management Studies, Noida–[A Constituent of Symbiosis International (Deemed University), Pune, Maharashtra, India] Plot–47 & 48, Block A, Industrial Area, Sector-62, Noida, Uttar Pradesh–201301, India, Tel.: (+91) 8447815224, E-mail: rubysengar@gmail.com (R. Sengar), Tel.: (+91) 9997771938, E-mail: narendraiet15@gmail.com (N. S. Chaudhary), Tel.: (+91) 9810182974, E-mail: bb.bhushan@gmail.com (B. Bhushan)*

[2]*Department of Management, IIT Roorkee, Uttarakhand, India, Tel.: (+91) 9410543454, E-mail: srangnekar1@gmail.com*

ABSTRACT

It is fairly prominent in a country like India that the acceptance of lesbians, gays, bisexual, and transgender (LGBT) community is hitting a setback. The present study aims to propose a linkage between workplace diversity, engagement, and workplace culture so as to propel the acceptance of the LGBT community in Indian corporations. This chapter can be revolutionary in bringing a major change in the mind-sets of the people in India and help in accepting the sexual minorities, which are becoming an important part of the workplaces nowadays and appreciating their contribution is to the overall organization betterment. Apparently, there is a dearth of studies on LGBT's sexual identity which remains negotiated in the workplace and acts as an obstacle to impartiality in the workplace organizations need to gear up to match the standards set by other organizations around the globe who have maximized their efforts, i.e., workplace diversity by including LGBT people. For a business to grow, flourish, and augment performance, they must recognize the significance of deciphering LGBT issues by focusing on LGBT inclusive and diverse workplace environments.

1.1 INTRODUCTION

This study aims to propose a linkage between workplace diversity, engagement, and workplace culture so as to propel the acceptance of the LGBT community in Indian corporations. Apparently, there is a dearth of studies on LGBT's sexual identity which remains negotiated in the workplace and acts as an obstacle to impartiality in the workplace (Woodruffe-Burton and Bairstow, 2013). The time has come when Indian organizations need to gear up to match the standards set by other organizations around the globe who have maximized their forte, i.e., workplace diversity by including LGBT people. For a business to grow, flourish, and augment performance they must recognize the significance of deciphering LGBT issues by focusing on LGBT inclusive and diverse workplace environments. An organization can become an employee's first choice as well as can gain a competitive advantage if it is able to hold talent from assorted backgrounds and give them a milieu which is best suited for making them feel their whole selves. Conversely, the businesses will lose upon their LGBT talents if they don't stay up-to-date with the diversity rules and exhibit superior inclusiveness for the LGBT employees. People in organizations usually follow an attitude of "don't ask, don't tell" when it comes to employee's gender identity or sexual orientation. Most of the leaders at the corporate do not consider LGBT as an important corporate agenda and thus neglect it. In reality, most of the problems faced by the LGBT community with respect to gender identity or sexual orientation are usually not considered as a part of the organizational issue and therefore do not receive much corporate help. However, an LGBT role model or leader in a workplace can help in reducing the social stigmas associated with their community. In a country like India, there is a dearth of role models especially at managerial levels who can support a workplace culture of frankness and acceptance from the rest of the employees in the organization towards the LGBT employees. These managers can help in propelling the overall culture of the organization which in turn will enhance their commitment, engagement as well as will help in reaping the benefits of their diverse talents.

Apparently, India is a country which is deliberated as one of the probable budding superpowers of the globe. Though the nation is extensively recognized for its diversity in languages, religion, ethnicity, etc. but the acceptance of the LGBT community in the nation is hitting a setback. LGB refers to sexual orientation whereas T refers to gender identity. LGBT people are sometimes referred to as LGBTQ where Q refers to Queer. Whether it is civil society or organizations, the acceptability of these sexual minorities is very low in the country. In fact, India is deficient in certain inspirational role models whose presence can augment the acceptance of these sexual minorities both in society and organizations. Indeed, there is an enormous need to have various sensitive national camaraderie networks besides the inspirational role models who can together reinforce the support towards the sexual minorities.

Lesbian is defined as "a woman who is attracted emotionally and/or sexually to people of the same sex/gender" whereas gay is "a man who is attracted emotionally and/or sexually to people of the same sex/gender" (Srivastava, 2014). Bisexuals are "those people who are attracted towards both male and female, though there may be a preference for one gender over others" (Srivastava, 2014). Transgender, as defined by Connell refers to "individuals who deliberately reject their original gender assignment" (2010). Queer is a term used to slander LGBT people. Queer acts as "an umbrella to accommodate those who identify as LGBT as well as intersex and questioning" (Rumens, 2016). In fact, the queer theory focuses on "deconstructing and disrupting binary formations that structure heteronormative knowledge about gender and sexuality" (Sullivan, 2003). In this chapter, sexual orientation refers to "an individual's enduring physical, romantic, emotional, and/or spiritual attraction to members of the same and/or opposite sex" and gender identity means that birth-assigned sex and the internal sense of gender identity do not match which is meant only for transgender (Community Business, 2017). Also, here sexual minorities are those who recognize themselves as lesbian, gay, bisexual, or transgender which are not part of the heterosexual consortium.

Numerous studies on sexual minorities have found that the LGBT employees face workplace chauvinism and discrimination which eventually brings down the level of decent workplace experience, intensifies depression and psychological distress (Smith and Ingram, 2004; Leppel, 2014). Most of the leaders at the corporates do not consider LGBT as an important corporate agenda and thus neglect it. In reality, most of the problems faced by LGBT community with respect to gender identity or sexual orientation are usually not considered as a part of the organizational issue and therefore do not receive much corporate help. Indeed, employee engagement can help in plummeting humiliations associated with LGBT employees. Employee engagement in organization helps in satisfying the individual via providing companionship with co-workers in addition to gratification of employee's social needs (Howell and Costley, 2006). Better employee engagement enhances motivation level, job satisfaction, over, and above, feelings of empowerment (Macey and Schneider, 2008). But in a country like India, there is a dearth of such role models especially at managerial levels who can support a workplace culture of frankness and acceptance from rest of the employees in the organization. Apparently, there is a paucity of studies on LGBT's sexual identity which remains negotiated in the workplace and acts as an obstacle to impartiality in the workplace (Woodruffe-Burton and Bairstow, 2013).

1.2 NEED FOR DISCUSSING THIS THEME

With the passage of time, it has become quite evident in the Indian organizations that they are struggling with both lawful and socio-cultural constraints (Horton et al., 2015). The past studies have clearly stated that workforce diversity has received a slice of attention from various HR specialists and organizational leaders in both developed and developing economies (Priola et al., 2014). Nowadays both academicians and practitioners are outpouring their consideration towards diversity and inclusion, primarily focusing on gender, race, nationality, sexual orientation, etc., due to the fluctuations in the economy and statutory worldwide (Oswick and Noon, 2013). In reality, extremely prosperous multi-national companies (MNCs) globally had started investing hugely towards workplace diversity (Collins, 2012). Today workplace diversity is a well-thought-out human resource (HR) practice used to remain competitive in a very turbulent business environment (Thomas, 1990). This attention is basically attributed towards the generation of higher revenues in addition to the development of good image for the organizations in the market globally through workplace diversity.

For a business to grow, flourish, and augment performance they must recognize the significance of deciphering LGBT issues by focusing on LGBT inclusive, diverse, and engaged workplace environment. An organization can become employee's first choice as well as can gain a competitive advantage if it is able to hold talent from assorted backgrounds and give them a workplace culture or milieu which is best suited for making them feel their whole selves. Conversely, the businesses will lose their LGBT talents if they don't stay up-to-date with the diversity rules and exhibit superior inclusiveness for the LGBT employees. Recently, it has been seen that many organizations have entered the first phase of their diversity expedition and are hopefully going to have sustainability in their decisions that can make the workplace a better place to work for especially the sexual minorities.

A noticeable trend was seen by Indian corporations in targeting LGBT customers by an expenditure of huge amounts. The logic behind this was an attempt to lure 'pink money,' i.e., the money earned by LGBT people, through various marketing techniques. A report generated by Forbes India reported that India had approximately 30 million people as the LGBT adult population who had annual earnings of 1.5 lakh crores (Forbes India, 2017). This evidently states that companies want to portray themselves as LGBT friendly so that they can earn a maximum share of this community's disposable income. This clearly indicates the importance of amicable workplace culture for LGBT employees and their engagement so that the organization can boost their market image as more LGBT friendly that takes care of both LGBT employees and consumers.

It is quite visible that there is a shortage of researches on discrimination and acceptance of LGBT employees in Indian organizations. So, for that reason, this chapter is purely dedicated to the acceptance of LGBT employees via workplace diversity, workplace culture, and employee engagement in Indian organizations.

1.3 REVIEW OF THIS THEME

1.3.1 EVOLUTION OF THE THEME

The concept of LGBT has been investigated using certain variables namely workplace culture, workplace diversity, and employee engagement specifically in organizations which requires a comprehensive review of the literature on LGBT employees. For this purpose, the LGBT literature has been studied keeping in mind the challenges and issues faced by the LGBT employees in various organizations. Now, the studies related to LGBT employees in the organizations are discussed detail in Table 1.1.

According to the laws made in 1861 during Queen Victoria's rule, homosexuality was a felonious act for which a person was put behind the bars (The Hindu, 2017). As per Section 377 of the Indian Penal Code (IPC), homosexual acts were considered a criminal offense. In 2009, the Delhi High Court benched by two-judges gave a view that banning concordant homosexual sex between adults was violating the fundamental rights assured by the Constitution of India (The Hindu, 2017).

TABLE 1.1 Studies on LGBT Employees (2001–2015)

Year	Variables	Challenges and Issues	References
2001	Employee Engagement	Discrimination at workplace led to lower employee engagement in lesbian and gay employees.	Ragins and Cornwell, 2001; Button, 2001
2002	Workplace Culture	Heterosexual employees were earning 11–27% high in comparison to their gay colleagues in the US nation-state.	Berg and Lien, 2002
2002	Workplace Culture	Lesbian employees were facing huge interpersonal negativity at the workplace.	Hebl et al., 2002
2003	Workplace Culture	Lesbian employees were getting inferior performance appraisals than their heterosexual counterparts.	Horvath and Ryan, 2003
2007	Workplace Culture	Around 68% of the sexual minorities felt discriminated on promotions, performance assessment, etc. in the workplace.	Badgett et al., 2007
2008	Workplace Culture	Workplace biasness happened on the basis of sexual orientation.	Heilman and Eagly, 2008
2009	Workplace Culture	Heterosexuals were earning less in comparison to their lesbian colleagues.	Daneshvary et al., 2009
2015	Workplace Diversity and Workplace Culture	Few progressive changes had been seen in the acceptance of the LGBT community both in corporate and society in India.	Vohra et al., 2015

Judgment was passed by Supreme Court in the year 2013 stating that "a person cannot be booked under Section 377 based on his/her identity, i.e., his/her sexual orientation" (The Hindu, 2017). Indeed not even IPC Section 377 criminalizes people who are homosexual. Being a gay or lesbian couple is not a crime in India. Only homosexual acts are deliberated as illegal. In 2013, Supreme Court of India, which is deliberated as the crown court of the petition under the Constitution of India, upturned the 2009 judgment and declared that Section 377 of IPC was legitimate and applicable to sexual acts not taking into consideration the age or the accord of the people involved. Furthermore, in 2014, a review bench agreed to the fact that history should be left behind and recognized the judgment of the Delhi High Court (The Hindu, 2017). In fact, Mr. Shashi Tharoor made an individual effort and presented a private member's bill in the parliament with a hope to legalize homosexuality in December 2015 (The Hindu, 2017). Through the bill, it was demanded that certain changes should be brought in IPC by supporting the fact that Section 377 should be replaced. Although the bill got rejected by the majority of Parliament members still a ray of hope was aroused by the Supreme Court in February 2016 after listening to the therapeutic appeal (The Hindu, 2017). They further had referred the substitution of Section 377 to a new five-judge bench who will look at the matter from the scratch with no prejudices. So, it was a kind of temporary relief to the LGBT community all over India that some positive judgment can be expected in the future.

In recent times, the Supreme Court upheld the Right to Privacy. The judgment clearly stated that "individual privacy is a guaranteed fundamental right" (Supreme Court, 2017). It concluded privacy as "preservation of personal intimacies, the sanctity of family life, marriage, and sexual orientation" (Supreme Court, 2017). It also implies a right to be left unaccompanied. Certainly, it shields the individual choices leading a way of life. This judgment on the Right to Privacy further propelled the hopes of the LGBT community in the direction of making homosexual acts legal in India. Again with this, they have started dreaming of a society which will accept them as normal as heterosexual.

In India, as such no legal compliance has been designed that can safeguard employees belonging to this community from discrimination in the workplace. Due to international influences, companies involved in global businesses have been strictly following the anti-discrimination law, unlike India.

1.3.2 WORKPLACE DIVERSITY

For more than a decade, companies have acknowledged the fact that it is imperative to have diverse viewpoints in an attempt to land at prodigious solutions. Companies have begun to realize that if they accept people from assorted genders, religious values, or nationalities with equality they can build a working milieu that instigates paramount notions and talent.

Workplace diversity has been defined as "acceptance and respect for differences in an individual's race, gender, physical abilities, and sexual orientation (Majumdar and Adams, 2015). When it is presumed that sexual orientation" of the employee is known, few of the perceptions of them are made on stereotyping, disgrace, and organizational demography (Shore et al., 2009; Ozturk and Rumens, 2014). It has been seen that the sexual minorities in the organization had to face a lot of discrimination by the heterosexual employees in terms of progression in the career, more pressure, low rewards, and appraisals, etc. (Berg and Lien, 2002; Blandford, 2003). Thus, LGBT employees are basically left with two options either to pick a low-paid job wherein they have the independence to reveal their sexual orientation or to go for a job where they have to hide their sexual preferences but will be paid a handsome salary.

At the present time, for companies like Infosys, Godrej, etc., sexual orientation is given the least consideration, in fact, the focus has shifted to ensuring a harmless and harassment-free work environment. More importantly, the company should realize that sexual orientation has got nothing to do with their performance at work; it is their talent and diversity that brings best for the company in terms of profit and growth. Cisco has a subsidiary in India which has been focusing on reverse mentoring wherein the willing LGBT employees and leaders on monthly basis sit together for 2–3 hours to discuss the former's work and life experiences with the latter. Then the company offers a forum where the leaders get an opportunity to share their learning's from their relationship with the LGBT employees with the whole set of employees at the company. This technique has helped in changing mindsets of the employees towards the sexual minorities as well as has produced an open culture that made feel the LGBT employees more secure and increased their linkages to the LGBT employee network.

All-inclusive practices are imperative from an organization's point of view to maintain a positive status in the society so that attraction and retention of the workforce don't become a task. It is quite obvious that some percentage of the working population would be belonging to LGBT community and when all-inclusive practices are being followed, it not only appeals this section of the working population, but it also attracts the attention of those unidentified employees who have some known LGBT person in their friends circle or in relation and wish to be allied with an organization that illustrates support and esteem.

1.3.3 WORKPLACE CULTURE

Slowly and steadily, organizations have started comprehending the significance of workplace culture for a diverse workforce because, in the long run, it is the organization that is able to entice and retains the best talent in addition to the development of creative and innovative workforce which will sequentially escalate commitment and engagement of its diverse workforce. Workplace culture is described as "a set of assumptions, values, norms, symbols, and artifacts within the organization, which convey meaning to employees regarding what is expected and shape individual and group behavior" (Hatch, 1993).

Transgender had been facing more gender inequality in the workplace as seen in the past studies (Connel, 2010). Directors or Managers usually say that they believe in equality and at no cost discrimination are done on 'sexual orientation' of the employee. On the contrary, the hard-core reality is that the LGBT employees feel just the opposite of it. In India, a study was piloted by Mingle on "LGBT workforce and their workplace culture" taking a sample of 455 LGBT employees (Mingle, 2017). This study reported that gays who had openly disclosed their sexual orientation trusted their employers more in comparison to their closeted counterparts. The former felt more faithful towards their employers and were additionally expected to stay with the company for a longer period of time than the latter. Google has identified that workplace culture ought to include diversity which is deeply engraved into their core principles of doing business. Furthermore, the results from the study revealed that when LGBT employees were asked about any sort of discrimination happening to them from higher authorities and bosses only 11% and 9% of them responded as 'Often' and 'Sometimes.'

It is desirable that the society accepts the person willingly for who they are not for who they love. If it is established in our work culture, the society will follow. Ashok Row Kavi, Editor – 'Bombay Dost' India's first gay magazine said that *"Your sexuality is just another facet of your personality. It is like the color of your skin or the color of your hair. It tells you who you are, but does not tell you what you are capable of doing"* (Banerji et al., 2012). This means that belonging to an LGBT community does not mean that you are not talented the only thing that is required is acceptance from the peers which comes through amicable and flexible workplace culture. He even quoted that "the social stigma will remain and it is still a long struggle." But he also said that the change has to come from the people around who shall understand their LGBT fellows and accept them the way they are.

It is quite obvious that the change cannot happen over a night but companies like Infosys, Godrej, etc. have been promoting diversity for such a long time that this change is no longer impossible. The work culture should be such that it puts an affirmative impression on the happiness and comfort of LGBT employees and this is possible through the LGBT-friendly policies and procedures as being followed at Infosys. These policies should include diversity training on sexual orientation to all the employees, benefits offered to LGBT along with welcoming their partners at organizational social events, identification of exclusivity and worth of every employee, etc. An ideal work culture is characterized by a good understanding between the colleagues wherein differences are accepted easily and co-workers experience a reduced amount of tension, discrimination, and pestering. So, instead of having separate policies and practices for LGBT employees it's better to have general company policies applicable to all which removes discrimination, over, and above, support from colleagues can work wonders for this community especially from the retention point of view. Consequently, LGBT employees look for an organizational culture that supports diversity and calls out loud to all the people employed in the organization that they have no tolerance for any sort of discrimination. Tracy Ann Curtis who is a principal consultant at TAC Global clearly stated that "To ensure high performance, organizations need to cultivate environments which allow each of us, regardless of our diversity-to contribute our best selves-and this is what leads to highly productive and innovative workplaces" (The Hindu, 2017). Implementation of LGBT-friendly HR practices was found to be having a strong affirmative relationship with the organizational performance due to the initiation of indiscrimination or equality acts in various nations but the implementation of these acts did not happen globally (Wang and Schwarz, 2010) including India.

Since there is no indiscrimination act in India that can safeguard LGBT people in the organization, so the corporations can themselves take a foot forward to build an environment where in these employees feel safe and secure. With the intention of further cultivating a better workplace culture numerous suggestions have been given for Indian managers to stimulate enclosure of the LGBT employees by an information guide developed by Community Business (Community Business, 2017) which are as follows:

- **Benefits:** The advantages that other employees enjoy if they are married such as medical coverage, retirement schemes, compassionate leave, repositioning aid, etc. should also be provided to LGBT employee's partners giving no consideration to their marital status.
- **Communicate Corporate Culture:** This communication is important as this will tell the LGBT employees that the company cares, supports, and values them and can be done through meetings with diversity as an agenda, posters, intranet to highlight LGBT-friendly policies.
- **Community and Advocacy:** The companies should try to become a noticeable exemplar for equality of LGBT employees at the workplace through activities like recruitment, charity, public support, etc. for lawful LGBT equality. Not only becoming role models would do, companies should also share their prominent practices with other companies and leaders who can also endorse LGBT equality at their workplace.

- **Diversity Structure:** Develop a group of employees who will be held accountable for addressing LGBT hitches.
- **Diversity Training:** Various whys and wherefores due to which the society is unaware of the LGBT terminologies are negligible learning at schools and colleges, scarce role models in their community, fear of disclosure of their sexual orientation and gender identity in public, etc. So, this training will enhance awareness of this community's concerns.
- **Equal Opportunity Policies:** A policy that treats all employees as equals irrespective of their race, gender, sexual orientation or gender identity, forbid disparity, over, and above, make rules that can solve LGBT associated pestering, grumbles, etc. at the workplace.
- **Market Positioning:** The businesses should involve in a courteous and suitable promotion to the LGBT community. Though the communal approval of the LGBT is, however, in the bourgeoning stage yet there is a need to further make their presence visible as they have a huge purchasing power. This is why the inclusion of LGBT people in the workplace is very important.
- **Monitoring:** Companies should accumulate data on LGBT employee's recruitment, employee engagement, job satisfaction, performance appraisals, career progression, grievance handling, etc. to make sure that they are being treated as equals or not.

Through these practices, the corporate segment can play a critical role in changing the mindsets of Indians towards the acceptance of LGBT people as a whole. Businesses can provide a harmless and open culture that accepts LGBT people as normal human beings, which can change the disgraces, preconceptions, etc. associated with the LGBT community.

1.3.4 EMPLOYEE ENGAGEMENT: A TOOL TO ENHANCE WORKPLACE CULTURE

In reality, a number of researches have clearly indicated that the LGBT community which forms the minority group in the organization faces countless prejudgments in addition to discrimination which has adversely affected their work experience, motivation, engagement, and most importantly their retention. Employee engagement as defined by Schaufeli et al. (2002) is "a positive, fulfilling, work-related state of mind that is characterized by vigor, dedication, and absorption." In many organizations, LGBT individuals are considered as some sort of untouchables who are belittled on the grounds of their sexual orientation. As per their opinion, only heterosexuals have the right to work in an organizational setting where they can uphold their own individualities and none others. This hetero-normativity is considered as normal in most of the companies which have adverse repercussions for the sexual minorities at the workplace (Fielden and Jepson, 2016). In reality, the sexual minority in a workplace often feels the possibility of societal rejection leading to a negative impact on their perceptions of righteousness which eventually lessens their work engagement and propels the turnover ratio of the company.

Employee engagement can be amplified with the help of three tools namely, psychological meaningfulness, availability, and safety (Kahn, 1990). According to Kahn, Psychological meaningfulness is "a feeling that one is receiving a return on investment of one's self in a currency of physical, cognitive or emotional energy" (1990). This meaningfulness at the job occurs when an individual gets respected by their workplace (May et al., 2004). Psychological availability refers to "the sense of having the physical, emotional, or psychological resources to personally engage at a particular moment" (Kahn, 1990). This availability is focusing towards the ability of the individuals to handle numerous needs of personal and work facets of their lives. There should be defined reliable set of rules that ascertains worldwide protection of LGBT employees (Norman-Major and Becker, 2013). Psychological safety refers to "a feeling of being able to show and employ one's self without fear of negative consequences to self-image, career, or status" (Kahn, 1990). It is often seen that workplace eccentricities and maltreatment are mostly directed at employees belonging to LGBT community (Lim and Cortina, 2005). Therefore, for employee engagement to escalate it is essential that the employee identifies the job as meaningful, psychologically harmless along with the feeling of being resourceful at all times. Workplace culture and the assumption that the heterosexual traits should be rivaled by all the employees further makes it more emotionally difficult for the LGBT employees to engage themselves to the core (Banihani et al., 2013). Numerous researches on LGBT demonstrates that they as an individual in a workplace setting are valued less and viewed as less useful in comparison to heterosexuals. This, in turn, sinks their engagement level since this organizational setting doesn't allow them to be more of their selves in their job. To avoid this situation, organizations should try to emphasize more on strengthening the employee engagement process by making it a further rational, psychological, and physically harmless experience for the LGBT individuals in the organization.

When an employee works in an organization, he or she has to become a part of any workplace talk which is more of personal in nature as it helps in developing cordial relationships among each other and thus make them feel as they are truly engaged with each other. Conversely, there are employees those who wish not to become part of such talks due to reasons like they want to stay secluded, they feel they are being judged or they feel awkward. The research by Mingle also identified that 17% of the LGBT employees who wished to stay closeted believed that their productivity was highly affected by their confinement (2017). This would further impact their career progression (Hewlett and Sumberg, 2011). So, it can be clinched that if openness is compromised upon when diversity and inclusion of LGBT employees are concerned, then it will have an adverse effect over their engagement in addition to their retention in the organization. So, the workplace culture plays an imperative role in propelling engagement of LGBT employees in the organization via bridging the gap through more openness and independence in the work environment.

Similarly, Infosys has realized that if they offer a good working culture that supports their employees equally, giving no consideration to their color, race, gender or sexual orientation, they will be a step ahead of their competitors by the way of creating a feeling of empowerment among the employees. This is how they have set a benchmark for other companies which can help in bringing a transformation across the society. Gradually, companies are now on the track of joining hands in bringing LGBT community as close as possible.

1.3.5 ACCEPTANCE OF LGBT IN INDIAN CORPORATIONS

In the present scenario, however, the acceptance of LGBT community may be very low in the Indian society but still, there are major Indian corporate giants such as Infosys, Godrej, etc. who believed in breaking the league and have incorporated LGBT rights in their diversity policy and thereby encouraging it in their organizations. For example, Godrej has come up with talks, plays, etc. to inspire candor towards their LGBT employees (Banerji et al., 2012). Similarly, they have sponsored and hosted a number of LGBT events to propel their level of motivation and commitment. In fact, they have included 'others' as an option under the category of gender in their company forms. Parmesh Shahani, Head-Godrej India Culture Lab expressed his views by stating that "LGBT people are a talent to be pursued and they don't like working with companies that discriminate" (Banerji et al., 2012). Infosys moreover followed the same league and has devised a diversity structure which includes a working group that focuses on LGBT inclusion and is accountable for solving LGBT problems (Infosys, 2014). Furthermore, the company has developed a group known as 'Infosys Gays Lesbians and You' directed towards bringing LGBT employees together to discuss on the matters related to policy changes which further made their working environment more supportive and pleasant (Infosys, 2014). This diversity structure strengthens the company's principles of equality and veracity. Correspondingly, Goldman Sachs Indian branch too has created a diversity structure with more than 300 LGBT members enjoying their rights at the workplace.

IBM India had developed an effective employee resource group (ERG) named as The Employee Alliance for Gay, Lesbian, Bisexual, and Transgender Empowerment (Community Business, 2017). The purpose behind the establishment of this group was to give a safe workplace culture to its LGBT employees. In Kolkata, Krishnagar women's college had appointed it's first-ever transgender Principal Prof. Manobi Bandopadhyay in the history of India (The Hindu, 2017). Another transgender namely Padmini Prakash was employed by Lotus news channel as television news anchor. She is continuously contributing by becoming a voice for her community on a larger scale (Times of India, 2017). Likewise, in the year 2015 'Madhu Kinnar' was elected India's first transgender mayor of Chhattisgarh (Reuters, 2017). All these employments, particularly from the transgender community, happened in the country after Supreme Court accepted them as the third gender legally and the government was instructed to make certain that they had been receiving equal treatment everywhere in India. The transgender employments were successfully accepted in the above organizations as legality was involved and society to a certain extent has started accepting the community without biases. Similarly, if the government of India establishes certain employment laws on inclusion and equality of the LGBT community, their life might become more serene like heterosexuals in the workplace.

LGBT community supporters and the organizations that fight for their rights have put a lot of pressure on the corporate world for their acceptance. It actually resulted in setting up of certain workshops by various top companies for their managers. The main focus of these workshops was the conception of responsiveness and preparing them for the ways in which they should deal with the LGBT employees in their organization (DNA-Essel Group, 2017).

1.3.6 INTERNATIONAL PRACTICES

In 2013, a survey by Human Rights Campaign was carried out on Fortune 500 firms which highlighted that 88% of the firms safeguard their LGBT employees from discrimination by establishing corporate policies (Community Business, 2017). One of Fortune 500 companies is Google which has established a group known as Gayglers who voluntarily endorse and makes responsiveness about LGBT co-workers in an organization (Google, 2017). This is how Google is able to communicate to the world that they provide a work culture that works for one and all. Correspondingly, Accenture has developed a meritocracy policy which is communicated to LGBT employees during orientation, diversity training, etc. (Accenture, 2017a). They introduced a global non-discrimination policy to promote equality (Stonewall, 2018). They have further developed an international network of LGBT employees aimed at fetching all LGBT employees collectively for sharing data as well as mentoring. The companies like Accenture and British Council have developed an equal opportunity policy that is spread globally to all the subsidiaries all over the world (Accenture, 2017b). Similarly, Bank of America has been awarded by 'The Human Rights Campaign' as one of the Best places to work for LGBT equality for providing equal opportunities in the workplace (Glassdoor, 2018). Likewise, the company named Buzzfeed is known for welcoming the members of the LGBT community as employees with open arms (Glassdoor, 2018).

Apple, the company known worldwide for its revolutionary innovations, has a gay as its CEO highlighting the fact that the company believes in LGBT inclusion showing support to LGBT community (Glaad, 2018). One of the renowned universities, Lehigh is known for its methods to bring about cognizance and support for its LGBT employees and students (Glassdoor, 2018). The university has initiated a pride center for LGBT employees and students so as to form a fair and impartial world by community building. It clearly states that the employees are treated equally and their self-respect will always be given the first priority. IBM has involved itself in diversity training, teaching their workforces to be more unprejudiced, unbiased, and change their mindset on LGBT employees at the workplace (Banerji et al., 2012).

1.4 IMPLICATIONS

Williams Institute piloted a survey on "the business impact of LGBT-supportive workplace policies" and identified that the employees belonging to LGBT community who hide their sexual orientation and identity face huge levels of hassle and nervousness leading to a number of health issues as well as job associated glitches (Glaad, 2018). Accordingly, if the companies focus on inculcating a workplace culture that supports LGBT inclusion they can reduce the insurance cost associated with the lower health of the LGBT employees. In fact, more customers, specifically the LGBT consumers, are added just by showcasing themselves as a socially responsible company which has established a good public image in the market via LGBT inclusion. This inclusion will enhance diversity and thus ultimately propel innovation throughout the company. Promoting an LGBT-inclusive workplace culture benefits an organization in (Hewlett and Yoshino, 2016).

- **Attracting and Retaining Top Talent:** Majority of the LGBT employees want to get their true selves at the job which is possible through inclusion policies. Companies are now focused on hiring talented people and also put efforts in the direction of making them stay engaged and committed. They are ready to "go the extra mile" to help the company in succeeding. Further, it allows them to attain their supreme potential.
- **Captivating the Fidelity of Discerning Customers:** Typically, the LGBT community prefers buying from an organization that treats LGBT as equals and supports inclusivity. Many companies such as American Express have started working in the direction of tapping the LGBT market.
- **Hitching the Awareness of LGBT Employees to Motivate Market Innovation:** When the company has team members sharing the sexual orientation with the target customers they are in a better situation to comprehend the market desires and wants. This is how a company can be benefitted with LGBT inclusivity.

As the world is becoming progressive and moving beyond its conventional boundaries bringing entire world together under one roof. It is imperative for present organizations in India to adopt global practices to realize their full potential and sustain in the near future. This can only be done when the organizations can embrace the wide diversity available in their surroundings. They need to come out of their preconceived notions and have to accept the diversity in its original form. The onus lies greatly on the management and leadership by leading with examples like done by many business entities starting from big giants like Google, IBM, etc. and many Indian companies such as Godrej, Infosys, etc. which are capitalizing the gains of diversity inclusion in their ecosystem. LGBT community is a very much part of our society and organization works in the same society. Being a part of the same environment, it's important to include LGBT in their main workforce to utilize their talents for the sake of organizational interests. Organizations need to do away with the prejudices to have an amicable and cordial work environment which fits all giving way to "Unity in Diversity" to reap the benefits of the diverse talent existing in our workplace.

Managers need to ensure that cultural DNA is not disturbed but is open to accept all without any reservations based upon their diverse traits ranging from culture, caste, color, ethnicity, or sexual orientation. The acceptance of the LGBT or the sexual minority in the organizations will pave a new way and set a benchmark for others. This will not only help management to gain increased employee motivation, commitment, and engagement but also enhance the employer brand. Indian organizations/employers need to uplift their standards to match the international benchmark. This will improve the overall goodwill of the company and also attract the top talent which helps in placing the organizations in the top list. The changing strata of the workforce with rising millennials who desire to work with the employer of their choice. They are more inclined to the workplace which focuses more on their performance and work-related behavior and gives space to live with their individual differences. The managers need to rework their strategies to live up to the expectation of the employees to sustain and retain their best talent. We can summarize the entire gist in the words of Max de Pree (1990) as *"We need to give each other the space to grow, to be ourselves, to exercise our diversity. We need to give each other space so that we may both give and receive such beautiful things as ideas, openness, dignity, joy, healing, and inclusion."*[1]

1.5 CONCLUSION

In India, the LGBT community has been facing harassment and inequality for a very long time. It will take many more years or even decades for Indians to easily swallow this taboo. Though social acceptance to this community is still questionable, at least the corporate sector is doing a bit on its part by showing huge acceptance to these sexual minorities by offering secure and open workplace culture that allows these people to be demonstrating their original identity at the workplace. This has helped in plummeting

---

[1]DePree, M., (1990). What is leadership? *Planning Review, 18,* 14–41.

chauvinisms and changed the mindsets of many. By doing this, companies have earned a good reputation in the global market and set apart themselves from the competitors in terms of a workplace culture that is all-inclusive of workplace diversity with an equal opportunity policy in place.

Be the company in any part of the globe, today's employees desire to be a part of that organization whose working culture supports the entire workforce without giving weightage to their sexual orientation or gender identity. Moreover, it is the moral responsibility of the government to introduce certain laws that protect the rights of the sexual minorities in the workplace, i.e., an equal opportunity policy shall be developed to ensure a ban on discrimination on the basis of gender identity and sexual orientation. As well as these policies should confirm that there is no inequality in recruitment, training, compensation, etc. of sexual minorities. But then again the laws alone cannot remove this discrimination, it is essential that the co-workers, peers, supervisors, bosses, etc. understand, support, and accept LGBT employees as one-of-their-kinds in the organization. Apart from policies, diversity at the workplace should be included in terms of the training and structure of the organization. Subsequently, the workplace culture too plays a very important role in shaping the organization's employees friendly and co-operative towards their LGBT colleagues.

From the past few years only, Indian corporations have started focusing more on the inclusion of the LGBT community. This has become relatively obvious that workplace diversity can never happen in the organizations if the advocates of diversity emphasize on tags of gender, caste, ethnicity, sexual orientation, etc. Indian Corporate world has the potential to build an environment wherein employees from the LGBT community exhibit pride, confidence, and work without any fear in their minds. Therefore, to an extent, LGBT issues at the workplace can be lessened if both government and corporations work hand-in-hand. In the subsequent section, a case covering the above concept has been provided (refer Appendix 1.1 for plausible case discussion).

1.6 CASE: DISCRIMINATION OR PERCEPTUAL BIAS AT IMPULSE

1.6.1 INTRODUCTION

"This is so inhumane, how they can do so? They never walk their talk. They say something and do something. Why are they, such hypocrites? How can they fire an employee just because she is a lesbian"? Karan Thapar, the software

engineer at Impulse India Private Limited, Gurgaon, branch of Impulse Private Limited a reputed MNC in the corporate world, thought to him.

Karan was feeling uneasy and unhappy after returning from the office party. He was disturbed by the discussions and recent incidents which took place in the company. Karan remembered how happy he was? When he joined Impulse India Private Limited, an MNC's six months back in January 2017. He was delighted with a fact that he would be part of the system; where people appreciate and respect diversity and individual differences in the workplace. He was under impression that he was working for an organization, which took care of their people and respected them for "who they are" and respect diverse talents and person's individuality.

His expectation of landing into the right organization was short-lived. It only lasted, till his illusion was blown away after an encounter with harsh reality in the company. People were judgmental and don't hold respect for anything which was beyond their own liking and preferences. The questions were hovering in his mind, making him restless and indecisive. Am I at the right place? Should I stay or leave?

1.6.2 ABOUT THE ORGANIZATION

Impulse Private Limited is an American multinational technology company headquartered in California, United States with global income: $115.2 million with operations in over 160 countries and employing more than 300,000 people worldwide. The company originated in 1991 and was listed on major stock exchanges of the world with total assets: $ 10.5 billion. It provides services to various industries ranging sectors like banking, retail, telecommunications, and government and addresses wide spectrum from providing software, analytics, cloud computing, cognitive solutions, internet solutions, extending IT infrastructure along with mobile technology and providing security solutions to the companies.

Impulse India Private Limited is a fully owned Indian subsidiary of Impulse Private Limited. It was operating in the industrial hub of Gurugram; Haryana in a spread of 20 acres of land and employs around 1500 people.

It was one of the reputed and well-known service providers in its field and catering needs of industrial houses operating both in northern and southern Indian territories. It was accredited with ISO 9001: 2000 certification to ensure the quality standards in the company. Impulse India Private Limited was a reliable name in its field and became one of the most profitable subsidiaries of Impulse Private Limited in India.

1.6.3 KARAN THAPAR

Karan Thapar was a 25 year's old, single who graduated from a reputed college in Delhi. He got selected through recruitment drive and had been working for Impulse India Private Limited Company for the past six months. Karan hails from Chandigarh and belongs to a reputed family. Karan was ambitious, hardworking, and focused on having liberal thoughts and believes in non-discrimination and equality for all. He always stands as a firm believer in promoting equality and treating everyone in a dignified manner without judging or discriminating anyone, irrespective of their gender, caste, religion, or sexuality, etc. He always remained appreciative of the progressive societies and wanted to work in an eco-system which can work without getting biased. Karan was inclined towards the same gender and identifies him as part of the LGBT community. But, he never disclosed his interest and his personal liking to anyone for a fear of inviting unpleasant surprises for himself.

1.6.4 INCIDENT 1

In the office on late Friday evening in July 2017 while going through the news feed on Facebook. He read that an e-commerce startup company named 'Zefo,' which is into buying and selling of revamped and second-hand furniture. The company has fired a 25-year-old woman, after the news aired by Karnataka media on the lesbian relationship of two young ladies belonging to Bengaluru. On the basis of the blurred picture of the couple telecasted by Kannada news channel, Company's HR has asked the lady employee on phone to leave the organization without having any further conversation over the matter. On the contrary, Zefo's culture talks about zero tolerance for discrimination at the workplace.

Karan discussed this news with his superior and his colleagues in the party the same evening. He was disappointed after listening to their viewpoints, who rather hailed the decision taken by 'Zefo.' The boss said that unnatural behavior is not acceptable in the civilized society. It is against our culture and societal norms and should be highly condemned in our society. His colleagues said that such people should be kept away as they are spoiling the environment. A bad egg can spoil the environment of the entire organization. So, elimination of the bad egg was a wise decision of Zefo. In our company also, we have few such persons, they should be also thrown out. We don't understand how a company can hire even such people? They are a curse to the humanity and spoils overall workplace culture.

Karan was really upset after knowing such viewpoints of the people in the organization. On one hand, they speak about the rich culture of the organization which talks about the inclusion of all and promotes equality despite their diverse traits. While on another hand, hold biased opinions and subjective views driven by their own cropped norms and outdated value system. Karan felt that the Indian colonial mindset is yet to come out of the criminality of gay relations, although the countries from where such restrictions originated have reformed into acceptance of such relations.

1.6.5 INCIDENT 2

It was July 2017, the month of performance appraisal in the organization where people hope to get rewarded for all their hard work and efforts after year-long wait and struggle. Mohit Sanyal was a promising young aspiring software engineer presently working in the position of System Analyst, operations in Impulse India Private Limited. He after serving the organization for almost 5 years was awaiting his next promotion to be a Project Manager. He was a performer with a focused and result-oriented approach which was visible from his past year performance. His efforts helped the company gain a lot of business by soothing out the processes handled by him with accuracy and diligence. This led to the customer satisfaction which was only made possible by the strategies implemented by Mohit Sanyal.

Mohit was openly straight acting openly gay and held his strong opinions about the LGBT community and its associated issues. He was vocal about his viewpoints and was a straightforward approach which was not liked by many people in the organization. But, they can't say anything to him openly as per company's policy. This is because no one can be denied job on the basis of his/her sexual orientation. During the time of performance evaluation, despite his credentials, he was not promoted and some other guy from his department with comparatively low badges got a promotion. On inquiry, he was told that he will be considered next year. However, he learnt unofficially that company hesitates to promote gays and lesbians to senior positions as they allegedly spoil the work culture of the organization and would promote their LGBT agenda over here.

Mohit was possibly denied promotion for being gay in spite of his remarkable work performance and was sidelined from the getting fair chance. Although, there is nowhere written that being gay is a crime and a fair chance should be denied to the concerned person. The company policy nowhere talked about this discrimination that LGBT people will not be promoted but the people within the organization and their skewed mindsets discriminated him because of their discriminatory attitude and ignorance.

1.6.6 CONCLUSION

Karan was really upset and felt disengaged after seeing these incidences in the organization. He felt annoyed and helpless at the same time that how an MNC operating in today's global world which preaches that we have LGBT friendly policies and works for their welfare and talks about diversity inclusion and respect of all etc. can deviate from their norms and discriminate anyone one on the basis of their sexuality or individual choices. He was devastated with a feeling of being cheated and thinking whether I have landed in a right place? What will be my fate, if they came to know about my choices/orientation?

APPENDIX 1.1: CASE DISCUSSION

Case Synopsis

In July 2017, Karan Thapar, a software engineer at the XYZ, Gurugram, Haryana subsidiary of ABC Private Limited, is faced with a dilemma. He was upset by the series of incidents which happened in the organization. First, he was shocked to know the reaction of his superiors and colleagues towards people of the LGBT community in their organization. Secondly, he witnessed Mohit Sanyal who was denied promotion for just being gay despite his credentials and performance. Karan who was secretly gay was moved by these incidents making him restless and indecisive. Thinking of whether he is at the right place? Should he stay or leave? As he joined the organization with an expectation that it respects diversity and individual identity, but found an almost different scenario which crushed his hopes.

Theoretical Base

The primary teaching objective of this case is to explore how organizations can address the diversity issues in their organizations with special reference to LGBT employees. The organizations have to understand that the workplace culture and the approach of the people make a lot of difference, which can

Engaging an LGBT Workforce

oust out good talent from the system. The organization needs to understand the changes happening in the global world and should accommodate the same. They need to address the issues which can impact the morale of their employees and should prefer merit over personal biases during performance evaluations rather than focusing on their individual interest and differences.

Main Questions

- Are we ready for global environment in our organizations in spite of our traditional values and thinking?
- Is it alright for the organization to focus only on the work-related behavior of the employees and ignore his private life activities?
- Was the company ethically correct to deny promotion to Mohit Sanyal?
- Should Karan continue in Impulse India Private Limited despite his apprehensions about the company's Hippocratic treatment of the LGBT community?

Intended Audience

The case has covered various issues related to the subject matter of organizational behavior and human resource (HR) management. It is intended to develop logical and analytical thinking among the students to critically appraise and apply their learned concepts. This case can be used for different levels of the management education practitioners and the students of undergraduates, postgraduates, and the executives working in the organization.

Teaching Plan (2 Hours Class)

- 0–10: Introduction and sharing of previous experiences.
- 10–30: Listing out options and rationale.
- 30–50: Basic issues in the case.
- 50–80: Decision.
- 80–100: Action Plan.
- 100–120: Wrap up.

Instructors can start the class by asking students if they had heard or been in a similar situation. Hopefully, few students will be there in the class who will narrate their experiences. This is meant to illustrate to the rest of the class how difficult it is for the people from the LGBT community to survive in the company and help students to appreciate their viewpoint and issues faced by them rather than jumping to preconceived notions. Once a few students have shared their experiences, the instructor can then pose the question, "As Karan, what are their options?" All the options can be put on the board as separate headings. Once all the options are listed, then instructors can take a poll as to who supports which option. Instructors can then push the students on the rationale behind their thought process. Instructors can act as devil's advocate by asking students why such issues should be addressed. This should bring out the need and benefits of diversity for the organizations. After a discussion of the issues involved in the case and the various options, instructors should select one option and push students for a detailed action plan. What should he do? What would be the repercussions of this approach to the organization? How should organizations proceed to resolve diversity issues in the organization? What could be the associated benefits for the organization? How could acceptance of the LGBT people be increased within the organization?

- The class can be concluded by outlining the key learning points.
- This case is majorly based on the approach of the people in the organization towards LGBT community. It will highlight the perceptual biasedness which people hold towards the sexual minority in the organization. The case will also throw light on the organizational practices and approaches towards diversity in their systems. The discomfort and level anxiety people have been part of the organization system, which talks about equality but didn't practice it. The hypocritical character of the organizations and people therein who never walk their talk.

Teaching Experience of this Case

The case was tested in the class of Organization Behavior and human resource management (HRM) for undergraduate students. They were able to identify the issues like diversity issues, inequality, and perceptual biases, demotivation, discontentment, career advancement, job dissatisfaction. The students came up with various options based on their perceptions and understanding and debated putting their views in front of each other. The students discussed about the approaches of Karan Thapar, Superior, and employees in the organization based on their viewpoint in light of its merits and demerits.

KEYWORDS

- employee engagement
- lesbians, gays, bisexual, and transgender (LGBT)
- pink money
- sexual minorities
- workplace culture
- workplace diversity

REFERENCES

Accenture, (2017a). *Global-Meritocracy*. https://www.accenture.com/us-en/global-meritocracy (accessed on 24 February 2020).

Accenture, (2017b). *Inclusion and Diversity*. https://www.accenture.com/in-en/careers/team-culture-diversity (accessed on 24 February 2020).

Badgett, M. V. L., Lau, H., Sears, B., & Ho, D., (2007). *Bias in the Workplace: Consistent Evidence of Sexual Orientation and Gender Identity Discrimination* (pp. 1–27). The Williams Institute. [Online]. https://escholarship.org/uc/item/5h3731xr (accessed on 24 February 2020).

Banerji, A., Burns, K., & Vernon, K., (2012). *Creating Inclusive Workplaces for LGBT Employees in India* (pp. 1–54). Hong Kong: Community Business. [Online]. https://www.communitybusiness.org/latest-news-publications/creating-inclusive-workplaces-lgbt-employees-india-resource-guide-employers (accessed on 24 February 2020).

Banihani, M., Lewis, P., & Syed, J., (2013). Is work engagement gendered? *Gender in Management: An International Journal, 28*, 400–423. [Online]. https://www.emeraldinsight.com/doi/full/10.1108/GM-01-2013-0005 (accessed on 24 February 2020).

Berg, N., & Lien, D., (2002). Measuring the effect of sexual orientation on income: Evidence of discrimination. *Contemporary Economic Policy, 20*, 394–414. [Online]. https://onlinelibrary.wiley.com/doi/abs/10.1093/cep/20.4.394 (accessed on 24 February 2020).

Blandford, J. M., (2003). The nexus of sexual orientation and gender in the determination of earnings. *Industrial and Labor Relations Review, 56*, 622–642. [Online]. https://journals.sagepub.com/doi/abs/10.1177/001979390305600405 (accessed on 24 February 2020).

Button, S. B., (2001). Organizational efforts to affirm sexual diversity: A cross-level examination. *Journal of Applied Psychology, 86*, 17–28. [Online]. https://www.ncbi.nlm.nih.gov/pubmed/11302229 (accessed on 24 February 2020).

Collins, E. C., (2012). Global diversity initiatives. *The International Lawyer, 46*, 987–1006. [Online]. https://www.jstor.org/stable/23643948?seq=1#page_scan_tab_contents (accessed on 24 February 2020).

Community Business, Creating Inclusive Workplaces for LGBT Employees in India-A Resource Guide for Employers. http://www.outandequal.org/wp-content/uploads/2015/05/Community-Business-Inclusive-Workplace-India-2012.pdf (accessed on 24 February 2020).

Connell, C., (2010). Doing, undoing, or redoing gender? Learning from the workplace experiences of transpeople. *Gender and Society*, *24*, 31–55. [Online]. https://journals.sagepub.com/doi/abs/10.1177/0891243209356429 (accessed on 24 February 2020).

Daneshvary, N., Waddoups, C. J., & Wimmer, B. S., (2009). Previous marriage and the lesbian wage premium. *Industrial Relations: A Journal of Economy and Society*, *48*, 432–453. [Online]. https://onlinelibrary.wiley.com/doi/abs/10.1111/j.1468-232X.2009.00567.x (accessed on 24 February 2020).

DNA-Essel Group. *Bangalore IT Forms Wake up to LGBTI Rights, Finally!* http://www.dnaindia.com/bangalore/report-bangalore-it-firms-wake-up-to-LGBTi-rights-finally-1853316 (accessed on 24 February 2020).

Fielden, S. L., & Jepson, H., (2016). An exploration into the career experiences of lesbians in the UK. *Gender in Management: An International Journal*, *31*, 281–296. [Online]. https://www.emeraldinsight.com/doi/full/10.1108/GM-03-2016-0037 (accessed on 24 February 2020).

Forbes India. *The Lure of the Pink Rupee*. http://forbesindia.com/article/briefing/the-lure-of-the-pink-rupee/6652/1 (accessed on 24 February 2020).

Glaad. *The Value of LGBT Equality in the Workplace*. https://www.glaad.org/blog/value-lgbt-equality-workplace (accessed on 24 February 2020).

Glassdoor. 5 *Companies with LGBT-Friendly Offices*. https://www.glassdoor.com/employers/blog/5-companies-with-lgbt-friendly-offices/ (accessed on 24 February 2020).

Google. *Gayglers*. https://www.google.com/diversity/at-google.html (accessed on 24 February 2020).

Hatch, M., (1993). The dynamics of organizational culture. *Academy of Management Review*, *18*, 657–693. [Online]h ttps://journals.aom.org/doi/abs/10.5465/amr.1993.9402210154 (accessed on 24 February 2020).

Hebl, M. R., Foster, J. B., Mannix, L. M., & Dovidio, J. F., (2002). Formal and interpersonal discrimination: A field study of bias toward homosexual applicants. *Personality and Social Psychology Bulletin*, *28*, 815–825. [Online]. https://journals.sagepub.com/doi/abs/10.1177/0146167202289010 (accessed on 24 February 2020).

Heilman, M. E., & Eagly, A. H., (2008). Gender stereotypes are alive, well, and busy producing workplace discrimination. *Industrial and Organizational Psychology*, *1*, 393–398. [Online]. https://www.cambridge.org/core/journals/industrial-and-organizational-psychology/article/gender-stereotypes-are-alive-well-and-busy-producing-workplace-discrimination/79B1C722310ABBBE8D0EA90157689B97 (accessed on 24 February 2020).

Hewlett, S. A., & Sumberg, K., (2011). *For LGBT Workers, Being 'Out' Brings Advantages* (pp. 1, 2). Harvard Business Review. [Online]. https://hbr.org/2011/07/for-lgbt-workers-being-out-brings-advantages (accessed on 24 February 2020).

Hewlett, S. A., & Yoshino, K., (2016). *LGBT-Inclusive Companies are Better at 3 Big Things*. https://hbr.org/2016/02/lgbt-inclusive-companies-are-better-at-3-big-things (accessed on 24 February 2020).

Horton, P., Rydstrøm, H., & Tonini, M., (2015). Contesting heteronormativity: The fight for lesbian, gay, bisexual and transgender recognition in India and Vietnam. *Culture, Health and Sexuality*, *17*, 1059–1073. [Online]. https://www.tandfonline.com/doi/abs/10.1080/13691058.2015.1031181 (accessed on 24 February 2020).

Horvath, M., & Ryan, A. M., (2003). Antecedents and potential moderators of the relationship between attitudes and hiring discrimination on the basis of sexual orientation. *Sex Roles*, *48*, 115–130. [Online]. https://link.springer.com/article/10.1023/A:1022499121222 (accessed on 24 February 2020).

Howell, J. P., & Costley, D. L., (2006). *Understanding Behaviors for Effective Leadership* (2nd edn.). Pearson: NJ.

Infosys. *Creating a Diverse Workplace.* https://www.infosys.com/careers/culture/diversity-inclusivity/Pages/index.aspx (accessed on 24 February 2020).

Kahn, W. A., (1990). Psychological conditions of personal engagement and disengagement at work. *Academy of Management Journal, 33,* 692–724. [Online]. https://journals.aom.org/doi/abs/10.5465/256287 (accessed on 24 February 2020).

Leppel, K., (2014). Does job satisfaction vary with sexual orientation? *Industrial Relations, 53,* 169–198. [Online]. https://onlinelibrary.wiley.com/doi/abs/10.1111/irel.12053 (accessed on 24 February 2020).

Lim, S., & Cortina, L. M., (2005). Interpersonal mistreatment in the workplace: The interface of impact of general incivility and sexual harassment. *Journal of Applied Psychology, 90,* 483–496. [Online]. https://www.ncbi.nlm.nih.gov/pubmed/15910144 (accessed on 24 February 2020).

Macey, W. H., & Schneider, B., (2008). The meaning of employee engagement. *Industrial and Organizational Psychology, 1,* 3–30. [Online]. https://www.cambridge.org/core/journals/industrial-and-organizational-psychology/article/meaning-of-employee-engagement/0517A938DBEDA2E0BE2FBE27A9DDC4DB (accessed on 24 February 2020).

Majumdar, S. R., & Adams, M. O., (2015). Diversity in the master of public administration programs at minority-serving institutions. *Journal of Public Affairs Education, 21,* 215–228. [Online]. https://www.tandfonline.com/doi/abs/10.1080/15236803.2015.12001829 (accessed on 24 February 2020).

May, D. R., Gilson, R. L., & Harter, L. M., (2004). The psychological conditions of meaningfulness, safety and availability and the engagement of the human spirit at work. *Journal of Occupational and Organizational Psychology, 77,* 11–37. [Online]. https://onlinelibrary.wiley.com/doi/abs/10.1348/096317904322915892 (accessed on 24 February 2020).

Mingle. *Annual LGBT Workplace Diversity and Inclusion Survey.* http://mingle.org.in/pdf/LGBTWorkplace-Survey-2012-MINGLE.pdf (accessed on 24 February 2020).

Norman-Major, K., & Becker, C., (2013). Walking the talk: Do public systems have the infrastructure necessary to implement and enforce LGBT and gender identity rights? *Journal of Public Management and Social Policy, 19,* 31–44. [Online]. https://search.proquest.com/openview/d1841d16d684526e00720f5cb98184f9/1?pq-origsite=gscholar&cbl=2026673 (accessed on 24 February 2020).

Oswick, C., & Noon, M., (2013). Discourses of diversity, equality, and inclusion: Trenchant formulations or transient fashions? *British Journal of Management, 25,* 23–39. [Online]. https://onlinelibrary.wiley.com/doi/abs/10.1111/j.1467-8551.2012.00830.x (accessed on 24 February 2020).

Ozturk, M. B., & Rumens, N., (2014). Gay male academics in UK business and management schools: Negotiating heteronormativities in everyday work life. *British Journal of Management, 25,* 503–517. [Online]. https://onlinelibrary.wiley.com/doi/full/10.1111/1467-8551.12061 (accessed on 24 February 2020).

Priola, C., Lasio, D., Simone, S. D., & Serri, F., (2014). The sound of silence: Lesbian, gay, bisexual and transgender discrimination in "inclusive organizations. *British Journal of Management, 25,* 488–502. [Online]. https://onlinelibrary.wiley.com/doi/abs/10.1111/1467-8551.12043 (accessed on 24 February 2020).

Ragins, B. R., & Cornwell, J. M., (2001). Pink triangles: Antecedents and consequences of perceived workplace discrimination against gay and lesbian employees. *Journal of Applied*

Psychology, 86, 1244–1261. [Online]. https://www.ncbi.nlm.nih.gov/pubmed/11768065 (accessed on 24 February 2020).

Reuters. *First Transgender Mayor Elected in Chhattisgarh-Media.* http://in.reuters.com/article/india-transgender-idINKBN0KE0MO20150105 (accessed on 24 February 2020).

Rumens, N., (2016). Sexualities and accounting: A queer theory perspective. *Critical Perspectives on Accounting, 35*, 111–120. [Online]. https://www.sciencedirect.com/science/article/pii/S104523541500060X (accessed on 24 February 2020).

Schaufeli, W. B., Salanova, M., Gonzalez-Roma, V., & Bakker, A. B., (2002). The measurement of engagement and burnout: A two sample confirmatory factor analytic approach. *Journal of Happiness Studies.* [Online] *3*, 71–92 https://link.springer.com/article/10.1023/A:1015630930326 (accessed on 24 February 2020).

Shore, L. M., Chung-Herrera, B. G., Dean, M. A., Holcombe, E. K., Jung, D. I., Randel, A. E., & Singh, G., (2009). Diversity in organizations: Where are we now and where are we going? *Human Resource Management Review, 19*, 117–133. [Online]. https://www.sciencedirect.com/science/article/pii/S1053482208000855 (accessed on 24 February 2020).

Smith, N. G., & Ingram, K. M., (2004). Workplace heterosexism and adjustment among lesbian, gay, and bisexual individuals: The role of unsupportive interactions. *Journal of Counseling Psychology, 51*, 57–67. [Online]. https://www.researchgate.net/publication/232557288_Workplace_Heterosexism_and_Adjustment_Among_Lesbian_Gay_and_Bisexual_Individuals_The_Role_of_Unsupportive_Social_Interactions (accessed on 24 February 2020).

Srivastava, S. S., (2014). Disciplining the 'desire': 'Straight' state and LGBT activism in India. *Sociological Bulletin, 63*, 368–385. [Online]. https://journals.sagepub.com/doi/abs/10.1177/0038022920140303?journalCode=soba (accessed on 24 February 2020).

Stonewall. *Engaging the Majority to Create an LGBT Inclusive Workplace.* https://www.accenture.com/t20170201t214109__w__/gr-en/_acnmedia/pdf-42/accenture-stonewall-uk-star-performer-guide.pdf (accessed on 24 February 2020).

Sullivan, N., (2003). *A Critical Introduction to Queer Theory.* Edinburgh University Press: Edinburgh.

Supreme Court. *In the Supreme Court of India Civil Original Jurisdiction.* http://supremecourtofindia.nic.in/pdf/jud/ALL%20WP(C)%20No.494%20of%202012%20Right%20to%20Privacy.pdf (accessed on 24 February 2020).

The Hindu. *India's First Transgender College Principal Takes Charge.* http://www.thehindu.com/news/national/other-states/manobi-bandopadhyay-indias-first-transgender-college-principal-takes-charge/article7298708.ece# (accessed on 24 February 2020).

The Hindu. *LGBT Rights: The Journey till Now.* http://www.thehindu.com/specials/in-depth/LGBT-rights-the-journey-till-now/article14056726.ece (accessed on 24 February 2020).

Thomas, R. R. Jr., (1990). From affirmative action to affirming diversity. *Harvard Business Review, 1*, 107–117. [Online]. http://letr.org.uk/references/storage/S49MGQI3/Thomas%20-%201990%20-%20From%20affirmative%20action%20to%20affirming%20diversity.pdf (accessed on 24 February 2020).

Times of India. *First Transgender News Anchor Says She's Rid of Her Demons.* https://timesofindia.indiatimes.com/india/First-transgender-news-anchor-says-shes-rid-of-her-demons/articleshow/42851729.cms (accessed on 24 February 2020).

Vohra, N., Chari, V., Mathur, P., Sudarshan, P., Verma, N., Mathur, N., & Dasmahapatra, V., (2015). Inclusive workplaces: Lessons from theory and practice. *Vikalpa, 40*, 324–362. [Online]. https://journals.sagepub.com/doi/full/10.1177/0256090915601515 (accessed on 24 February 2020).

Wang, P., & Schwarz, J. L., (2010). Stock price reactions to GLBT non-discrimination policies. *Human Resource Management, 49*, 195–216. [Online]. https://onlinelibrary.wiley.com/doi/abs/10.1002/hrm.20341 (accessed on 24 February 2020).

Woodruffe-Burton, H., & Bairstow, S., (2013). Countering heteronormativity: Exploring the negotiation of butch lesbian identity in the organizational setting. *Gender in Management: An International Journal, 28*, 359–374. [Online]. https://www.emeraldinsight.com/doi/full/10.1108/GM-01-2013-0015 (accessed on 24 February 2020).

FURTHER READING

Acharya, A., & Gupta, M., (2016a). An application of brand personality to green consumers: A thematic analysis. *Qualitative Report, 21*(8), 1531–1545.

Acharya, A., & Gupta, M., (2016b). Self-image enhancement through branded accessories among youths: A phenomenological study in India. *Qualitative Report, 21*(7), 1203–1215.

Aslan, F., Şahin, N. E., & Emiroğlu, O. N., (2019). Turkish nurse educator's knowledge regarding LGBT health and their level of homophobia: A descriptive-cross sectional study. *Nurse Education Today, 76*, 216–221.

Barnett, A. P., Molock, S. D., Nieves-Lugo, K., & Zea, M. C., (2019). Anti-LGBT victimization, fear of violence at school, and suicide risk among adolescents. *Psychology of Sexual Orientation and Gender Diversity, 6*(1), 88–95.

Bryan, A., (2019). Kuchu activism, queer sex-work, and "lavender marriages," in Uganda's virtual LGBT safe(r) spaces. *Journal of Eastern African Studies, 13*(1), 90–105.

Donaldson, W., Smith, H. M., & Parrish, B. P., (2019). Serving all who served: Piloting an online tool to support cultural competency with LGBT U. S. military veterans in long-term care. *Clinical Gerontologist, 42*(2), 185–191.

Goodin, D., (2019). Wearing gay history: A digital archive of historical LGBT T-shirts. *Dress, 45*(1), 103–105.

Gupta, M., & Kumar, Y., (2015). Justice and employee engagement: Examining the mediating role of trust in Indian B-schools. *Asia-Pacific Journal of Business Administration, 7*(1), 89–103.

Gupta, M., & Pandey, J., (2018). Impact of student engagement on affective learning: Evidence from a large Indian University. *Current Psychology, 37*(1), 414–421.

Gupta, M., & Ravindranath, S., (2018). Managing physically challenged workers at micro sign. *South Asian Journal of Business and Management Cases, 7*(1), 34–40.

Gupta, M., & Sayeed, O., (2016). Social responsibility and commitment in management institutes: Mediation by engagement. *Business: Theory and Practice, 17*(3), 280–287.

Gupta, M., & Shaheen, M., (2017a). Impact of work engagement on turnover intention: Moderation by psychological capital in India. *Business: Theory and Practice, 18*, 136–143.

Gupta, M., & Shaheen, M., (2017b). The relationship between psychological capital and turnover intention: Work engagement as mediator and work experience as moderator. *Journal Pengurusan, 49*, 117–126.

Gupta, M., & Shaheen, M., (2018a). Does work engagement enhance general well-being and control at work? Mediating role of psychological capital. *Evidence-Based HRM, 6*(3), 272–286.

Gupta, M., & Shukla, K., (2018b). An empirical clarification on the assessment of engagement at work. *Advances in Developing Human Resources, 20*(1), 44–57.

Gupta, M., (2017). Corporate social responsibility, employee-company identification, and organizational commitment: Mediation by employee engagement. *Current Psychology, 36*(1), 101–109.

Gupta, M., (2018). Engaging employees at work: Insights from India. *Advances in Developing Human Resources, 20*(1), 3–10.

Gupta, M., Acharya, A., & Gupta, R., (2015). Impact of work engagement on performance in Indian higher education system. *Review of European Studies, 7*(3), 192–201.

Gupta, M., Ganguli, S., & Ponnam, A., (2015). Factors affecting employee engagement in India: A study on offshoring of financial services. *Qualitative Report, 20*(4), 498–515.

Gupta, M., Ravindranath, S., & Kumar, Y. L. N., (2018). Voicing concerns for greater engagement: Does a supervisor's job insecurity and organizational culture matter? *Evidence-Based HRM, 6*(1), 54–65.

Gupta, M., Shaheen, M., & Das, M., (2019). Engaging employees for quality of life: Mediation by psychological capital [Involving employees in improving quality of life: mediation of psychological capital]. *Service Industries Journal, 39*(5/6), 403–419.

Gupta, M., Shaheen, M., & Reddy, P. K., (2017). Impact of psychological capital on organizational citizenship behavior: Mediation by work engagement. *Journal of Management Development, 36*(7), 973–983.

Han, X., Han, W., Qu, J., Li, B., & Zhu, Q., (2019). What happens online stays online?: Social media dependency, online support behavior and offline effects for LGBT. *Computers in Human Behavior, 93*, 91–98.

Hinrichs, K. L. M., & Christie, K. M., (2019). Focus on the family: A case example of end-of-life care for an older LGBT veteran. *Clinical Gerontologist, 42*(2), 204–211.

Hirschtritt, M. E., Noy, G., Haller, E., & Forstein, M., (2019). LGBT-specific education in general psychiatry residency programs: A survey of program directors. *Academic Psychiatry, 43*(1), 41–45.

Hoare, J. P., (2019). Narratives of exclusion: Observations on a youth-led LGBT rights group in Kyrgyzstan. *Soviet and Post-Soviet Sexualities*, 171–191.

Hull, K. E., & Ortyl, T. A., (2019). Conventional and cutting-edge: Definitions of family in LGBT communities. *Sexuality Research and Social Policy, 16*(1), 31–43.

Khatua, A., Cambria, E., Ghosh, K., Chaki, N., & Khatua, A., (2019). Tweeting in support of LGBT? A deep learning approach. *ACM International Conference Proceeding Series*, 342–345.

Kortes-Miller, K., Wilson, K., & Stinchcombe, A., (2019). Care and LGBT aging in Canada: A focus group study on the educational gaps among care workers. *Clinical Gerontologist, 42*(2), 192–197.

Logie, C. H., Dias, L. V., Jenkinson, J., Newman, P. A., MacKenzie, R. K., Mothopeng, T., Madau, V., Ranotsi, A., Nhlengethwa, W., & Baral, S. D., (2019). Exploring the potential of participatory theatre to reduce stigma and promote health equity for lesbian, gay, bisexual, and transgender (LGBT) people in Swaziland and Lesotho. *Health Education and Behavior, 46*(1), 146–156.

Marston, K., (2019). Researching LGBT+ youth intimacies and social media: The strengths and limitations of participant-led visual methods. *Qualitative Inquiry, 25*(3), 278–288.

Mikulak, M., (2019). Godly homonormativity: Christian LGBT organizing in contemporary Poland. *Journal of Homosexuality, 66*(4), 487–509.

Mongelli, F., Perrone, D., Balducci, J., Sacchetti, A., Ferrari, S., Mattei, G., & Galeazzi, G. M., (2019). Minority stress and mental health among LGBT populations: An update on the evidence. *Minerva Psichiatrica, 60*(1), 27–50.

Morrison, M. A., Bishop, C. J., & Morrison, T. G., (2019). A Systematic review of the psychometric properties of composite LGBT prejudice and discrimination scales. *Journal of Homosexuality, 66*(4), 549–570.

Naal, H., Abboud, S., & Mahmoud, H., (2019). Developing an LGBT-affirming healthcare provider directory in Lebanon. *Journal of Gay and Lesbian Mental Health, 23*(1), 107–110.

Ní, & Mhaoileoin, N., (2019). The ironic gay spectator: The impacts of centering western subjects in international LGBT rights campaigns. *Sexualities, 22*(43467), 148–164.

Pandey, J., Gupta, M., & Naqvi, F., (2016). Developing decision making measure a mixed method approach to operationalize Sankhya philosophy. *European Journal of Science and Theology, 12*(2), 177–189.

Salkind, J., Bevan, R., Drage, G., Samuels, D., & Hann, G., (2019). Safeguarding LGBT+ adolescents. *BMJ (Online), 364*.

Smith, R. W., Altman, J. K., Meeks, S., & Hinrichs, K. L. M., (2019). Mental health care for LGBT older adults in long-term care settings: Competency, training, and barriers for mental health providers. *Clinical Gerontologist, 42*(2), 198–203.

The Lancet Public Health LGBT: The vital fight for the right to health. *The Lancet Public Health, 4*(3), e116.

Tobin, T. W., & Moon, D., (2019). The politics of shame in the motivation to virtue: Lessons from the shame, pride, and humility experiences of LGBT conservative Christians and their allies. *Journal of Moral Education, 48*(1), 109–125.

Toft, A., Franklin, A., & Langley, E., (2019). Young disabled and LGBT+: Negotiating identity. *Journal of LGBT Youth, 16*(2), 157–172.

Waite, S., & Denier, N., (2019). A research note on Canada's LGBT data landscape: Where we are and what the future holds. *Canadian Review of Sociology, 56*(1), 93–117.

Wilson, B. M., & Gianella-Malca, C., (2019). Overcoming the limits of legal opportunity structures: LGBT rights' divergent paths in Costa Rica and Colombia. *Latin American Politics and Society, 61*(2), 138–163.

CHAPTER 2

Active Aging: Engaging the Aged

RITU GUPTA,[1] SHRUTHI J. MAYUR,[1] and SWATI HANS[2]

[1]T A Pai Management Institute, Manipal, Karnataka, India,
Tel.: (+91) 9652081180, E-mail: rgritu@gmail.com (R. Gupta),
Tel.: (+91) 9447703833, E-mail: shruthi.mayur@tapmi.edu.in (S. J. Mayur)

[2]ICFAI Business School (IBS), Hyderabad, The ICFAI Foundation for Higher Education (IFHE), Hyderabad, Telangana–501203, India,
E-mail: swati.hans@ibsindia.org

ABSTRACT

With changing demographics around the world, it is imperative that older employees are given equal opportunities and kept engaged in the workplace. More initiatives need to be taken by organizations to consciously include older people in the workforce. The initiatives address various domains of individual resources such as physical, financial, social, emotional, cognitive, and motivational. Optimizing opportunities related to health, participation, and security enhances quality of life post-retirement and psychological well-being as people age.

2.1 INTRODUCTION

Greying of the world, the era we are living in has a more grey population than ever before, and it is expected to double in the years to come. The population is aging; it is not just a segment of the population, but a significantly larger proportion. Hence, it becomes important to include the aging population in social and economic development. How can we promote this, what are the challenges of including the aged population in socio-economic development? These are some of the questions this chapter answers.

Graceful aging is a process of optimizing opportunities for physical, social, and mental well-being during the later years of one's life, in order to ensure a healthy, independent, and quality life. The World Health Organization (WHO) has termed this process as 'active aging' since the late 1990s, WHO defined it as "the process of optimizing opportunities for health, participation, and security in order to enhance quality of life as people age."

This definition focuses on policy action primarily on three areas:

- **"Health":** which is understood to be physical health as well as mental and social well-being, following the WHO recommended definition.
- **"Participation":** which in turn is understood as a multifaceted array of activities by older persons in social, economic, cultural, spiritual, and civic affairs, in addition to their participation in the labor force.
- **"Security":** which is concerned with the access of older persons to safe and secure physical and social environment, income security, and (when applicable) the securing of a rewarding employment (refer Figure 2.1).

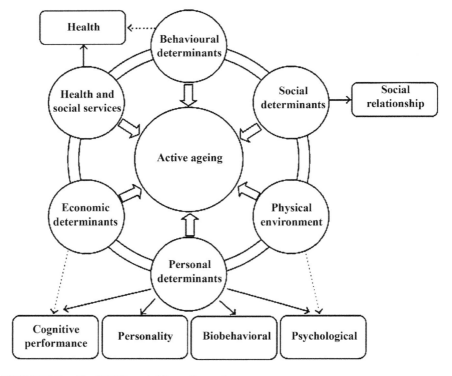

FIGURE 2.1 The WHO model for active aging.
Source: Reprinted with permission from Paúl, Ribeiro, and Teixeira, (2012). https://creativecommons.org/licenses/by/3.0/

Hence, the focus of active aging is to create greater opportunities for labor market engagement. The expectation from the aging cohort is that they should be engaged with their community and family, whilst the work they do maybe unpaid, yet it is considered productive for the society. They can collectively enrich the society by maintaining and optimizing their health, not only physically but by preserving mental well-being as well (Sidorenko and Zaidi, 2013). This may be successful aging, but it should not be necessary to be busy or active all the time to age successfully (Ekerdt, 2018). The Second Global Summit on Aging in Madrid in 2002 looked at active aging as 'a model for planning the future.' Here, older people's contribution to society is values, and their contribution in all facets of life in society is actively encouraged.

Aging has been explored in several other ways; some of them are discussed here. Productive Aging has been defined by several scholars differently, but what really counts is the promotion and economic contributions made by the older adults in the labor market (Davey, 2002). A more inclusive definition of productive aging was given by Rowe and Kahn (1997), who include the social contribution made by older adults in the productivity. It is seen that active aging includes healthy and productive aging at large (Boudiny, 2013). Finally, some of the subjective measures which are missed in productive aging are included in positive aging (Paul, Teixeira, and Ribeiro, 2015).

'Social aging' explores the expectations and constraints of older people's work life. Their age, health, life expectancy, cognitive ability, and other factors are considered (Sidorenko and Zaidi, 2013). The current stage of life is as important as the remaining years and plans for making it wholesome.

Active aging has a clear focus on the social aging phenomenon; the increase in life expectancy requires tapping the potential of the older workforce. The participation of older individuals in the workforce has become crucial and organizations need to enable their continued participation.

2.2 NEED FOR DISCUSSING THIS THEME

People are aging successfully. Today, they enjoy a high life expectancy rate than individuals two decades ago. Aging cohorts' number is on the rise across the world because of better health measures. In Europe, the demographic shift is even more evident with the high population of elder people and low birth rates. There are fewer active people to support the social cost of health care of elderly.

Organizations need to understand the importance of the aging cohort, their wisdom, dignity, productiveness, and self-control which comes with a lifetime of experience. The values they help infuse in the workforce cannot be neglected. The organizations view these employees as a drain on funds, while they do not understand the spiritual, cultural, and socio-economic contributions made by them (Chakraborti, 2004). Organizations need to create a framework in which individual and collective contribution of the aging cohort is captured and they are treated with respect and gratitude, dignity, and sensitivity. Government of Vietnam has an ordinance about the aged in the society and clearly describes the care and respect they deserve (Chakraborti, 2004). Also, the staggering number of individuals expected to be in this age group by 2050, as projected by the United Nations, needs our attention (refer to Table 2.1 and Figure 2.2).

2.2.1 EVOLUTION OF ACTIVE AGING

WHO has termed the process of graceful aging as 'active aging' since the late 1990s. Other international organizations, academic circles, and government groups (including G8, the Organization for Economic Co-operation and Development (OECD), the International Labor Organization (ILO), and the Commission of the European Communities (CEC)) are also working on the concept using 'active aging.' This movement towards 'active aging' is to ensure the idea of continuous involvement of aged persons in socially productive activities and meaningful work by the aging workforce. This is to ensure social engagement of the aged employees. OECD and ILO used this concept to help organizations address the challenges faced by labor market due to lengthening retirement.

2.2.2 AGING AND DEVELOPMENT

We have reached a consensus that active aging requires individuals to be healthy and contribute economically towards the development of the society. To achieve this, it is essential that the aging population should be included in the mainstream economic activities and provided adequate support to perform. In the following sections, we discuss some of the ways this integration can be successfully adopted and the challenges faced, which can also be seen in the elaborate model of active aging as adopted by WHO (refer Figure 2.3).

TABLE 2.1 Countries with the Largest Share of Aged Population (60 and above) from 1980 to 2050 (Projected)

Rank	1980 Country or Area	1980 Percentage Aged 60 Years or Over	2017 Country or Area	2017 Percentage Aged 60 Years or Over	2050 Country or Area	2050 Percentage Aged 60 Years or Over
1	Sweden	22.0	Japan	33.4	Japan	42.4
2	Norway	20.2	Italy	29.4	Spain	41.9
3	Channel Islands	20.1	Germany	28.0	Portugal	41.7
4	United Kingdom	20.0	Portugal	27.9	Greece	41.6
5	Denmark	19.5	Finland	27.8	Republic of Korea	41.6
6	Germany	19.3	Bulgaria	27.7	China, Taiwan Province of China	41.3
7	Austria	19.0	Croatia	26.8	China, Hong Kong SAR	40.6
8	Belgium	18.4	Greece	26.5	Italy	40.3
9	Switzerland	18.2	Slovenia	26.3	Singapore	40.1
10	Luxembourg	17.8	Latvia	26.2	Poland	39.5

Source: Adapted from United Nations (2017). World Population Prospects.

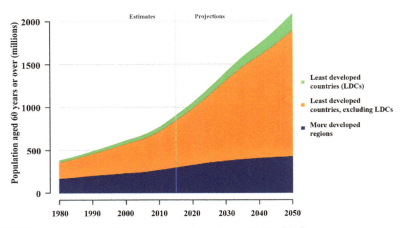

FIGURE 2.2   Projection of the aged population from 1980 to 2050.
Source: Reprinted from United Nations (2017). World Population Prospects.

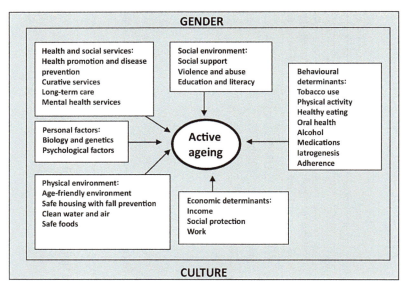

FIGURE 2.3   Active aging framework, adapted from WHO.
Source: https://apps.who.int/iris/bitstream/handle/10665/67215/WHO_NMH_NPH_02.8.pdf
;jsessionid=30F5701F715CAFEDAF13EA949E6F5E87?sequence=1

2.3   REVIEW OF THIS THEME

The concept of active aging was originally inspired by the work of Robert Havighurst on activity theory, arguing that people's well-being relies on them staying active as they age. Active aging is described as the process of optimizing opportunities for maintaining positive subjective well-being, enhancing the quality of life, and continued engagement in the labor force and community as one age.

Drawing on the theoretical model of the resource-based dynamic model (Wang et al., 2011), Leung, and Earl (2012) developed an assessment tool called the retirement resources inventory (RRI) to measure various domains of individual resources relevant to retirement well-being and retirement adjustment quality. These resources include physical, financial, social, emotional, cognitive, and motivational resources discussed in the paper of Wang et al. (2011). The results of Leung and Earl's (2012) demonstrate six resource domains that further categorized into three major groups: (1) Tangible resources include physical (e.g., perceived health and physical strength and illness) and financial resources (e.g., savings, investment, and perceived income adequacy); (2) Mental resources describe resources that contribute to one's mental capacity, including emotional (e.g., positive affect and emotional stability), cognitive (e.g., perceived control and memory capabilities), and motivational resources (e.g., perceived adaptability and flexibility in goal pursuit); and (3) Social resources refers to types of social support and quality of social interaction.

Today, retirement has become an active stage of life and people associate positive ideas with retirement. Globally, people expect their retirement income to come from government benefits, employer, and personal savings. It is critically important for older adults to have financial security for successful aging. It could be in the form of defined-contribution, defined-benefit pension plan, savings, portfolio of investments, medical expenses, etc. Defined-contribution and defined-benefit plans are often known as pension plans. In a defined-benefit plan, the employer maintains a retirement account, ponies up all the money and promises you to pay once you retire, whereas a defined-contribution plan requires you to put in your own money. People should take a personal responsibility of drafting their long-term financial security. Because of the widespread concern over the sustainability of government benefits, they should understand the importance of starting to save early in retirement savings and developing a written retirement strategy.

2.3.1   HEALTH

Cognition, health, and well-being are central to the idea of active aging, and the ways in which they change with age is of equal importance. Achieving successful aging depends on planning, saving, investment, and it also depends on staying in good health. To facilitate active aging and help older adults, their families, and health care staff improve quality of life in later years, it is important to study the within-person changes that occur on a daily basis across one's whole life. In 10 of the 15 countries surveyed globally, 43% respondents said that health in older age is a primary concern and 39% said that it is a minor concern. However, there is a mismatch found between concern about future health and current healthy behaviors. The retirees' advice younger generations to take current healthy measures to prepare for their older age. These advices include maintaining a balance between private life and professional life, exercise regularly, save money for future, bye old-age insurance, minimize stress, seek for leisure opportunities, spend more time on health prevention, etc.

2.3.2 BRIDGE EMPLOYMENT

Health is one of the most salient issues related to aging. Many factors have been investigated to determine health during retirements such as social support, socio-economic status, engagement in leisure activities, and employment status. Research highlights the increasing proportion of retirees who choose to engage in some form of employment rather than full retirement. The retirement trend has evolved from one time permanent process to a gradual exit. The continuation of employment after retirement is known as bridge employment. Bridge employment is defined as the pattern of labor force participation exhibited by older workers as they leave their career jobs and move toward complete labor force withdrawal (Shultz, 2003). It could be a self-employment, part-time job, activities involved in contribution to society, or temporary employment. Studied argued that financial pressure and good health status are the two most prominent motivators for older people to participate in bridge employment. Researchers contended that bridge employment (career field or different field) as compared to full retirement, is associated with better mental health and functional limitations (Zhan et al., 2009).

2.3.3 SOCIAL RESOURCES

Social resources can be broadly classified into three categories-sources of social interactions, quality of social interactions, and type of social support. The sources of social interactions are relationships with family, friends, spouse, and activities (leisure, group affiliations). The quality of values relationships and supportive interactions with spouse, family, and friends produce greater retirement satisfaction and positive psychological effect. The type of social support can be in the form of emotional, informational, and instrumental.

2.3.4 GOALS

Goals are an internal representation of desired states (Austin and Vancouver, 1996, p. 338). Individuals need to construct something positive for themselves, something they look forward to. Clearly indicative that having goals for later life is important, and having a path to achieve these goals is equally crucial. The research on goal is primarily focused on three qualitative aspects-works that focus on: *structure*, *process*, and *content*. Structural research, according to Austin and Vancouver, is concerned about categories of goals and how those goal categories are interrelated. In the retirement goals arena, a study by Hershey and Jacobs-Lawson (2009) provides support for the notion that retirement goals can be decomposed into higher-order categories of self-related retirement goals (be happy; be relaxed) and other-related goals (spend time with family, friends; volunteering). Process research describes how certain types of actions and behaviors are directed and motivated by goals. This line of work is inspired by studies conducted by Brougham and Walsh (2005, 2009), who demonstrated how goals for work and retirement determine the age at which one plans to leave the workforce. Content research, in turn, refers to the specific types of goals what an individual (or group of individuals) values.

There was an ethnographic study conducted by Savishinsky (2004) in which he compared and contrasted archetypes of retirement among American and Indian individuals. The findings revealed that Americans view retirement as a form of entitlement, or in other words, a reward for decades of hard work. After withdrawal from the workplace, they are interested in contributing back to the society (e.g., volunteerism). On the contrary, Indians believe that they have fulfilled all their obligations to family and society till retirement, and retirement is the final phase of their life. Post-retirement, they wish to pursue either spiritual life or a path of staying at home. The author further argued that beyond fundamentally different views of retirement in India and the U.S., the difference exists in the financial matters as well across countries. In India, older adults depend upon their children to support them economically during retirement, whereas in the U.S. retirees primarily rely on social security benefits, employer pensions, and personal savings as financial sources of support (World Economic Forum, 2013). In another study, it was found that Americans have more goals which are clear and concrete than their Indian counterparts (Gupta and Hershey, 2016) (refer to Table 2.2).

TABLE 2.2 Nine Coding Categories Used to Sort Retirement Goals

Exploration	Go on a world tour; See the whole of India; Travel around the world with my wife.
Attainment of Possessions	Buy a holiday home; Buy a big house; Build a home in Goa.
Leisure	Be a part of recreational activities; Pursue my hobbies; Read.
Self-related	Be happy; Have fun; Keep working to stay busy.
Contact with Others	Spend time with family; Ensure children are settled; Stay happy with family and friends.
Contributions to Others	Teach part-time; Work towards the education of poor children; Be philanthropic like Warren Buffet.
Spiritual-Transcendental	Get closer to God; Have a spiritual connection; Improve my knowledge of Hinduism.
Financial Stability	Be financially independent; Have sufficient money to retire; Have a big bank balance.
Other	Open my own restaurant; Retire at age 45; Fulfill my family dream.

Examples Shown are Drawn from the Indian Data Set.
Source: Reprinted with permission from Gupta and Hershey (2016). ©2016 Springer.

2.3.5 RETIREMENT ASPIRATIONS

Retirement has become an important phase of life about which people associate positive ideas and thoughts. They aspire to be socially connected, participate in the societal work, and remain secure and healthy. Globally, a large proportion of older people have positive associations with retirement than younger people aged 18 to 24. The most widely held retirement aspirations among respondents are traveling, spending quality time with friends and family, contributing to the community, leisure, freedom, paid work, enjoyment to mention a few.

2.3.6 WORKPLACE INTERGENERATIONAL CLIMATE (WIC)

WIC is an evolving phenomenon at the workplace that assesses and captures the shared belief of employees about employees from other generations. A burgeoning age-diverse workforce presents both opportunities and challenges for the companies. With Traditionalists (born before 1945), Baby Boomers (1946–1964), Generation X (1965–1980), and Generation Y or Millennials (born after 1980), members of four different generations work together in some companies. People from different generations carry different perspectives, social trends, style of communication, culture, and when they interact, these differences may resort to stereotyping of their co-workers (Bal, Reiss, Rudolph, and Baltes, 2011; Fritzsche and Marcus, 2013; Ng and Feldman, 2012; Walker, 1999). Workers of an age group can be found to interact more frequently with people of their generation set and disassociate themselves from others that recourse to the formation of subgroups at the workplace. Palmore (1972, p. 4) calls it ageism which is prejudice or discrimination against or in favor of an age group. Racism, sexism, and ageism are the three primary interpersonal and socio-cultural categorizations (Fiske, 1998; Kite, Deaux, and Miele, 1991). It is quite atypical that the research on age-based prejudice remains surprisingly scantly-examined in comparison to the other two "isms" of society, despite the salience of age in business-related and strategic matters.

Research on the effects of ageism on workers demonstrates the fact that people who encounter negative age-related stereotypes experience detrimental psychological and physiological changes (Levy, 1996, 2000, 2003; Levy, Ashman, and Dror, 1999). The increasing plausibility for multiple generations of workers working together on a daily basis, calls for an urgent need for companies to lessen the impact of ageism in its immediate environment. Traditionally, the literature on ageism has focused on stigmatization of old people and its devastating effects on their attitudes at work (Gaillard and Desmette, 2010). This does not circumscribe the dynamics of modern workplace climate where people of multiple generations contact and work with each other.

Diversity, if managed appropriately, may facilitate direct positive benefits to individuals and at large, to organizations. There is handful of socio-contextual factors introduced in the literature for age-diverse work environments fostering improved organizational performances (Iweins et al., 2013). More understanding is required in this domain; workplace intergenerational climate (WIC) can affect job attitudes and behavior, which should be explored.

2.3.7 ACTIVE AGING INDEX (AAI)

Active aging is described as a situation where people wish to continue working in the formal labor market and contribute through other unpaid productive activities (such as care provision to family members and volunteering) and live healthy, independent, and secure lives as they age (refer Figure 2.4). The AAI measurement would fall within the following four domains:

- Contributions through paid activities: Employment;
- Contributions through unpaid productive activities: participation in society;
- Independent, healthy, and secure living;
- Capacity and enabling environment for active aging.

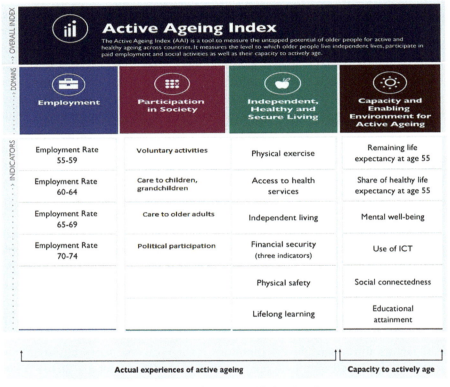

FIGURE 2.4 Domains and indicators of aggregated index (AAI).
Source: Reprinted from "Active Aging Index project. https://statswiki.unece.org/display/AAI/Active+Ageing+Index+Home."

This recommendation is inspired by Sen's capability focused conceptual framework, in which capabilities are defined as substantive opportunities and empowerments to enhance well-being and quality of life, such as life expectancy, health, education, social participation and so forth (see, e.g., Sen,

Active Aging: Engaging the Aged 45

1985, 1993, 2009) (refer Figure 2.5). This domain is therefore considered as measuring:

- Human assets by outcome indicators such as remaining life expectancy;
- Health capital with the healthy life expectancy and mental well-being indicators;
- Human capital aspects by educational attainment indicator.

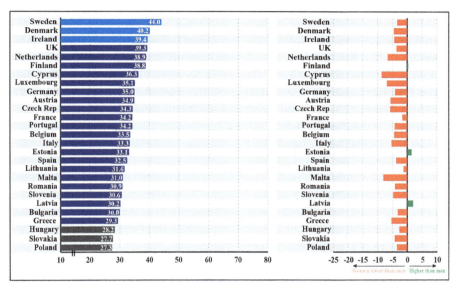

FIGURE 2.5 Active aging index (all domains together: overall) ranking of countries for total population and for differences between men and women.

Source: Reprinted from Active Aging Index Report, (2012). https://www.unece.org/fileadmin/DAM/pau/age/wg6/Presentations/ECV_AAI.pdf

2.3.8 CROSS-CULTURAL ASPECT

The phenomenon of active aging has been endorsed differently across cultures and countries. It is dependent upon prevalent demographic trends, data sources, and indicators. The framework used in one country is not directly generalizable to other contexts. Active aging index (AAI) developed by the experts from the European Center Vienna is a multifaceted global index comprising of 22 indicators and captures the potential of older people on four crucial aspects: (1) participation in society, (2) employment, (3) independent, healthy, and secure life, (4) capacity for active aging

(Varlamova et al., 2017). Spain is ranked 17th among the 37 EU countries in the AAI (Zaidi et al., 2012). Data sources used to arrive at Spanish AAI are broadly similar to those used to calculate AAI in Europe (Rodriguez-Rodriguez et al., 2017). Russia has a comparatively higher score on AAI results with other countries and stands 18th in the ranking. As conditions and individual well-being improves in reality by developing policies and practices, there remains sufficient scope to improve the active aging of the adult population (Rodriguez-Rodriguez et al., 2017). For instance, the Maltese government in the recent past has reformed the public policy related to aging and has initiated the implementation process of the National Strategic Policy for Active Aging, the Minimum Standards for Care Homes for Older Persons, and National Dementia Strategy (Formosa, 2017). Some of the European countries emphasize on the importance of addressing the gendered nature of aging and strategize different range of policies for men and women (Foster and Walker, 2013).

The EU-Commission has fixed two targets for 2010:

1. The Stockholm targets of 2001, which intend to reach 50% higher employment of elder people between 55 and 64; and
2. The Barcelona targets of 2002, which intend to progressively augment of 5 years the age of retirement in order to prolong the professional life of elder people.

2.3.9 ENGAGEMENT OF AGING WORKFORCE

Diversity in the workforce, longer life-spans, lower retirement age are just few of the aspects that have led industry to view the concept of employee engagement in the context of the aging workforce with greater interest.

Engaging employees is defined as when employees feel positive emotions toward work find their work personally meaningful, consider their workload to be manageable, and have hope about the future of their work (Nelson and Simmons, 2003). Engaging employees is one of the top five most important challenges for management, according to a survey of 656 CEOs across the globe (Wah, 1999). Organizations with higher engagement levels tend to have lower employee turnover, higher productivity, shareholder returns, and better financial performance (Baumruk, 2006).

While the above definition may hold good in general, the aims of engaging an aging workforce differ from the regular engagement activities a firm undertakes. To understand why the difference exists, it is necessary to first define 'aging workforce.' An aging workforce consists of people aged 45 and above. Much against the myth that the aging workforce maybe unfit to work, research has shown that people in their 60s and 70s have a desire to remain in the labor force and make meaningful contributions (Smyer and Pitt-Catsouphes, 2007). Many employers consider their older workers to be a valuable asset: positive aspects attributed to them include a strong work ethic, reliability, loyalty, business experience, institutional memory, and specialized skills (Banner, 2011).

To facilitate engagement in aging we need to recognize certain aspects of the aging workforce like changes in their abilities, beliefs, skills, and goals. Let's take for example 'goals.' On an average, with increasing age, people focus less on growth and development goals, and more on maintenance goals, prevention of losses, and emotional meaningfulness (Zacher et al., 2017). So, does this mean that the employee is not or cannot be engaged?

2.3.10 IMPORTANCE OF ENGAGEMENT IN ACTIVE AGING

In the work context, active aging means that as workers age they maintain or improve:

1. Their physical, mental, and social well-being;
2. Continue to show high levels of work engagement and performance; and
3. Experience fair treatment and employment security.

Therefore, older workers are not only able and motivated to work past traditional retirement ages; they also continue to be happy and productive members of the workforce (Zacher et al., 2017). The organization must acknowledge that every employee is a valuable and rare resource that cannot be replicated. The strength of the aging workforce lies in the fact that they have years of experience that the younger generation do not have. This experience has added skills, knowledge, and abilities that are not available in books or company manuals. These "silver skills," as they are often referred to, can be harnessed for the growth of the organization.

Thus, an organization must clearly articulate engagement strategies for the aging workforce to ensure their contribution towards organizational growth. One definition that contributes to this line of thought is given by Macey and Schneider (2008) who split engagement into three areas: *'trait engagement'* as an inclination to see the world from a vantage point, and this is reflected in the individual's *'state engagement'* which leads to *'behavioral engagement.'*

When it comes to engaging the aging workforce, the HR strategy may take into consideration the following:

- Ensure that the work is meaningful and purposeful so that the employee may take pride in what they do-true for any employees, how is it different for aging employees.
- Provide equal opportunities for training and development-though training needs would be different.
- Building an age-friendly organizational culture that values age diversity and the strengths of all workers independent of their age (Zacher et al., 2017).
- Age-inclusive career management interventions where workers of all ages have equal opportunity to develop their knowledge and skills.

2.3.11   CHALLENGES OF ACTIVE AGING

Aging poses many challenges for individuals, employers, and society at large. Some of them are discussed below:

1. **Disability Issue:** With increase in proportion of older adults, disability becomes a major challenge. Disability impacts the quality of life and makes older adults dependent. Its life expectancy is complemented with decreased disability, and then older people will be able to live an autonomous healthy life and put less pressure on the social system.
2. **Age Discrimination-Happens in Promotion and Training, Appraisals:** The complexity of the age discrimination issue is undeniable. As the young population is facing with higher unemployment, there is an unseen pressure on older workers to retire from the workplace. There has been considerable research done on ageism against older workers and the physiological and psychological consequences of the same. Age discrimination influences functional processes such as promotions, performance appraisals, training, and is evident across different cultures and contexts.
3. **Mandatory Retirement Age:** Mandatory retirement age restricts the direct participation of older adults in the labor force. The US organizations abide by Age Discrimination in Employment Act (ADEA) that eliminated mandatory retirement age options.

4. **Pension Crisis:** Mandatory retirement age limits the time of long-term savings as people age. Moreover, increasing concerns regarding sustainability of employer-related retirement benefits adds to the pension issue. This economic crisis may compel people to engage in paid employment whether they wish to continue working or not.
5. **Gender Issues:** Gendered structures—in society, in working life, and in the home—expose women to many obstacles in their career course. Though, there are evidences of men's ascending participation in domestic work in most countries since the 1960s, yet women carry the majority of household responsibility ranging from childcare to elderly care. The implications of gender discrimination have some visible consequences in the career patterns of male and females. During their lives, women are seen more often interrupting their careers than do men to tend to family care responsibilities. The notion that female employees have a greater inherent tendency to give priority to family over work shapes the professional lives of women and affects women's career scopes and salaries. Thus, it is important to study how women plan for retirement as well as ensure adequate support during retirement, which may differ due to certain life decisions they make. As women are more vulnerable to bad health, poor financial security, it becomes more crucial to understand the issues they face and take corrective measures.
6. **Trade Union Perspective:** Trade unions have provided their invariable support to help people work longer, but several factors like organizational policies, health condition, and infrastructure do not facilitate longer working. Trade unions have reported that organizations adopt different means to drive older workforce out. By empowering trade unions and taking actions on their reports, we may be able to support active aging workforce participation more.

2.3.12 RETIREMENT ADJUSTMENT

Retirement, as defined by the dictionary, refers to the practice of leaving one's job or ceasing to work after reaching a certain age. It is those 'golden years' one has worked towards through their lifetime. While this practice has been around for at least two centuries, when the eventuality of retirement comes, many of the individuals are plagued with a feeling of anxiety and insecurity. Why does this happen?

Many individuals feel that financial planning is the panacea to the apprehension towards retirement. While financial security is very important, equally significant is the emotional change that retirement brings with it. The change is in relationships, socializing, and having ample time on one's hands, just to name a few.

Post-retirement individuals who have not planned well may find it burdensome to adjust to the aspects mentioned above and hence comes the area of retirement adjustment. Retirement adjustment in simple terms refers to how one will cope with his/her life after retiring from active work life. Even though retirement is termed as a transition phase in adult life, the question of how well individuals adjust to retirement has been a focus of interest to researchers as well as to the popular media (Wang and Shultz, 2010). Understanding retirement adjustment gives us insight into how individuals simultaneously adjust to internal (i.e., physical, and psychological aging) and external (i.e., lifestyle, and societal norms) challenges in their life after retirement. Over the years, retirement adjustment has moved to being studied as a crisis event to having beneficial effects also.

While retirement adjustment can be challenging, having knowledge of some of the underlying theories is vital. The theories are as stated below:

- **Role Theory:** Understanding the role an individual has played in their working career as this would help predict certain behaviors. For example, A Security Guard. He will always be on the alert and any small disturbance will immediately grab his attention.
- **Continuity Theory:** Recognizing life patterns such does the individual get stressed easily or have they had a turbulent career and so on and how did they cope with the situation. This will help understand how they will cope with the adjustment to a retired life
- **Stage Theory:** Individuals must recognize and ease themselves through the various stages of retirement as put forth by this theory. The stages are as follows: early in the transition to retirement, retirees experience a honeymoon stage, in which they may feel more energetic and satisfied as they pursue new activities and roles. Later on, retirees may experience a disenchantment stage, in which they realize that they have fewer resources and/or had unrealistic expectations about retirement. As time passes, retirees enter a reorientation stage, during which they re-evaluate their life status, accept limitations, and focus on further adjustment to retirement. Eventually, retirees enter the stability stage, settling into a predictable daily life pattern.

While research is on in this area to understand what factors impact (positively and negatively) retirement adjustment, a comprehensive table depicted by Wang et al. (2011) gives us a snapshot of the factors that impact retirement adjustment. An excerpt of the same is given in Table 2.3.

TABLE 2.3 Summary of Variables That Influence Retirement Adjust Quality

Predictor Category	Variable	Effect
Individual attributes	Physical health	+
	Mental health	+
	Financial status	+
	Physical health decline	–
Preretirement job-related variables	Work stress	+
	Job demands	+
	Job challenges	+
	Job dissatisfaction	+
	Unemployment before retirement	+
	Work role identity	–
Family-related variables	Marital status (married vs. single/widowed)	+
	Spouse working status (working vs. not)	–
	Marital quality	+
	Number of dependent children	–
	Losing a partner during the transition	–
Retirement transition-related variables	Voluntariness of the retirement	+
	Retirement planning	+
	Retiring earlier than expected	–
	Retiring for health care reasons	–
	Retiring to do other things	+
	Retiring to receive financial incentives	+
Postretirement activities	Bridge employment	+
	Volunteer work	+
	Leisure activities	+
	Anxiety associated with social activities	–

Note: Plus sign (+) denotes a positive effect on retirement adjustment quality, and minus sign (–) denotes a negative effect on retirement adjustment quality.

Source: Reprinted with permission from Wang, Henkens, and Van Solinge, © American Psychology Association.

Another side of the picture is that with the growing life expectancies and better medical facilities, the working span of an individual has increased. Looking at the scenario from the employer's perspective, it can be a challenge.

The dilemma is whether to allow the aging workforce to continue working or let them go and be replaced by younger blood. This has many a time led to discrimination and disharmony in the workforce. Many employees feel compelled to retire earlier than expected. This leads to a different type of stress that impacts the retirement adjustment. What can employers do to alleviate this issue? As quoted by OECD in 2011: To deal with the dilemmas of a welfare state in an aging society, the consensus view among policymakers appears to be that credible and sustainable pension plans can only be attained by raising the statutory retirement age, reducing benefits, and shifting a considerable amount of financial risk onto the shoulders of individual citizens (Gerontologist, 2017).

The factors identified are expected to aid and hamper retirement adjustment. There are certain challenges and issues also, faced by the retiring working force. It would be prudent to plan for 'the golden years' keeping these factors in mind, thus leading to the well-being of the individual post-retirement.

2.4 IMPLICATIONS

It is clear from the above discussion that demographic trend is changing and the upsurge in older population is affecting the labor market. Organizations need to revisit their policies and practices not only to accommodate older people but also to ensure their contribution in socio-economic outputs.

Few activities adopted by different countries and companies around the world are discussed below which organizations may adopt to encourage active aging.

More than 72% of Danish companies realize the labor market concerns with shrinking population and increased aged population. But a small number of companies make policy amendments regarding the same. Having a written or unwritten policy for older personnel is a need of the hour, and in 2005, this was reported by approximately 45% of companies (Jensen and Møberg, 2012). Some of the most popular practices adopted by organizations in Denmark are part-time retirement, reduction of working time before retirement, training plans for older workers, early retirement schemes, possibilities of extra leave for older workers, decreasing the workload for older workers, reduction in task and salary (demotion), ergonomic measures, an age limit for irregular work/shift work, promoting internal job mobility, continuous career development and flexible working hours.

Taiwan is conscious of its aging population and has launched a series of senior service programs and senior health promotion measures since

1998. Subsequently, they have worked on the new labor pension system, senior education, and community care centers. Further, they adopted the 10-year Long-term Care Plan as its flagship program and promoted national pension legislation. Amendments to the Labor Insurance Act annuitized old-age benefits. The government is also working on friendly care services for the elderly and age-friendly cities (Lin, Chen, and Cheng, 2014). In Taiwan, people ranked active aging activities on basis of their importance: physical health, dignity, mental health, social interaction, education, environmental accessibility, volunteering, religious activities, being respected, economic security, community participation, senior employment, personal safety, sense of belonging to society and political participation.

In a study in Indonesia, they found that there are three pillars of active aging: participation, security, and, functional health. It was found that Indonesian older people actively participate in community building. The source of funds for financial security are mostly earned by men and transferred to women. They found that even though women outlive their male counterpart, they do not enjoy better health in the later stage. The economic dependence of women on men, along with poor health makes women vulnerable in old age. In Indonesia, the joint family structure supports older people who continue to stay with their children. The intergenerational transfer of financial and domestic support remains strong in the society, which also helps in reducing economic hardship and social isolation for older people (Arifin, Braun, and Hogervorst, 2012).

There is a need to make changes in the organizational policies to help the aged workforce remain active, some of the suggestions based on the readings include:

- Creating an ergonomically friendly work environment.
- Managing health and offering preventative healthcare:
  - Yoga sessions, or health programs, nutritionist, mindfulness (meditation).
- Changing working patterns and responding to the needs of associates:
  - Flexi-time, more breaks, work from home, reduced work hours, sick leaves.
- Training to keep up the skills of the workforce.
- Supporting associates who want to retire.
- Workplace intergenerational.
- Collegiality.

2.5 CONCLUSION

Individuals should be able to maintain their independence at all ages and stages, for this, planning is extremely important in regard to housing during retirement, medical decision making, and being conscious about health. Planning should also include activities that individuals would want to indulge in or novel or adventurous things they have always wanted to do. At large, the organizations need to emphasize on some core issues like employment opportunities for elderly, and avoiding age discrimination, abandoning mandatory retirement age, opportunities for part-time or partial employment. Organizations should strive for social engagement for the aging cohort and involve them in decision making. There is also a grave requirement for intergenerational transmission and inheritance rather than generational transfers or replacement (Walker, 2002).

"Aging is not 'lost youth,' but a new stage of opportunity and strength."

In the subsequent section, cases covering the above concept have been provided (refer Appendix 2.1 and 2.2 for plausible case discussion).

2.6 CASE 1: BRINGING BACK THE WORKING ALLURE

Susan, the HR Head at Creative Minds, an advertising agency, was putting together her thoughts for the exit interview she was about to conduct. She thought to herself that she must somehow convince Roger to reconsider his decision about leaving the firm.

Roger, the Creative Head, had joined the firm a decade ago. He was 44 years then. A '*go getter*' and very passionate about his work, Roger had enabled the firm grab and retain many big clients. Susan had always admired the way he worked and motivated his team to push their creativity way beyond their limits. Though not into client servicing, Roger always managed to build a good rapport with every client. The clock struck 10 and prompt as always Roger stepped into Susan's cabin.

Roger: "Good Morning Su, shoot away, I'm ready to answer."

Susan smiles: "Hi Roger, high spirited as always."

After an exchange of pleasantries, the exit interview got underway. Once the formalities were completed, Roger said: "Susan I would like to talk to you more about certain aspects that I have noticed here at the firm."

Susan: "Tell me Roger. Even I have been meaning to ask you about a few things."

Roger: "I have been in this firm for over 10 years. When I first joined, the atmosphere was so electrifying and stimulating, so much so that even a left-brainer would become creative. Over the past two years, I have noticed that "stimulation' is lacking. Though we are put through many training programs, given freedom to express our ideas and, not to forget, we are paid well compared to our counterparts, something is amiss."

Susan nodded: "Yes, go on Roger."

Roger: "I don't want to take names at this point, but Susan you really need to consider this ASAP. Let me help you a little by pointing out a few issues.

- Deadlines are not being met, inaccuracies have increased, there is little creativity and lots of procrastination.
- People are not contributing new ideas during meetings or even asking questions. They no longer seek ways to improve. Gone are the creativity, initiative, and willingness to help team members.
- Clock watching is a common phenomenon that has emerged. No one is ready to work late anymore.
- I don't mean to begrudge Jessi from the Market Research team, but when is the last time she shared an article of interest about the company or marketplace trends? When is the last time she shared anything at all? Curiosity is a good sign that an employee cares about her work and the organization. If you encourage learning and growth and employees don't share your enthusiasm, it's time to take a closer look.
- The coffee room that used to be abuzz with enthusiasm has now become more of a complaining cell. My colleagues are constantly complaining about something or the other. I no longer go to the lounge. It propagates negativity.

Roger looked at his watch and realized he had taken up too much time. He looked at Susan and said:

"Su I spoke my mind because I care. Most of the people here joined around the same time I did. The best part then was that we were all of the same age group give or take a year or two. We did so well then but something is lacking now. Maybe I should have informed you earlier …. But like they say *'Better late than never.'* Give it some thought and see what you can do to reignite Creative Minds."

After Roger left, Susan sat down and began to wonder where she should start and how to take the process forward. Roger had given her some good inputs and it was time for some real in-house research and strategy planning. She began making notes of the meeting with Roger.

Note: The employee engagement plans, across the board, at creative minds were:

- Long service awards to engage and retain loyal employees.
- Generate referrals and gratify.
- Birthday/anniversary gifts, festival gifts, referral gifts.
- Rewards and recognition programs to engage employees.
- Employee feedback surveys-to get information and act on what employees are thinking and feeling.
- Involving employees in company updates.

Critical Questions for the Case:

- Do you think the issues mentioned by Roger are related to employee engagement? Discuss.
- What are your thoughts on the employee engagement strategies being employed at Creative Minds?

2.7 CASE 2: LEVERAGING INTERGENERATIONAL TALENT

Anytime Bank, began its operations 2 decades ago, and had a well-established reputation in banking and insurance industry. The firm had spared no expense and had hired the "best and the brightest," graduates from top universities in the country and other experienced people for higher positions. They had very clear hiring policies and would settle for nothing less than the best. The average age of the 30 full-time employees, at present, ranged between 45 years to 55 years.

After having been in business for 20 years, the top brass decided to expand and automate operations. This was the result of rapid technological changes and the need to beat the ever-rising competition. This move called for hiring tech-savvy employees and training old hands with the new system.

Sam, the HR Head of the firm, knew that this operational change and new hiring would bring with it a spate of challenges, especially as the firm's workforce would undergo a generational shift from having primarily Baby Boomers and Gen X cohort to including Gen Y and/or even Gen Z cohort. Before calling in the consultant trainers, Sam, and his team needed to profile the new and old recruits and ensure that the training, in all aspects, from imparting the vision, mission, culture of Anytime Bank right down to the operational level, be delivered in the best possible manner. Some of the requirements that the top management had stressed that they would not short change on were: Loyalty, dedication, education, competitiveness, and service.

When the expansion and automation plans were informed to the employees, an increasing number of bankers in the Baby Boomers generation began to show signs of leaving the firm. Their reasons were that they had interests and commitments outside of work and were unwilling to forego having a personal life as they felt that the changes would bring in more workload. But this group had been with the bank since its inception and was a vital part of the future.

One day during tea, Sam overheard a conversation that made him take a closer look at the profile of the current and prospective employees. Tom, the 40-year-old Marketing Head, full of enthusiasm, says, "Hi Larry. How are things going?"

Larry, replies, "What can get worse for a 58-year-old like me? Bad enough, I have had to deal with your so-called tech-savvy generation, now I hear that the selfie clan seems to be coming in too." He went on cribbing saying how much he and his peers had to not only forgo but also adjust to the whims and fancies of the younger generation. Larry remarked vehemently, "If the company brings in a bunch of tweeters, bloggers, and the like and expect us to imbibe loyalty into them, they can forget about it. Cannot deal with the ruffians today."

Tom listened wondering why Larry was so upset. Tom, in fact, liked the working environment, though at times he faced communication problems with his team that comprised of five executives in the ages of 25 to 27 years of age. Many a time Tom was conservative in his decisions but noticed that his team was ready to take risks.

Soon after listening to this conversation, Sam called for a meeting with his team. He informed his team of the training program for the employees, the requirements of the management and hinted about the conversation he had just overheard. The 170 new recruits would be joining in a month's time. They needed to do something and do it fast. He was aware that the intergenerational cohorts needed and expected tailored packages in terms of training, benefits, shifts, pay packages and the list went on. Sam realized that the firm needed to make changes to address the generational issues that would otherwise pose a threat to its long-term viability. Sam and his team began to collate data to present to the trainers. Before assigning the roles, all employees had to go through a common induction program. The data at his disposal is given below:

Age (in Years)	Old Employees	New Recruits
55 and above	5	16
40–54	9	50
30–39	10	60
24–29	6	44

Critical Questions for the Case:

- How do you think Susan should go about understanding and engaging the aging workforce?
- How do you think Sam should prioritize his work?
- What engagement interventions would you suggest for the current set of employees in Case 2?

APPENDIX 2.1: CASE DISCUSSION FOR CASE 1

Synopsis

The case is a description of what Roger feels as he leaves Creative Minds. After his exit interview with Susan, the Head of HR, Roger apprises Susan of the aspects that he feels are unfavorable to the organization. During his discussion, he refers to the age group of the people he is referring to. Roger mentions that the issues need to be addressed ASAP. Susan begins making notes of her meeting with Roger and wonders how the issues that were brought to the front need to be addressed.

Teaching Objectives

The objective of this case is:

- Introduce the students to some of the behaviors that result when the employee is not engaged or actively disengaged.
- Identify the crux of the problem-is the cohort being discussed an aging workforce?
- Relook the engagement activities being adopted by the firm.
- Discuss what are the interventions required keeping in mind the age group and needs of the cohort.

Teaching Strategy

After the introductory remarks, the students are requested to form groups of five or six sitting in the same area of the classroom and asked to discuss the answers to the assignment questions. The discussions should last for 10 minutes. During this period, the instructor should walk around and listen to the discussions without getting involved. The instructor may identify two or three groups who may be called on later.

The instructor then reconvenes the class and leads the discussion that centers around the assignment questions. This discussion can take up to 55 minutes.

Discussion

Do you think the issues mentioned by Roger are related to employee engagement? Discuss. (10 minutes).

To begin the discussion, ask for a volunteer from one of the groups identified earlier or pick a volunteer at random. Ask the following two questions for which the answer will hopefully be "yes" if not you may want to ask why.

- Did everyone in the group get a chance to contribute to the discussion and share ideas?
- Did you learn anything from the group discussion which you had not thought of earlier?

The instructor can probably ask these questions to the other groups identified earlier. The purpose of the questions is to make participants understand the importance of learning in small groups and why it is important to participate in the same.

After the short opening, the instructor should ask one of those who participated earlier whether the issues mentioned by Roger are related to employee engagement. The answers may fall under one of the following options:

A. Yes, all are engagement related issues.
B. No. Only a few are engagement related issues.

If the answer is 'A,' ask the class if they too feel the same way and go ahead. Take up each of the issues mentioned by Roger and have a brief discussion on the same.

If the answer is 'B,' then ask which issues are related to engagement and which are not and why.

Ideally, the answer must be 'A.'

Start making a list of the points discussed on the board.

What are your thoughts on the employee engagement strategies being employed at Creative Minds? (10 minutes)

The instructor must now call upon one of the groups identified earlier to discuss the employee engagement strategies being employed at Creative Minds. During the discussion, ask if the strategies are fit for all employees "across the board." The reason for this trigger question is for the following reasons:

- To form the setting for the third assignment question.
- To see if the students could figure the point at which the age of the employees was mentioned (As this is crucial to the case).
- To relook whether the engagement strategies must change for an aging workforce.

How do you think Susan should go about understanding and engaging the aging workforce? (15 Minutes)

This can be a rather open-ended discussion. The goal of this discussion is to get the students to be able to identify the age group of the cohort that Roger is referring to (aging workforce). Students are required to discuss whether the current engagement practices are relevant to the workforce under review. The instructor may call at random any student/s to voice their opinion what should be the course of action to be adopted by Susan. The discussion may take on the following points:

- Discuss whether the current engagement practices are good enough across the board.
- What are the points to be considered while creating strategies for engaging an aging workforce?
- List the steps that Susan must take to deal with the issues on hand.

Summary (10 Minutes)

At this point, it is necessary to highlight the following and wind up the discussion:

- Understanding the engagement issues involving the aging workforce.
- Understand the change in motivation triggers and age-related goals.

- Stressing on how Susan can overcome the problems by employing some of the following guidelines:
  - Ensure that the work is meaningful and purposeful so that the employee may take pride in what they do-true for any employees, how is it different for aging employees.
  - Provide equal opportunities for training and development-though training needs would be different.
  - Building an age-friendly organizational culture that values age diversity and the strengths of all workers independent of their age.
  - Age-inclusive career management interventions where workers of all ages have equal opportunity to develop their knowledge and skills.

APPENDIX 2.2: CASE DISCUSSION FOR CASE 2

Synopsis

The case is about a 20-year-old bank that is planning to go in for expansion and automation of its operations. Sam, the HR Head of the firm, knew that this operational change and new hiring would bring with it a spate of challenges, especially as the firm's workforce would undergo a generational shift. On overhearing a conversation between a 58-year-old and a 40-year-old employee, Sam got a clear idea of the challenges ahead. Sam realized that the firm needed to make changes to address the generational issues that would otherwise pose a threat to its long-term viability. Sam and his team began to collate data to present to the trainers. Before assigning the roles, all employees had to go through a common induction program.

Teaching Objectives

The objective of this case is:
- Introduce the students to different generations present in the workforce.
- Identify the crux of the problem-challenges posed by the intergenerational workforce.
- Discuss what are the interventions required keeping in mind the traits and needs of each generation.

Teaching Strategy

- After the introductory remarks, the students are requested to form groups of 5 or 6 sitting in the same area of the classroom and asked to discuss the answers to the assignment questions. The discussions should last for 10 minutes. During this period, the instructor should walk around and listen to the discussions without getting involved. The instructor may identify two or three groups who may be called on later.
- The instructor then reconvenes the class and leads the discussion that centers around the assignment questions. This discussion can take up to 45 minutes.

Discussion

Q.1: How do you think Sam should prioritize his work?

Sam's first step would be to make a broad classification of the employees (30 old employees and 170 new recruits) in terms of characteristics (Note 3), and by examining the organization's demographics, culture, and management practices.

1. **Generational Demographics:**
   - What is the generational mix?
   - Are there many Traditionalists and Boomers versus Gen-X, Gen-Y, and Gen-Z?
   - What is the distribution in management ranks and in core business functions such as engineering or sales?

   Whatever the mix, creating a generational profile of the organization is a critical first step.

2. **Culture:** Do the words that employees and customers use to describe the organization match the ones it uses to describe itself or is there a mismatch. For example, the firm may describe itself as an "innovative, industry leader and standard-setter" but the employees describe it as "risk adverse and bureaucratic," there is an obvious disconnect. There needs to be a common culture across all generations as the generational dynamics will affect teamwork, productivity, and the firm's growth.

3. **Management Practices and Leadership:** Managing a multi-generational workforce requires the ability to lead change and manage individual and organizational performance. The emerging leadership model is one that focuses on competence, character, and collaboration. It reflects a concern for people and the community as well as a concern for profit. In this model, the leader's behavior is consistent with the organization's stated vision, values, and mission. This leader shares power, collaborates, and works to build and maintain trust in ways that are different from past approaches that emphasized position, power, and authority. *(Recent studies by the Gallup Group and the Leadership Council document that employees do not leave companies; they leave managers. This requires a renewed emphasis on management training and development).*

Next, he needs to come up with a strong engagement strategy at the earliest as the firm could not afford to lose the employees who were the building blocks of the organization.

Third, he had to ensure the training would address the age gap issues of the workforce as their values, drivers, and needs of these different generations were worlds apart. Essentials that a multi-generational training program must include are:

- Defining characteristics of generations including generational values and behaviors *(Note 1, 2 and 3)*.
- Identifying common stereotypes and unconscious biases.
- Recognizing and responding to cross-generational dynamics.
- Understanding emerging trends in demographics.
- Using tools and techniques designed to improve communication, maintain respect, and enhance working relationships.

Q.2: What engagement interventions would you suggest for the current set of employees?

Answer: Note 4-Engagement options

Notes: The notes given below are prepared to assist in better understanding and discussion with reference to case 2.

Note 1: The birth years for each generation

- Gen Z, iGen, or Centennials: Born 1996 and later;
- Millennials or Gen Y: Born 1977 to 1995;
- Generation X: Born 1965 to 1976;
- Baby Boomers: Born 1946 to 1964;
- Traditionalists or Silent Generation: Born 1945 and before.

Note 2: How to leverage the multigenerational workforce

The older workers can teach younger workers about:

- Coping with difficult times;
- Loyalty;
- Experience;
- Interpersonal skills;
- Regrets and how to get over them;
- Independence.

The younger workers can also contribute by the teaching the older workers about:

- New technology;
- Diversity;
- Risk-taking;
- Balancing work/life issues;
- Fulfilling dreams.

Note 3: Characteristics of each generation

The table given below gives an insight into the characteristics, needs, motivators, and engagement intervention requirements specific to each generation.

ACKNOWLEDGMENT

The authors wish to thank Dr. Douglas A. Hershey, Professor, Oklahoma State University for his constructive feedback on this chapter.

	Traditionalists	Baby Boomers	Generation X	Millennials	Generation Z
Traits	– Patriotic and loyal – Fiscally conservative – Faith in institutions	– Nuclear families – Equal rights – Optimism, anything is possible	– Latchkey kids – Diverse, non-traditional	– Flexible and self-reliant – Personal freedom and equality	– Achievement-oriented – Civic duty – Diversity and socially conscious
What they want from work	– Established rules – Strong leadership authority – Employer loyalty	– Employer loyalty – Company commitment – Documented feedback	– Immediate and continuous feedback – Work/life balance – People (not Company) loyalty	– Continuous learning – Fun at work – Workplace location and flexible hours – Feedback – Dedicated cause	– High expectations of leadership – Expect to work for success – Return to employer loyalty
How to motivate them	– Recognize experience as valued and respected – Acknowledge tacit and historical knowledge	– Show appreciation for the quality of work – Recognize work as valued and needed	– Empower and encourage innovation – Provide challenges and opportunity	– Provide opportunities to network – Provide opportunities to work with people outside the team	– Show "dream" positions of aspiration – Help them work toward these opportunities
How to keep them engaged	– Provide ways to get results daily – Take time to provide chances to build skills – Give loyalty and support	– Help become comfortable with technology – Specific points when providing feedback – Be direct in messaging	– Provide competitive pay – Don't micromanage – Establish strong vision and good reputation	– Embrace technology – Provide rapid advancement opportunities – Offer flexibility in work locations and hours	– Embrace diversity – Establish active community in and around workplace – Support environment sustainability

KEYWORDS

- active aging
- cross-cultural
- elder's well-being
- engagement
- engagement of aging workforce
- retirement-planning
- silver skills
- social aging phenomenon

REFERENCES

Active Ageing, 2019, Center Services. Retrieved from: https://activeaging.org/special-events-programs/center-services/ (accessed on 24 February 2020).

Arifin, E. N., Braun, K. L., & Hogervorst, E., (2012). Three pillars of active aging in Indonesia. *Asian Population Studies*, *8*(2), 207–230.

Asghar, Z., Katrin, G., Maria, M. H., Orsolya, L., Bernd, M., Ricardo, R., Andrea, S., Pieter, V., & Eszter, Z., (2012). *Active Aging Index Report: Concept, Methodology, and Final Results*. European Center Vienna, UNECE Grant No: ECE/GC/2012/003.

Austin, J. T., & Vancouver, J. B., (1996). Goal constructs in psychology: Structure, process, and content. *Psychological Bulletin*, *120*(3), 338.

Bal, A. C., Reiss, A. E., Rudolph, C. W., & Baltes, B. B., (2011). Examining positive and negative perceptions of older workers: A meta-analysis. *Journals of Gerontology Series B: Psychological Sciences and Social Sciences*, *66*(6), 687–698.

Banner, N., (2011). '*An Aging Workforce*', POST note, 391.

Baumruk, R., (2006). '*Why Managers are Crucial to Increasing Engagement*. Strategic HR Review.

Boudiny, K., (2013). 'Active aging': From empty rhetoric to effective policy tool. *Aging and Society*, *33*(6), 1077–1098.

Brougham, R. R., & Walsh, D. A., (2005). Goal expectations as predictors of retirement intentions. *The International Journal of Aging and Human Development*, *61*(2), 141–160.

Brougham, R. R., & Walsh, D. A., (2009). Early and late retirement exits. *The International Journal of Aging and Human Development*, *69*(4), 267–286.

Chakraborti, R. D., (2004). *The Greying of India: Population Aging in the Context of Asia*. Sage.

Davey, J. A., (2002). Active aging and education in mid and later life. *Aging and Society*, *22*(1), 95–113.

Ekerdt, D., (2018). In defense of the not-so-busy retirement. *Wall Street Journal*. Extracted from: https://www.wsj.com/articles/in-defense-of-the-not-so-busy-retirement-1524449520 (accessed on 24 February 2020).

Fiske, S. T., (1998). Stereotyping, prejudice and discrimination at the seam between the centuries: Evolution, culture, mind and brain. *European Journal of Social Psychology, 30*, 299–322.

Formosa, M., (2017). Responding to the active aging index: Innovations in active aging policies in Malta. *Journal of Population Aging, 10*(1), 87–99.

Foster, L., & Walker, A., (2013). Gender and active aging in Europe. *European Journal of Aging, 10*(1), 3–10.

Fritzsche, B., & Marcus, J., (2013). The senior discount: Biases against older career changers. *Journal of Applied Social Psychology, 43*(2), 350–362.

Gaillard, M., & Desmette, D., (2010). (In)validating stereotypes about older workers influences their intentions to retire early and to learn and develop. *Basic and Applied Social Psychology, 32*(1), 86–98.

Gupta, R., & Hershey, D. A., (2016). Cross-national differences in goals for retirement: The case of India and the United States. *Journal of Cross-Cultural Gerontology, 31*(3), 221–236.

Havighurst, R. J., (1963). Successful aging. *Processes of Aging: Social and Psychological Perspectives, 1*, 299–320.

Henkens, K., van Dalen, H. P., Ekerdt, D. J., Hershey, D. A., Hyde, M., Radl, J., ... & Zacher, H. (2018). What we need to know about retirement: Pressing issues for the coming decade. *The Gerontologist, 58*(5), 805–812.

Hershey, D. A., & Jacobs-Lawson, J. M., (2009). Goals for retirement: Content, structure and process. *New Directions in Aging Research*, 167–186.

International Labor Organization, 2019, Retirement age, overtime top lawmakers' discussion on revised Labor Code, Retrieved from: https://www.ilo.org/hanoi/Informationresources/Publicinformation/WCMS_712016/lang–en/index.htm (accessed on 24 February 2020).

Iweins, C., Desmette, D., Yzerbyt, V., & Stinglhamber, F., (2013). Ageism at work: The impact of intergenerational contact and organizational multi-age perspective. *European Journal of Work and Organizational Psychology, 22*(3), 331–346.

Jensen, P. H., & Møberg, R. J., (2012). Age management in Danish companies: What, how, and how much? *Old site of Nordic Journal of Working Life Studies, 2*(3), 49–66.

Kite, M., Deaux, K., & Miele, M., (1991). Stereotypes of young and old: Does age outweigh gender? *Psychology and Aging, 6*, 19–27.

Leung, C. S., & Earl, J. K., (2012). Retirement resources inventory: Construction, factor structure and psychometric properties. *Journal of Vocational Behavior, 81*(2), 171–182.

Levy, B. R., (1999). The inner self of the Japanese elderly: A defense against negative stereotypes of aging. *The International Journal of Aging and Human Development, 48*, 131–144.

Levy, B. R., (2000). Handwriting as a reflection of aging self-stereotypes. *Journal of Geriatric Psychiatry, 33*, 81–94.

Levy, B. R., (2003). Mind matters: Cognitive and physical effects of aging self-stereotypes. *The Journals of Gerontology Series B: Psychological Sciences and Social Sciences, 58*(4), P203–P211.

Levy, B. R., Ashman, O., & Dror, I., (1999–2000). To be or not to be: The effects of aging self-stereotypes on the will-to-live. *Omega: Journal of Death and Dying, 40*, 409–420.

Levy, B., (1996). Improving memory in old age by implicit self-stereotyping. *Journal of Personality and Social Psychology, 71*, 1092–1107.

Lin, W. I., Chen, M. L., & Cheng, J. C., (2014). The promotion of active aging in Taiwan. *Aging International,* *39*(2), 81–96.

Macey, W. H., & Schneider, B., (2008). The meaning of employee engagement. *Industrial and Organizational Psychology,* *1,* 3–30.

Manfredi, S., & Vickers, L. (2013). Meeting the challenges of active aging in the workplace: is the abolition of retirement the answer?. *European Labour Law Journal,* *4*(4), 251–271.

Nelson, D. L., & Simmons, B. L., (2003). Health psychology and work stress: A more positive approach. *Handbook of Occupational Health Psychology* (pp. 97–119) Washington DC: American Psychological Association. New York: Springer.

Ng, T. W., & Feldman, D. C., (2012). Employee voice behavior: A meta-analytic test of the conservation of resources framework. *Journal of Organizational Behavior,* *33*(2), 216–234.

Organization for Economic Co-Operation and Development (OECD), (2020), Data retrieved from: http://www.oecd.org/employment/ (accessed on 24 February 2020).

Palmore, E., (1972). Compulsory versus flexible retirement: Issues and facts. *The Gerontologist,* *12*(4), 343–348.

Paúl, C., Ribeiro, O., & Teixeira, L., (2012). Active aging: An empirical approach to the WHO model. *Current Gerontology and Geriatrics Research, 2012,* pp. 1–10 doi:10.1155/ 2012/382972.

Paúl, C., Teixeira, L., & Ribeiro, O., (2015). Positive aging beyond "success": Towards a more inclusive perspective of high level functioning in old age. *Educational Gerontology,* *41*(12), 930–941.

Rodriguez-Rodriguez, V., Rojo-Perez, F., Fernandez-Mayoralas, G., Morillo-Tomas, R., Forjaz, J., & Prieto-Flores, M. E., (2017). Active aging index: Application to Spanish regions. *Journal of Population Aging,* *10*(1), 25–40.

Rowe, J. W., & Kahn, R. L., (1997). Successful aging. *The Gerontologist,* *37*(4), 433–440.

Savishinsky, J., (2004). The volunteer and the Sannyāsin: Archetypes of retirement in America and India. *The International Journal of Aging and Human Development,* *59*(1), 25–41.

Sen, A., (1985). 'Well-being, agency and freedom.' *The Journal of Philosophy, LXXXII*(4), 169–221.

Sen, A., (1993). 'Capability and well-being.' In: Nussbaum, M., & Sen, A., (eds.), *The Quality of Life.* Clarendon Press, Oxford.

Shultz, K. S., (2003). Bridge employment: Work after retirement. In: Adams, G. A., & Beehr, T. A., (eds.), *Retirement: Reasons, Processes, and Results* (pp. 215–241).

Sidorenko, A., & Zaidi, A. (2013). Active aging in CIS countries: Semantics, challenges, and responses. *Current gerontology and geriatrics research, 2013.* pp. 1–17 http://dx.doi.org/10.1155/2013/261819.

Smyer, M., & Pitt-Catsouphes, M. (2007). The meanings of work for older workers. *Generations,* *31*(1), 23–30.

The Aegon Retirement Readiness Survey, (2017). *Successful Retirement-Healthy Aging and Financial Security.* https://www.aegon.com/en/Home/Research/Aegon-Retirement-Readiness-Survey-2014/ (accessed on 24 February 2020).

Varlamova, M., Ermolina, A., & Sinyavskaya, O., (2017). Active aging index as an evidence base for developing a comprehensive active aging policy in Russia. *Journal of Population Aging,* *10*(1), 41–71.

Wah, L., (1999). Engaging employee's: A big challenge. *Management Review,* *88*(9), 10.

Walker, A., (1999). The future of pensions and retirement in Europe: Towards productive aging. *The Geneva Papers on Risk and Insurance: Issues and Practice,* *24*(4), 448–460.

Walker, A., (2002). The politics of intergenerational relations. *Zeitschrift für Gerontologie und Geriatrie, 35*(4), 297–303.
Wang, M., & Shultz, K. S., (2010). Employee retirement: A review and recommendations for future investigation. *Journal of Management 36*, 172–206. doi: 10.1177/0149206309347957.
Wang, M., Henkens, K., & Van Solinge, H., (2011). Retirement adjustment: A review of theoretical and empirical advancements. *American Psychologist, 66*(3), 204–213.
WHO Active Aging: A Policy Framework, (2018). Extracted from: http://www.who.int/ageing/publications/active_ageing/en/as (accessed on 27 February 2020).
Zacher, H., Dirkers, B. T., Korek, S., & Hughes, B., (2017). Age-differential effects of job characteristics on job attraction: A policy-capturing study. *Frontiers in Psychology, 8*, 1124. http://doi.org/10.3389/fpsyg.2017.01124 (accessed on 24 February 2020).
Zaidi, A., Gasior, K., Hofmarcher, M. H., Lelkes, O., Marin, B., Rodrigues, R., Schmidt, A., Vanhuysse, P., & Zolyomi, E., (2012). Active aging index 2012 for 27 EU member states. *Vienna: European Center for Social Welfare Policy and Research*.
Zhan, Y., Wang, M., Liu, S., & Shultz, K. S., (2009). Bridge employment and retirees' health: A longitudinal investigation. *Journal of Occupational Health Psychology, 14*(4), 374.

FURTHER READING

Acharya, A., & Gupta, M., (2016a). An application of brand personality to green consumers: A thematic analysis. *Qualitative Report, 21*(8), 1531–1545.
Acharya, A., & Gupta, M., (2016b). Self-image enhancement through branded accessories among youths: A phenomenological study in India. *Qualitative Report, 21*(7), 1203–1215.
Arpino, B., & Solé-Auró, A., (2019). Education inequalities in health among older European men and women: The role of active aging. *Journal of Aging and Health, 31*(1), 185–208.
Barslund, M., Von Werder, M., & Zaidi, A., (2019). Inequality in active aging: Evidence from a new individual-level index for European countries. *Aging and Society, 39*(3), 541–567.
Carta, M. G., Atzeni, M., Perra, A., Mela, Q., Piras, M., Testa, G., Orrù, G., & Kirilov, I., (2019). Cost-effectiveness of US national institute of health and European Union FP7 projects on active aging and elderly quality of life-author's reply. *Clinical Practice and Epidemiology in Mental Health, 15*(1), 10–14.
Cesta, A., Cortellessa, G., De Benedictis, R., & Fracasso, F., (2019). ExPLoRAA: An intelligent tutoring system for active aging in (flexible) time and space. *CEUR Workshop Proceedings, 2333*, 92–109.
Cunha, P. M., Ribeiro, A. S., Nunes, J. P., Tomeleri, C. M., Nascimento, M. A., Moraes, G. K., Sugihara, P., et al., (2019). Resistance training performed with single-set is sufficient to reduce cardiovascular risk factors in untrained older women: The randomized clinical trial. Active aging longitudinal study. *Archives of Gerontology and Geriatrics, 81*, 171–175.
De Maio Nascimento, M., & Giannouli, E., (2019). Active aging through the university of the third age: The Brazilian model. *Educational Gerontology*.
Del Barrio, E., Marsillas, S., Buffel, T., Smetcoren, A. S., & Sancho, M., (2019). Erratum: From active aging to active citizenship: The role of (age) friendliness [*Social Sciences, 7*, (2018), 134]. *Social Sciences, 8*(1). doi: 10.3390/socsci7080134.
Desjardins, R., (2019). The relationship between attaining formal qualifications at older ages and outcomes related to active aging. *European Journal of Education, 54*(1), 30–47.
Desjardins, R., Olsen, D. S., & Midtsundstad, T., (2019). Active aging and older learners: Skills, employability and continued learning. *European Journal of Education, 54*(1), 1–4.

Gupta, M., & Kumar, Y., (2015). Justice and employee engagement: Examining the mediating role of trust in Indian B-schools. *Asia-Pacific Journal of Business Administration, 7*(1), 89–103.

Gupta, M., & Pandey, J., (2018). Impact of student engagement on affective learning: Evidence from a large Indian University. *Current Psychology, 37*(1), 414–421.

Gupta, M., & Ravindranath, S., (2018). Managing physically challenged workers at Microsign. *South Asian Journal of Business and Management Cases, 7*(1), 34–40.

Gupta, M., & Sayeed, O., (2016). Social responsibility and commitment in management institutes: Mediation by engagement. *Business: Theory and Practice, 17*(3), 280–287.

Gupta, M., & Shaheen, M., (2017a). Impact of work engagement on turnover intention: Moderation by psychological capital in India. *Business: Theory and Practice, 18*, 136–143.

Gupta, M., & Shaheen, M., (2017b). The relationship between psychological capital and turnover intention: Work engagement as mediator and work experience as moderator. *Journal Pengurusan, 49*, 117–126.

Gupta, M., & Shaheen, M., (2018). Does work engagement enhance general well-being and control at work? Mediating role of psychological capital. *Evidence-Based HRM, 6*(3), 272–286.

Gupta, M., & Shukla, K., (2018). An empirical clarification on the assessment of engagement at work. *Advances in Developing Human Resources, 20*(1), 44–57.

Gupta, M., (2017). Corporate social responsibility, employee-company identification, and organizational commitment: Mediation by employee engagement. *Current Psychology, 36*(1), 101–109.

Gupta, M., (2018). Engaging employees at work: Insights from India. *Advances in Developing Human Resources, 20*(1), 3–10.

Gupta, M., Acharya, A., & Gupta, R., (2015). Impact of work engagement on performance in the Indian higher education system. *Review of European Studies, 7*(3), 192–201.

Gupta, M., Ganguli, S., & Ponnam, A., (2015). Factors affecting employee engagement in India: A study on offshoring of financial services. *Qualitative Report, 20*(4), 498–515.

Gupta, M., Ravindranath, S., & Kumar, Y. L. N., (2018). Voicing concerns for greater engagement: Does a supervisor's job insecurity and organizational culture matter? *Evidence-Based HRM, 6*(1), 54–65.

Gupta, M., Shaheen, M., & Das, M. (2019). Engaging employees for quality of life: mediation by psychological capital. *The Service Industries Journal, 39*(5–6), 403–419.

Gupta, M., Shaheen, M., & Reddy, P. K., (2017). Impact of psychological capital on organizational citizenship behavior: Mediation by work engagement. *Journal of Management Development, 36*(7), 973–983.

Kenbubpha, K., Higgins, I., Wilson, A., & Chan, S. W. C., (2019). Testing psychometric properties of a new instrument 'promoting active aging in older people with mental disorders scale' from a cross-sectional study. *Psycho Geriatrics*.

Ko, P. C., & Yeung, W. J. J., (2018). An ecological framework for active aging in China. *Journal of Aging and Health, 30*(10), 1642–1676.

Leuzzi, R., Monopoli, V. G., Cupertino, F., & Zanchetta, P., (2018). Active aging control of winding insulation in high frequency electric drives. *2018 IEEE Energy Conversion Congress and Exposition, ECCE*, pp. 6852–6858.

Li, C., Li, J., Pham, T. P., Theng, Y. L., & Chia, B. X., (2018). Promoting healthy and active aging through exergames: Effects of exergames on senior adults' psychosocial well-being. *Proceedings-2018 International Conference on Cyberworlds, CW 2018*, pp. 288–291.

Meersohn, S. C., & Yang, K., (2019). Controlling active aging: A study of social imaginaries of older people in Chile. *Aging and Society.*

Mucci, N., Tommasi, E., Giorgi, G., Taddei, G., Traversini, V., Fioriti, M., & Arcangeli, G., (2019). The working environment as a platform for the promotion of active aging: An Italian overview. *Open Psychology Journal, 12*(1), 20–24.

Navarro, O. S. A., & Pilar, M. G. M., (2019). Accessibility and new technology MOOC-disability and active aging: Technological support. *Advances in Intelligent Systems and Computing, 880,* 992–1004.

Orte, S., Subías, P., Fernández, L., Mastropietro, A., Porcelli, S., Rizzo, G., Boqué, N., Guye, S., Röcke, C., Andreoni, G., Crivello, A., & Palumbo, F., (2019). Dynamic decision support system for personalized coaching to support active aging. *CEUR Workshop Proceedings, 2333,* 16–36.

Pandey, J., Gupta, M., & Naqvi, F., (2016). Developing decision-making measure a mixed-method approach to operationalize Sankhya philosophy. *European Journal of Science and Theology, 12*(2), 177–189.

Pistoia, M., Parata, C., Giugni, P., Urlini, G., Loi, S., & Borrelli, G., (2019). Virtual modeling of the elderly to improve health and wellbeing status: Experiences in the active aging at home project. *Lecture Notes in Electrical Engineering, 540,* 71–83.

Punyakaew, A., Lersilp, S., & Putthinoi, S., (2019). Active aging level and time use of elderly persons in a Thai suburban community. *Occupational Therapy International,* 7092695.

Sánchez, G., & Lobina, V., (2019). Programmed physical activity for the elderly as a motor of active aging. *Advances in Intelligent Systems and Computing, 801,* 444–450.

Schuller, T., (2019). Active aging and older learners: Trajectories and outcomes. *European Journal of Education, 54*(1), 5–8.

Um, J., Zaidi, A., & Choi, S. J., (2019). Active aging index in Korea-comparison with China and EU countries. *Asian Social Work and Policy Review, 13*(1), 87–99.

CHAPTER 3

Engaging Diverse Races at Work

ARTI SHARMA and RAJESH MOKALE

Indian Institute of Management, Indore, Madhya Pradesh, India, Tel.: (+91) 9001270997, E-mail: f15artis@iimidr.ac.in (A. Sharma), Tel.: (+91) 8108441844, E-mail: f17rajeshm@iimidr.ac.in (R. Mokale)

ABSTRACT

The scholarship in diversity research has advanced considerably in surface-level diversity reporting the influences of observable characteristics such as age, gender, and ethnicity at work. One of the strong and widely impacting factors in all observable diversity characteristics is race. The presence of the members of different races in an organization poses a greater challenge in the organizations due to the presence of different backgrounds. But, when engaged adequately, this racial diversity can provide a competitive advantage to the organization. In view of this, the chapter presents an account of research in racial diversity, engaging racial diversity at work, the importance of managing racial diversity, and its practical implications.

3.1 INTRODUCTION

The global business scenario is changing with a much faster pace with fading boundaries, increasing technological innovations, and intense competition. This has brought about a considerable change in the nature of workforce by the inclusion of diversity at the workplace. The fading physical boundaries across nations in the global business environment developed the need for the effective management of a globally diverse workforce to attain competitive advantage. The pressure for local product demand, availability of formidable labor, and other resources in recipient country emphasize on the need to attain global diversity goals (Florkowski, 1996). On the other hand, the advent of extended business avenues brought in wider social understanding in organizations to work with diminishing social differences and introduces diversity goals through broadening organizational culture. This understanding started the exchange of cultural and social practices in organizations through formal and informal means which helped to develop a wider cultural understanding interestingly called 'Glocal Culture' (Shi, 2013). The word 'Glocal' is formed by the combination of two words- Globalization and Localization. This has created the need to accommodate and assimilate the local and global workforce for the achievement of better business outcomes.

But the opportunities of wider business avenues also brought wider challenges to manage diversity, thus managing workforce diversity has been one of the biggest challenges faced by the organizations in the present times (Tsui and Gutek, 1999). With the world becoming a global village, diversity became an important element to sustain the business and the need for new work practices has taken effect. Paradoxically, it has created the problem of uneven growth and the need for 'Inclusive Growth' in society across the globe. BRICS nations faced opposition to many development-oriented projects from local citizens due to this uneven growth. It has raised the challenges for the organizations to maintain diversity in human resources (HRs) through educating and recruiting multiple races. This dynamism in the global market has created a greater need for businesses to include diverse races to compete at the global front.

The management of workforces comprising of local citizens and immigrants in different countries is defined as *Global Diversity Management* (Barak, 2016). One important aspect of Global Diversity Management is 'Racial diversity and engagement of diverse races in the workforce.' But, little has been said about engaging the racial diversity in the workforce. It is the prime concern of recent times of globalization and multination business operations. The trade across nations can't move forward without addressing the diversity and inequality in external society as business organizations recruit HRs from the society and serve its products to the wider societies. This calls for a quick attention to address and remove racial practices in organization and society for sustainable business practices.

Scholars have extensively studied diversity with respect to surface level and deep level characteristics (Harrison, Price, and Bell, 1998). The surface level diversity refers to the study of observable attributes such as age, gender, ethnicity, religion (Milliken and Martins, 1996) while deep level diversity comprises of less observable attributes (Riordan, 2000) which gets reflected over time such as, values (Jehn, Chadwick, and Thatcher, 1997), attitudes, preferences, and belief (Harrison et al., 1998) and affect (Barsade, Ward, Turner, and Sonnenfield, 2000). Across the globe, there lie apt differences in visible and non-visible features in the world community. This chapter is an attempt to contribute to the literature in surface level diversity based on engaging diverse races at work. In a growing world market, the need for assimilation of diverse races has been one of the greatest needs for any business to grow. With race as a feature of the global community, there are challenges inaptly implementing racial diversity, which can make a business flourish or die. The racial diversity is regarded as the "most challenging HR and organizational issues of our time" (Richard, 2000). The racial identity comes all the way in differentiating and assimilating HRs in organization (Pickett and Brewer, 2001). Thus, there lies a need to study engagement of racially diverse workforce in society. On the positive side, presence of diversity in the workforce brings a wider knowledge base that can foster the organization success (Conner and Prahalad, 1996). The racially diverse workforce and its effective management can help organization to reap the competitive advantage from the varied know-how that can act as a strategic asset (Barney and Wright, 1998). While on the negative side, the rise of China as 'the world's factory' (Zhang, 2005; Mess, 2016), India as 'consultancy services pioneer' and Philippines as 'world's call center'[1] couldn't resolve industry-people problems completely. Thus through this chapter, we intend to address issues of racial engagement in organizations.

Thus, our main focus will be on exploring the issues of racial diversity and engaging in racial diversity in literature. The chapter consists of four sections. The first section presents, the meaning of diversity. In the Second section, we put forth the literature on racial diversity. The third section identifies the ways of engaging in racial diversity. The fourth section intends to focus on the importance of managing racial diversity at workplaces. The fifth section presents the challenges for diversity at workplaces. In the penultimate section, we cover some cases of diversity. Finally, this chapter ends with managerial implication and conclusion.

3.2 NEED FOR DISCUSSING THIS THEME

Liberty at large provokes diversity and diversity preserves liberty (Parekh, 2011). Business units as private institutions of society seek liberty for

---

[1]https://www.bloomberg.com/news/articles/2017-10-04/call-center-nation-fights-to-keep-automation-from-stealing-jobs.

deriving profits. As organizations try to reduce labor, cost, and increase profit through widening product and service reach. This paradox is resolved through encroaching financially viable territories beyond the limits of states and nations. Many times organizations achieve this liberty through achieving economies of scopes and economies of scale. The economies of scale are achieved through expansion of the business horizons by concurring new territories in diverse structures of society, while economies of scope for the niche products/markets are achieved through selling the niche products and services to specialty groups in globally diverse societies. While outsourcing and offshoring mechanisms made the search of financially sustainable avenues easier for a greater business growth. This invokes social and geographical diversity in world trade. Similarly, exploitation, and exploration (March, 1991) paradigm suggests us, organizations essentially attract similar identities, when there is a need for learning through the exploitation of already trained similar HRs and their ideas, while in contrast newer divergent identities are attracted to satisfy the innovation and exploration needs. These peculiarities identify the need to address diversity in HRs.

Let's look into the definition of diversity. In designing the categorization-elaboration model (CEM) Kinnpenberg, Dreu, and Homan (2004) defines diversity as differences between individuals or any attribute that might lead to the perception that another person is different self. Diversity is defined at work unit level as well as individual level in the organizations. At organizational level, Organizational studies define diversity as 'as the structural property of organizational work units' (Ragins, 2017, p. 485). While at the individual level, diversity is addressed in literature through social as well as workplace context. In the social context, Kossek, and Lobel's (1996) defines of diversity as a range of differences in ethnicity/nationality, gender, function, ability, language, religion, lifestyle (Bassett-Jones, 2005). Cox (2001) defines Workplace diversity as the differences in social and cultural identities existing among people working together in a specified employment or market scenario. Taking it further, diversity is defined as the segregation of differences among the members of an organization unit bounded with a common characteristic such as task requirement, ethnicity, pay, or tenure (Harrison and Klein, 2007). We define, Workforce Diversity is defined as individual social differences in work attitudes presented by employees based on their social cognition (Greenwald and Banaji, 1995) and experiential cognition (Norman, 1993).

Overall literature shows that diversity in the workforce can be understood as an inclusive term addressing the similarities and differences found among the employees with respect to age, race, gender, culture, physical abilities and disabilities, and sexual orientation. In literature, there are two major streams for the conceptualization of diversity. The narrow perspective considers diversity with reference to age, gender, race, ethnicity, nationality, religion, and disability. Morrison (1992) classified diversity at four levels, namely, racial/ethnic/sexual and cultural diversity. The broader perspective comprises of personality traits, beliefs, values, attitudes, education, language marital status, background, tenure, economic status, marital status, and lifestyles, etc. Diversity looks into visible dimensions like gender, age, race, disability as well as hidden dimensions like beliefs, values, attitudes, and culture such as religious practices, sexual orientation, thinking styles, working styles, and personality traits (Ferdman and Sagiv, 2012; Hays-Thomas, 2004). A two-factor conceptualization of diversity segregates it as, the surface level diversity and the deep level diversity (Jackson, May, and Whitney, 1995). The surface-level diversity focuses on demographic characteristics of a population like race, gender, age, education, etc. Deep level diversity focuses on the deeper level of difference, mainly between diversity of values and attitudes in groups (Martins, Milliken, Wiesenfeld, and Salgado, 2003). The process of social categorization explains surface-level diversity (Riordan, 2000), wherein the surface level characteristics act as processes of depersonalization which develop in-group and out-group feelings leading to exclusion of out-groups (Brewer and Brown, 1998; Brown and Turner, 1981). This fosters in developing a more positive attitude towards each other among the in-group members and negative attitude towards out-group members (Egan, and O'Reilly, 1992). On the other hand, deep-level diversity comprises of values, beliefs, attitudes, skills, and other characteristics of individuals which are "subject to construal and more mutable" (Jackson et al., 1995, p. 217) in contrast to other characteristics. This relates to the heterogeneity in values, attitudes, and beliefs which are communicated through verbal and non-verbal patterns of behavior displayed during the process of individual interactions and information exchange (Harrison, Price, and Bell, 1998). Such a spread of literature invites researchers to study multiple socio-psychological factors like privileges, favoritism, advantages, and power relations. The roots of diversity management find their base in different theoretical perspectives borrowed from social psychology (Byrne, 1971; Newcomb, 1961), sociology (Amir, 1969; Berger, Rosenholtz, and Zelditch, 1980) and organization behavior (Schneider, 1987; Turner, 1987).

The field of diversity studies has taken a big leap from abiding by legal compliance and equality legislation to embracing differences at the workplace (Cassell, 2001) and harnessing diverse capabilities by efficient inclusion and empowerment (Cornelius, 2002). Diversity literature has extensively discussed the role of social identity and the identification process in forming the in-group and the out-group (Ashforth and Mael, 1989). The identity is a combined function of personal identity (Brewer and Gardner, 1996) and the social identity (Tajfel and Turner, 1986). The third aspect, i.e., organizational identity is studied minimal in literature from diversity perspective. The similarity-attraction theory explains the reason for the weakening of diversity mechanism as the group members try to reinforce qualities which are socially acceptable in developing the in-group feelings (Clore and Byrne, 1974), while derogating dissimilar out-group characteristics (Rosenbaum, 1986). The process of diversity studies has been shifted from cross-cultural differences to social cognitive mediation (Hong and Chiu, 2001). For organizations, the real depiction of workplace diversity is way beyond the words, it reflects through micro-level social and cultural practices as well as macro-level organizational policies woven in through organization culture, inclusionary HR policies and practices, respectful interpersonal communication. In organizations, the diversity orientation culture finds diversity as an important asset for organization while in contrast identity blind structure considers a compulsion from the larger legal and social discriminatory structure (Richard and Johnson, 2001). The respective accounts of diversity are studied using individual-level attributes such as age, gender, race, and ethnicity (Lawrence, 1995), which further reflects in the group, team, or departmental dynamics. This finally reflects at a larger organizational level and generates outcomes resembling with organizational goals and objectives, which is important for generating profitable returns to the organization.

One such important identity defined in society is 'Race.' Racial diversity studies are done in literature from inequality perspective. In organizations, races were saturated at each level creating problems of steepness for moving vertically to underrepresented races thus horizontally concentrating each race into each compartment of hierarchy. This motivates researchers to study problems of identity and cumulative accumulation of races at each level in the organizations. Bunderson and Vegt (2017) differentiate between the phenomenon of diversity and inequality. Diversity looks after the horizontal spread of racial identities at a specific level in the organization while inequality looks after vertical spread racial identities in the organization. But many times the horizontal spread and vertical spread can't be studied in silo. Limitations for the vertical spread of each race develop problems of the steepness of hierarchy while limited horizontal spread generates the problem of concentration. Thus, it is important to understand the problem from two-way analysis of the problem.

3.3 REVIEW OF THIS THEME

Looking at workplace setup, organizations work with boundaries that are implicit in organizational structures and positions. When organizational structure tries to replicate traditional social order of identities, it develops barriers to entry to minority social groups at upper echelons in the organization. This makes researchers important to study the domain identities. One such social identity is race. Race is defined in terms of physical characteristics which are common to an inbred based on geographically isolated population such as skin color, facial features, hair type, etc. (Betancourt and Lopez, 1993). Race is defined by Census Bureau in the USA as self-identified racial and national origin or sociocultural groups in the country[2]. Race is different from ethnicity but often considered synonymously in day-to-day use. Race is differentiated from ethnicity where race is often identified from physical factors while ethnicity is identified from geographical, ancestral origins, and more specifically culture, traditions, and language (Bhopal, 2004).

It is a form of surface-level diversity which makes oneself easy to understand the racial features through visible counts of human eyes. Though presently, the biological conception of race is unjustifiable and invalid for biological classification itself, another conceptualization identifies race as the product of nation's political and social history (Freeman, 2000). The racial slurs are routinized in the cognition of dominant racial individuals to oppress weakly placed racial individuals directly or indirectly make it disrespectful to accept, which can lead to ruthless conflict among racially diverse societies. Any organization situated in a racially diverse territory cannot sustain, if racially sound cultural and structural mechanism is absent. Thus, it is a big challenge to engage the racially diverse workforce to improve individual experiences at micro-individual level.

Feaign (2010) identifies the roots of racist practices in slavery and Jim Crow Laws in 19th century USA. The role of pro-white image, attitude, and emotions perpetuated the need for the parallel imagery of equal opportunity black system in US society. The thinking in terms of racial stereotypes inherited from the past and continuously reinforced in present led to social as well as workplace bias of whites towards non-whites (Feaign, 2010). This became the social norm in US society which led to the development of racially devoid workplaces in society. Thus federal government with the national motive started collecting data on race in order to draw policy decisions for maintenance of civil rights, assess racial disparities in health and

---

[2](Refer: US Census Bureau, https://www.census.gov/topics/population/race/about.html).

environmental risks and to promote equal employment opportunities towards all the races in the native country. This was the start of the diversity era in the democratic setup. Thus, Racial Diversity has taken a big leap starting from the traditional conceptualization of Races in equal employment opportunity Commission (EEOC) in USA to the attraction, retention, and inclusion of diverse workforce in business organizations.

In literature, Racial Diversity is defined as the degree of variation of racial categories in a group membership (Blau, 1977; Harrison and Klein, 2007). The major ideas of diversity are combined into two aspects mainly diversity orientated culture and identity blind literature (Richard and Johnson, 2001). Optimal distinctive theory (ODT) by Brewer (1991, p. 477) addresses the tensions faced by individuals in "human needs for belongingness and similarity to others (on the one hand) and a countervailing need for uniqueness and individuation (on the other)." Brewer (1991) argued that individuals strive for a balance between the two needs which they seek by attaining an optimal level of inclusion in the group. An individual looks for the harmony of both the need for uniqueness and belongingness to opt for a social identity associated with a particular group (Pickett, Bonner, and Coleman, 2002). Thus either of the need can become salient as per the demand of the situation (Correll and Park, 2005; Pickett and Brewer, 2001). According to the theoretical lens of ODT, we suggest that an individual in a racially diverse group experiences a tension between the need to get associate, bond, and form a strong relationship (need for belongingness) and the need to maintain the racial identity (need for uniqueness). To maintain an equalized position, people opt for different social identities to best suit respective different groups in order to gain wider acceptance (Baumeister and Leary, 1995). In return, of the group membership, individuals of different races expect loyalty, cooperation, and trustworthiness in the group thereby ensuring the security of all members (Brewer, 2007).

A racially diverse workforce should be managed by carefully understanding the individuals' need for belongingness and need for uniqueness. A proper care should also be taken to avoid too much of racial homogeneity as it may hit the individuals' need of uniqueness (Snyder and Fromkin, 1980). We argue that too much similarity will motivate individuals to work for enhancing their need for uniqueness for their race leading to the danger of comparison and competition within their own group. When we will look at the operationalization of diversity in the organization, we need to consider the composition of different races in the upper echelons of the organization as it helps in cultural enrichment of the workforce. It is also important to understand, a culturally diverse workforce helps in providing different opinions thereby facilitate quality decision making (Cox, 1994; McLeod et al., 1996). This improvement in decision quality and problem-solving due to racial diversity is termed as, "value-in-diversity hypothesis" or the "information/decision-making notion (Richard, 2000).

Multiculturalism in the workforce made it crucial to have an international perspective for an effective global diversity management. The US Labor Department made it compulsory to identify and achieve the targets of diversity through Employer Information Consolidated report[3]. EEO collects the information on many diversity aspects from the employers in the USA's territory. The prospective employees are protected in the US through EEOC which looks after discrimination against cases of discrimination with regard to many biased understandings of employers including races. This is a strong and responsible legislative arrangement, which is difficult to implement in the developing country with not fully developed infrastructure and low corruption perception (Transparency International, 2018) making it difficult to report and investigate all such situations. But there is a direct arrangement made in the public sector in India where the arbitrary privileges are extended to the minority population in society (Parekh, 2011). Initial quantitative research doesn't provide direct support for the relationship between racial diversity and firm performance (Richard, 2000), but it lacks 'how' to address diversity issues remain unanswered. The latter research found that increasing racial diversity congruence in two spaces community and organization increases store unit sales performance (Richard, Stewart, McKay, and Sackett, 2017). Here, our goal is to understand the solutions that resolve the problem of disengagement in a diverse workforce. The issues if addressed can lead to the diverse workforce to work in holistic ways and get the improved results. In the next section, we will see the ways of engaging racial diversity at workplaces.

3.3.1 ENGAGING RACIAL DIVERSITY AT WORK

Race is one of the important characteristic studied in diversity literature. The operationalization of racial diversity is possible only through practical recognition of diversity. Such recognition can be given through recruiting racially diverse workforce, developing policies for non-discrimination at workplaces, developing engaging environment through shared culture

---

[3]Learn more at equal employment opportunity Council about the regulation regarding anti-discrimination laws in the USA - https://www.eeoc.gov/employers/reporting.cfm.

and values. The symbolic inclusion of neglected races doesn't portray the culture of diversity, they may remain as torchbearers of the diversity, but it is important to devise ways to bring diverse practices in organizations from diverse racial locations that can make organizations feel more as home to the employees rather than an entity to get salaries. Denson and Chang (2009) identify three possible areas to address diversity in the educational institutions: Structural Diversity, curricular diversity, and interactional diversity. On the similar grounds organizations can address racial diversity through recruiting racially diverse workforce at each possible level in organizational structure, imbibing the cultural traits of diverse workforce in its culture and policy documents in order to nurture them, increase the bonding among the employees through formal organizational interaction and developing informal spaces to accompany racially diverse individual to develop amicable and comfortable climate to engage in trustworthy dialogue.

According to Hofstede (1994), the structure should follow culture. As the purpose of a structure is to coordinate activities, it is important that diverse culture should be able to coordinate activities. Even though bureaucratic organizations generate profits, they facilitate patriarchal management and dominant employee contracts in organizations, which direct member behavior through established rules (Block, 1987) and create impediments for organizational diversity. It is advisable to develop an egalitarian, non-bureaucratic organizational structure to sustain diverse workforce through a narrow hierarchy. The diversity acceptance in organizations thus is based on the attributes of workforce affect how people think, feel, and behave at work, and their acceptance, work performance, satisfaction, or progress in the organization.

There are following few approaches for racial diversity management mechanisms possible in the organization. In past, the studies related to organizational diversity explored various possible approaches like the compositional approach (Tsai and Gutek, 1999; Jackson, Joshi, and Erhardt, 2003) or the configurational approach (Moynihan and Peterson, 2001). The compositional approach is based on relational demography based on self-categorization theoretic model (Abrams et al., 1990; Turner, 1999) and configurational approach is based on socially defined groups (Jackson, Joshi, and Erhardt, 2003).

It is observed that high levels of racial representativeness increase quality of services (King et al., 2011) and productivity (Avery et al., 2008). The presence of a diverse workforce can be achieved through interaction method (Leonard et al., 2004). Interaction method looks for separate interaction terms for each racially diverse group based on proportion of employees present and customer from the particular race (Richard et al., 2015). It is important that such treatment should be positive and accepted in that race. The second approach is racial representation approach (King et al., 2011) which is also called as racial diversity congruence. Racial representation approach looks at the degree of match between employee and customer race-based numerical index. A proportionate congruence of employees with a particular race (from the external community where the business in located) improves racial diversity in organization (Richard, 2015). Thus, in operationalizing multiple ways are available to sustain the organization through the mentioned operationalization methods.

There are some functional approaches that are important to operationalize diversity. Thomas and Ely (1996) identify well-articulated and widely understood mission about diversity, leader's divergent understanding about the perspectives, approaches, learning opportunities, and challenges act as the preconditions for paradigm shift for diversity in organizations. The high standards performance culture and stimulus for personal development help organizations to sustain interpersonal dialogue in diverse workforce. The internal culture of firm remains congruent with local needs but the thirst for business expansion creates interaction among cultures through entry into newer geographies making the local practices remain settled locally and enforce organizations to adapt global range of practices. With differences in individual and group understanding, the spectrum of diversity becomes wider. If the employee is selected from the internal labor market, organizations look for closure of job vacancy from person-team fit perspective so that interaction should interactive part of job should be flawless. While, if the person is to be recruited from the external labor market organizations look for person-organization fit such that selected candidate should understand organizational culture, values, norms, and philosophy initially and sustain in the change s/he faced in organization. When designing a job role, organizations are unaware of diversity in labor market but when organization face external social and cultural contexts, they configure manpower requirements by understanding sociocultural aspects of internal culture and external organizational environment, present-day social identity of majority employees in organization as well as demographic dividend available in internal and external labor market.

Racial diversity is looked from in-group and out-group associations (Ashforth and Mael, 1989). The role identification bases the social identity in organization. It increases the in-group and out-group behavior and attitude in organization based on social identities. The out-group feeling can be reduced only when the specific implicit biases about the certain social groups are removed. While, some organizations adopt identity conscious practices to bring in more inclusiveness in organization for the excluded racial groups at different levels of hierarchy in organization. This in turn reduces the conflict of hierarchical division of social identities in organization which the hierarchical groups in organization are based on only the organizational structures. It reduces the gaps in social structures of organization and improves interpersonal communication in organization among the races due to increased vicinity among members.

As organizational culture becomes well known to diverse set of members, they adopt the practices of organization and develop shared understanding in society, in turn, satisfies the need for organizational sustainability through no external shocks due to racial differences. DiTomaso (2015) argues that, racial diversity can't be just improved by reducing out-group hostility but in-group favoritism also needs to be reduced. The distinction between the out-group hostility and in-group favoritism is difficult. It has been seen that, democratic setup has reduced out-group hostility but there is little evidence to understand and reduce in-group favoritism. The studies should now need to move from understanding and identifying allocation of emotional and material privileges to understanding the reasons for in-group favoritism and reducing it. The close contact with in-group member is based on lesser social distance which makes it easy for the employer to trust in-group members easily and thus put it on recognizable position in organization. Thus in in-group favoritism plays a vital role. This is the major disadvantage for the business organizations to better develop diversity in organization. For the development of ideal team for taking collective action, Jarzabkowski, and Searle (2000) defined team composition based on collective actions and diversity scale. They took diversity measures for demographic, behavioral, and informational diversity and designed four quadrant framework presenting four types of team based on strength of diversity and strength of collective action. A dysfunctional team is low on diversity and collective action, consensual team was low on diversity but high on collective action, a conflictual team was low on collective action and high on diversity while the effective team was high on both the dimensions (Jarzabkowski and Searle, 2000).

Moving on step ahead, the implementation of diversity finds hurdles in hierarchical structures of the organizations. The individual's socio-economic status in society is associated with individual employee's involvement in particular organizational processes (Kerckhoff, 1995) and positions held in organizations, which formulates boundaries among organizational groups.

When the social group identities are retained, the organizational positions are seen as scarce resources and social groups bring competition rather than diversity understanding into consideration to gain wider control organizational resources, organizational structures, and positions (Struch and Schwartz, 1989), which generate dynamism to maintain racial diversity and sustain racial representation in organization. The solution to this problem is suggested through the development of an egalitarian structure (Konrad, 2003) in the organization, which is highly impossible as organizations have prime motive is sustain and generate revenue by competing with other firms through asymmetric implementation of ideas. In recent times, the problems such as division in social and organizational spheres and their interdependence on each other have developed the need for racial diversity.

3.3.2 IMPORTANCE OF MANAGING RACIAL DIVERSITY AT WORK

As diversity has become emerging issue in organization, it is important to identify the takeaways from the organizational diversity. Diversity is an important construct which build a multicultural environment beyond the social boundaries. It brings diverse perspectives in the pallet of an organization. Diversity is important in developing transformational leadership (Judge and Piccolo, 2004) in organization. With the team heterogeneity and expanding global market, the need for transformational leadership that can sustain the diverse minds and bring in advantage for the organization is possible. The role of diversity management in organizations is helpful for managers to bring in the convergent thinking for the diverse social groups to create sense of belongingness in organizational settings. The development of sustainable organizational culture and bringing in diverse understanding for resolving organizational solutions helps managers to deal in distinctive ways to identify and associate with multiple identities in rightful manner.

At board level, diversity will help in getting the diverse perspectives and understand the executive proposal with reduced complacency and narrow mindedness (Kosnik, 1990). However, during strategic turbulence it may reduce the group efforts to take decisive actions (Goodstein, Gautam, and Boeker, 1994) as members bring in individual as well as constituencies' commitment and interest (Baysinger and Butler, 1985; Kosnik, 1990) with respect to occupational and professional associations (Powell, 1991; Thompson, 1967) that lead to divergent definitions of organizational goals and policies (Clegg, 1990; Mintzberg, 1983; Powell, 1991). While the study on Dutch firms shows that, diversity of age and socio-economic background shield organizations in the times of crisis (Van der Laan and Den Berg, 2012). Divergent thinking can be limited by doing dialectic inquiry that will lead to effective strategic decision making (Mitroff et al., 1977). It has been argued as double-edged sword from the productive/non-productive potential of the teams.

Organizations can attract and acquire from a diverse pool of prospective employees coming from varying racial backgrounds through increased resource pool. In longer run, if the labor market condenses, this result into sustain competitive advantage. Bertrand and Mullinathan (2004) have found that race plays selective role in resume callbacks sent to employers in Boston and Chicago newspapers. It is important for the companies to get rid of these prejudices and biases against anyone (DiTomaso, 2015). These recruitment practices based on racial differences can be reduced in labor market by designing diversity goals and guidelines (Bertrand and Mullainathan, 2004). Prejudice reduction, anti-racism, and anti-bias programs acts as the foundation global diversity management goals for the organizations (Dixon et al., 2012). One of the advantage ripe out of the diversity at employee level is that, the increased diversity increase the cost of integrating workers for the labor unions, thus it will give an edge to organizations who handle it well (Cox and Blake, 1991).

Marketing in the diverse population will be possible through diverse pool of employees coming from different racial groups that will, in turn, create sense of belongingness about the product company and build brand loyalty. Diverse thinking process will help in understanding creativity through wider pool of employees which reduces submissiveness and conformity biases with respect to dominant opinion. With wider approaches and heterogeneous thinking, organizations develop new perspectives. System Flexibility will increase with multicultural environment in organizations through flexible choice in the problem solving with divergent ways of looking at the systems. It makes the systems more flexible and non-standard making easy availability of optimal solutions in sudden crisis situations. Richard, Barnett, Dwyer, and Chadwick (2004, p. 264) suggest that for high risk-taking teams moderate diversity exhibit better performance in management groups.

3.3.3 CHALLENGES FOR ENGAGING RACIAL DIVERSITY AT WORK

Racial Diversity seems easy-going till this point. But is it that easy to address racial diversity in organizations? Can organization's internal structure and employees accept the racial diversity at similar/higher level? Do power relations play any role? Such questions are necessary to answer when we consider importance of diversity. There are some major challenges for identifying and accommodating diverse individuals at workplaces which are listed below.

Firstly, in the context of racial biases, priming effects are evident and based in discriminatory understanding about races. Dovidio et al. (2002) identify that, racial biases are either explicit or implicit. The racial attitudes of whites are biased towards black and friendly towards white. This attitude travels further when organizations use informal channels of recruitment. Such path-dependent recruitments develop exclusionary recruitment patterns far away from diversity. This restricts diverse employment pool and reduces the organizational sustainability in longer run in social sphere due to restrictive employment practices. Secondly, if the workplace is racially diverse, we find the differing non-verbal expression towards dissimilar individual. It is found that these racially based expressions are more spontaneous and controllable while verbal behaviors are implicitly influenced by traditional way of cognition (DePaulo and Friedman, 1998).

Thirdly, the results of diversity in organization are associated with team effectiveness, power conflict, and development of negative affective emotions. Richard, Barnett, Dwyer, and Chadwick (2004, p. 264) suggest that above higher level diversity can induce cognitive biasing and communication biases leading to communication problems and develop conflicts in organization. Kirkman et al. (2004) found that racially diverse teams feel less empowered and thus perform less effectively. These findings were supported by much research previously done on team effectiveness. This effective challenge for the organization occurs when there arises need include diverse races at same level in the hierarchy or weakly represented race at the higher level in hierarchy. Such a challenge to work in teams can lead to high turnover and reduced performance for the organizations. But if diversity is ignored, it may lead to higher levels of centralization of each race thus demographic homogeneity, power centralization, and singular race dominance through stratification based on hierarchy in organization (Benderson and Vegt, 2017). It can be resolved through giving the team more interactive and common goal oriented task rather than just cohabitating them in work spaces.

Finally, in this aspect there is responsibility on HR department to audit racial diversity for their organization annually to implement broader diversity practices and design diversity goals in every function that is void of racial recognition. Many times such processes become mechanical and conscious but are important to identify the skewness of racial population at each level in organizational hierarchy to develop acceptable environment for diverse races. This creates a need for the development of long term diversity acceptance culture through continuous training and generating organic feeling in the teams to overcome racial difference of each other. Similarly, top management team diversity is studied by Jarzabkowski and Searle, (2000). It shows that diversity can be valued through developing diversity evaluation mechanisms and regular process review, making team learn organizational and team goals and unlearn racial differences and hatred through developing super-ordinate goals through internally inclusive leadership and externally through diversity training. It has also been argued that perceived similarity creates cohesion in organizations (Hogg and Terry, 2000), thus it is important for the managers to maintain cohesion among workgroups through creating the narrative of work similarity and simultaneously maintaining the racial diversity. In this section we studied, challenges to maintain the focus of the workgroup more of their tasks beyond their understanding of prototypes of races constructed in their memory.

3.4 IMPLICATIONS

Pless and Maak (2004) identify that for the diversity management there is much attention given towards the strategic part through diversity policies, procedures, and systems, while lesser is understood about the norms and values in developing a diverse workforce. It is more of cultural question, i.e., associated with norms, beliefs, values, and expectations. The lesser attention given to the 'inclusionary' practices has impacted passively to the efforts taken by the organizations. Diversity has a critical role in workforce mobilization (Konrad, 2003), reaping business benefits for sustaining and gaining competitive advantage internationally (Florkowski, 1996). The diverse workforce serves as an asset for business organizations by contributing effectively to manage the diverse set of customers, markets, and bring new product solutions which can increase market share for the organization (Barak, 2016). Thus, diverse workforce attends diverse set of problems in the organization with its varied skills, knowledge, and experience parting from the status quo (Cox, 1994; McLeod et al., 1996; Priem, Harrison, and Muir, 1995; Schuler, and Jackson, 1987) by improving work and organizational processes (Gonzalez and Denisi, 2009; Homan et al., 2008), innovating new products and services (Insight Forbes, 2011) and more recently developing inclusionary practices (Daya, 2014) for yielding greater returns (Catalyst, 2004; Herring, 2009).

Many organizations came forward to defining their diversity goals and programs; McDonald is one such organization. It identifies diversity as inclusion and tries to engage all the employees with organization Global Diversity, inclusion, and community engagement team. It focuses on bias-free, inclusionary principles through their education portfolio 'Food For thought, Beyond Bias.' It signifies that human unconscious biases slow them down while moving them beyond biases speed them. It has also developed 'Employee business network,' which promotes an inclusive environment at workplace, foster relationships among people of color, support career development opportunities, and grow the business together. The initiatives at McDonald reduce not only horizontal information asymmetries but also it creates access to senior leadership, opens up information on career strategies and opportunities for advancement[4]. Humanity is plural, not singular. The best way the world works is 'Everybody In, Nobody Out'—is the philosophy at Apple Inc. Recently, Apple has increased the representation of people of color through more new hires coming from people of color and diverse races and ethnicities. It doesn't serve people with diverse races but also feel that, 'Technology for everyone should be made by everyone.' Apple has not just identified Diversity from inclusion and social disparity perspective, but it also showcases the importance of diverse workforce for business innovation, building future culture and work in common direction to achieve most for the organization[5]. It is important to notice that, organizations give more importance to education and considers it as important equalizer. Thus, we see diverse set of employees at all the levels of organization in the USA.

It is necessary to operationalize the diversity to reap these benefits else, it would be a barrier in generating business value. The four-step model of Pless and Maak (2004) identified: (a) the awareness generation programs, (b) development of diversity-oriented vision, (c) review of management concepts and principles to align with diversity goals, and finally, (d) operationalizing diversity by looking it as competency and fostering the inclusive behavior in organizations as needful areas for the rightful management of diversity in organization. Also, the selective operationalization of diversity is always harmful, as it is identified that organizational managers expect the minority population like people of color to assimilate in mainstream organizational culture abruptly rather than working on integration and inclusion of people of color in workforce (Thomos and Gabbaro, 1999). Thus, it is important to reason, respect, and routinize the diverse culture through better understanding and responding in rightful way to people of color, rather than dishonoring/disrespecting them by controlling or ignoring their voices in decision-making process. There should be intercultural

---

[4] The details of McDonald Global Diversity, Inclusion, and Community Engagement campaign are listed on their website accessible at: https://corporate.mcdonalds.com/corpmcd/about-us/diversity-and-inclusion.html.

[5] Managers can identify and operationalize diversity policies in at their workplaces on lines of global companies like: https://www.apple.com/diversity/.

morality maintained by the individual employees in the organization and quality of relationship maintained show completeness.

3.5 CONCLUSION

The research on diversity presents that it affects organizations both positively and negatively. The neatly addressed differences in organizational structures and social structures to develop egalitarian understanding can make organizations sustain in competitive environment. The engagement of racially diverse workforce has addressed the problems of racial differences. We put forth need for wider expansion of diversity in temporal and spatially wide dimension. As expansion helps organizations to bring new business to the organizations but organizations can't expand without understanding the local requirements to set up business. The localization of business into alternative cultural settings will be easy for the organizations with diverse employee base. It will help organizations to expand rapidly and sustain the challenges of multicultural, liaison with government offices in the respective business localities. Racial inclusion in specific helps organizations to diversify and expand businesses across geographies.

External world is changing stochastically, so is the role of organizations. The sense-making management with wider understanding of the world can help organizations to sustain the change. There is need for organization to adopt new insights not to look diversity as an impediment for sustaining scarce jobs in organizations, but expanding organizations to the new horizons through engaging in diversity with better management of diversity practices. We conclude that, racial diversity is one of the most important construct to sustain and develop organization in changing times. In the subsequent section, a case covering the above concept has been provided (refer Appendix 3.1 for plausible case discussion).

3.6 CASE: THREE SCENARIOS FOR RACIAL DIVERSITY AND ENGAGEMENT

3.6.1 CASE SCENARIO 1

Rohan works in an Information Technology organization and is quite popular among his officemates. Everyone is fond of his technical expertise, approachable, and ready to help attitude. He is the first choice to be picked up for any

Engaging Diverse Races at Work

project in the organization. Recently, he got selected for a big project with a well renowned Fortune 500 company. It was one of his dreams to work for this company. He reported to the team leader, George who introduced him with other team members. The project involved Michael, Ben, Suraj, and Chirstina as the other team members. George called for a brainstorming meeting followed by initial task allocation. Upon being asked on suggestion, Michael said, "George, I think the coding for the designs could be best handled by this bloody Indian, Rohan. He is terrific" This didn't go well with Rohan and other Indian Suraj in the team. Rohan turned to Michael and asked, "What do you mean by bloody Indian here" To which, Michael replied, "Yes! You, I believe you are very good at your coding work and hence, I pointed out your name. And, yes, I would hate to say this, but I admit that you are one of those bloody and creepy Indian, who are sucking all our jobs." The meeting turned into a heated altercation with serious verbal and racial references. George had to call off the meeting immediately only soon to realize the upcoming issues and poor team performance in this situation.

Assuming yourself as George, what would you do in this situation?

3.6.2 CASE SCENARIO 2

June 20, 2018

"It's a pleasure to inform you that you are selected for the post of Tele Executive at our organization" Mahek read this line twice. It was the best gift she could have received on her eighteenth birthday. Now she can easily manage home and fund her studies as she has got a job at a renowned BPO.

August 25, 2018

It's been twenty days now, after the one-month training. Mahek is enjoying her work and likes the workplace. The phone rings at her desk and she answers, "Hello! This is Daisy, How can I help you?" The caller asked, "Hey, Sheldon here, how do I book my flights with the cheapest fare from Ireland to Geneva" "Daisy checked on screen, the caller was a British national and as inquired she started giving her information and directions. Somehow, Sheldon wasn't able to follow the instructions and suddenly started abusing Mahek and said, "You don't know anything! You Asian bitch!! Just go to hell and get yourself another bitch for a company" As BPO etiquette taught during her training, Mahek had no option of responding to his abuses, has to listen to whatever he is saying and that too she can't hang up the phone by herself. After almost few minutes of verbal abuses and scolding, Sheldon hang up the phone and here, Mahek broke up with big tears in her eyes and felt pathetic with no fault of her own and decided to resign. Assuming you as her manager, how would you tackle this situation and avoid her resignation?

3.6.3 CASE SCENARIO 3

Mr. Swaminathan is the president for learning and development cell at ABC Infomatics in Bangalore (India). The cell offers a list of skills and courses that are being offered, and employees can select from and then study at respective institutes of world-class merit with a complete sponsorship from the organization. The most competitive among all the offered courses is the one-year course in business strategy at Harvard Business School. The selection is based on academic credentials and outstanding organizational performance for the past three years. This year too, Swaminathan has received 25 applications and among all only, one has to be selected. Swaminathan comes across the application of Krishnan Sarvapalli, a native of his place, and upon checking he feels that his application lacks in some selection criteria. But, he decides to select him over and above other eligible candidates thinking that selecting him is a kind of a way giving back to his own native place and community. The selection of Sarvapalli bring about an unrest among the other eligible candidates and they felt a racial preference in nominating Sarvapalli by Swaminathan and soon the Chief HR Officer, Mr. Gopinathan received four resignations from the organization's top talents. What would you do, if you were Gopinathan?

APPENDIX 3.1: CASE DISCUSSION

The case is a representation of real-life situations. The case does not offer an accurate answer. The objective of this is to sensitize the students with the issues pertinent to racial diversity. The cases put forth some scenarios which present an ideal situation to discuss in class to help the participants in understanding the concept of racial diversity, its implications and the effective way to manage it. It is advised that not to impose any solution as THE only solution to the case, nor completely rejecting or ignoring a solution suggested by the participants. The literature and discussion provided in this chapter can be used for class participation and reaching out for the solution.

Case Scenario 1

A similar situation is a common talk in any IT organization. The case is inspired from a real life situation wherein the team started facing trouble at the very inception due to racial comments by one of the team members. The racial comments add an atmosphere of superiority and inferiority in the work environment wherein the person reflecting superiority tries to dominate the one who is made to feel inferior. The later will experience oppression and won't be able to contribute productively and gradually will withdraw from work, thereby affecting the overall team performance.

Case Scenario 2

This situation narrates the typical problems being faced by executives who are engaged with dealing clientele with different cultures. There are some job profiles which deal with teleconsulting or tele calling which involves such instances. The company employees are trained to not to reply for any of such abuses but it affects their inner harmony and impact their work productivity.

Case Scenario 3

Another instance of regional and racial discrimination is presented, inspired from a true incident. The preferences for a particular race or a region are typical for some regions. Such kind of incidence loosens the faith and trust of employees in the top management, hampering their loyalty and overall performance in the organization. In an organization, the academic credentials and yearly performance report of all employees is open to all. It takes no time to diagnose such faulty nomination. In this case, the personal obligation of Swaminathan is affecting the organization by losing top bright talents. Gopinathan should call and ask for the grievances of the employees who offered resignation. After knowing their grievance, he can call for an inquiry and select the one with the most eligibility and merit.

KEYWORDS

- **diversity at work**
- **diversity engagement**
- **global diversity management**
- **globalization**
- **glocal**
- **optimal distinctive theory**
- **race**
- **races at work**
- **racial diversity**
- **surface-level diversity**

REFERENCES

Abrams, D., Wetherell, M., Cochrane, S., Hogg, M. A., & Turner, J. C., (1990). Knowing what to think by knowing who you are: Self-categorization and the nature of norm formation, conformity, and group polarization. *British Journal of Social Psychology*, *29*(2), 97–119.

Amir, Y., (1969). Contact hypothesis in ethnic relations. *Psychological Bulletin*, *71*(5), 319.

Ashforth, B. E., & Mael, F., (1989). Social identity theory and the organization. *Academy of Management Review*, *14*(1), 20–39.

Avery, D. R., McKay, P. F., & Wilson, D. C., (2008). What are the odds? How demographic similarity affects the prevalence of perceived employment discrimination. *Journal of Applied Psychology*, *93*(2), 235.

Banks, J. A., (1991). Teaching multicultural literacy to teachers. *Teaching Education*, *4*(1), 133–142.

Barak, M. E. M., (2016). *Managing Diversity: Toward a Globally Inclusive Workplace*. Sage Publications.

Barney, J. B., & Wright, P. M., (1998). On becoming a strategic partner: The role of human resources in gaining competitive advantage. *Human Resource Management*, *37*, 31–46.

Barsade, S. G., Ward, A. J., Turner, J. D., & Sonnenfeld, J. A., (2000). To your heart's content: A model of affective diversity in top management teams. *Administrative Science Quarterly*, *45*(4), 802–836.

Baumeister, R. F., & Leary, M. R., (1995). The need to belong: Desire for interpersonal attachments as a fundamental human motivation. *Psychological Bulletin*, *117*(3), 497.

Baysinger, B. D., & Butler, H. N., (1985). Corporate governance and the board of directors: Performance effects of changes in board composition. *Journal of Law, Economics, and Organization*, *1*(1), 101–124.

Bennett, D., (1998). *Multicultural States: Rethinking Difference and Identity*. Psychology Press.
Berger, J., Rosenholtz, S. J., & Zelditch, Jr. M., (1980). Status organizing processes. *Annual Review of Sociology*, 6(1), 479–508.
Bertrand, M., & Mullainathan, S., (2004). Are Emily and Greg more employable than Lakisha and Jamal? A field experiment on labor market discrimination. *American Economic Review*, 94(4), 991–1013.
Bhopal, R., (2004). Glossary of terms relating to ethnicity and race: For reflection and debate. *Journal of Epidemiology and Community Health*, 58(6), 441–445. doi: 10.1136/jech.2003.013466.
Block, P., (1987). *The Empowered Manager*. San Francisco: Jossey-Bass.
Bloomberg.com. Terms of Service Violation, (2018). [online] Available at: https://www.bloomberg.com/news/articles/2017-10-04/call-center-nation-fights-to-keep-automation-from-stealing-job (accessed on 24 February 2020)
Brewer, M. B., & Brown, R. J., (1998). Intergroup relations. In: Gilbert, D. T., Fiske, S. T., & Lindzey, G., (eds.), *Handbook of Social Psychology* (4th edn., pp. 554–594). Boston: McGraw Hill.
Brewer, M. B., (2007). The importance of being us: Human nature and intergroup relations. *American Psychologist*, 62(8), 728.
Bunderson, J. S., & Van Der Vegt, G. S., (2017). Diversity and inequality in management teams: A review and integration of research on vertical and horizontal member differences. *Annual Review of Organizational Psychology and Organizational Behavior*.
Byrne, D. E., (1971). *The Attraction Paradigm* (Vol. 11). Academic Pr.
Cassell, C., (2001). Managing diversity. In: Redman, T., & Wilkinson, A., (eds.), *Contemporary Human Resource Management: Text and Cases* (pp. 404–431). Harlow: Pearson Education.
Catalyst, (2004). *The Bottom Line: Connecting Corporate Performance and Gender Diversity*. Catalyst.
Clegg, S., (1990). *Modern Organizations: Organization Studies in the Postmodern World*. Sage.
Clore, G. L., & Byrne, D., (1974). A reinforcement-affect model of attraction. *Foundations of Interpersonal Attraction*, pp. 143–170.
Conner, K. R., & Prahalad, C. K., (1996). A resource-based theory of the firm: Knowledge versus opportunism. *Organization Science*, 7, 477–501.
Cornelius, N., (2002). *Building Workplace Equality: Ethics, Diversity, and Inclusion*. Cengage Learning EMEA.
Correll, J., & Park, B., (2005). A model of the in group as a social resource. *Personality and Social Psychology Review*, 9(4), 341–359.
Cox, Jr, T., (2001). *Creating the Multicultural Organization: A Strategy for Capturing the Power of Diversity*. Jossey-Bass.
Cox, T. H., & Blake, S., (1991). Managing cultural diversity: Implications for organizational competitiveness. *The Executive*, pp. 45–56.
Cox, T. H., Lobel, S. A., & McLeod, P. L., (1991). Effects of ethnic group cultural differences on cooperative and competitive behavior on a group task. *Academy of Management Journal*, 34, 827–847.
Cox, T., (1994). *Cultural Diversity in Organizations: Theory, Research, and Practice*. Berrett-Koehler Publishers.
Daya, P., (2014). Diversity and inclusion in an emerging market context. *Equality, Diversity, and Inclusion: An International Journal*, 33(3), 293–308.

DePaulo, B. M., & Friedman, H. S., (1998). *Nonverbal Communication*.
Dovidio, J. F., Kawakami, K., & Gaertner, S. L., (2002). Implicit and explicit prejudice and interracial interaction. *Journal of Personality and Social Psychology*, 82(1), 62.
Ferdman, B. M., & Sagiv, L., (2012). Diversity in organizations and cross-cultural work psychology: What if they were more connected? *Industrial and Organizational Psychology*, 5(3), 323–345.
Florkowski, G. W., (1996). Managing diversity within multinational firms for competitive advantage. *Managing Diversity* (pp. 337–364). Blackwell: Oxford.
Freeman, H. P., (1998). The meaning of race in science-considerations for cancer research. *Cancer*, 82(1), 219–225.
Global Diversity, (2018). Inclusion and Community Engagement | McDonald's. Retrieved from: https://corporate.mcdonalds.com/corpmcd/about-us/diversity-and-inclusion.html (accessed on 24 February 2020).
Gonzalez, J. A., & Denisi, A. S., (2009). Cross-level effects of demography and diversity climate on organizational attachment and firm effectiveness. *Journal of Organizational Behavior*, 30(1), 21–40.
Goodstein, J., Gautam, K., & Boeker, W., (1994). The effects of board size and diversity on strategic change. *Strategic Management Journal*, 15(3), 241–250.
Greenwald, A. G., & Banaji, M. R., (1995). Implicit social cognition: Attitudes, self-esteem, and stereotypes. *Psychological Review*, 102(1), 4.
Harrison, D. A., Price, K. H., & Bell, M. P., (1998). Beyond relational demography: Time and the effects of surface-and deep-level diversity on work group cohesion. *Academy of Management Journal*, 41(1), 96–107.
Hays-Thomas, R., (2004). Why now? *The Contemporary Focus on Managing Diversity*.
Herring, C., (2009). Does diversity pay? Race, gender, and the business case for diversity. *American Sociological Review*, 74(2), 208–224.
Hofstede, G., (1994). The business of international business is culture. *International Business Review*, 3(1), 1–14.
Hogg, M. A., & Terry, D. I., (2000). Social identity and self-categorization processes in organizational contexts. *Academy of Management Review*, 25(1), 121–140.
Homan, A. C., Hollenbeck, J. R., Humphrey, S. E., Van Knippenberg, D., Ilgen, D. R., & Van Kleef, G. A., (2008). Facing differences with an open mind: Openness to experience, salience of intragroup differences, and performance of diverse work groups. *Academy of Management Journal*, 51(6), 1204–1222.
Hong, Y. Y., & Chiu, C. Y., (2001). Toward a paradigm shift: From cross-cultural differences in social cognition to social-cognitive mediation of cultural differences. *Social Cognition*, 19(3), 181–196.
Inclusion and Diversity, (n.d.). Retrieved from: https://www.apple.com/diversity/ (accessed on 24 February 2020).
Insights, Forbes, (2001). *Global Diversity and Inclusion: Fostering Innovation through a Diverse Workforce*. Forbes Insight, New York.
Jackson, S. E., May, K. E., & Whitney, K., (1995). Understanding the dynamics of diversity in decision-making teams. *Team Effectiveness and Decision Making in Organizations*, 204, 261.
Jehn, K. A., Chadwick, C., & Thatcher, S. M., (1997). To agree or not to agree: The effects of value congruence, individual demographic dissimilarity, and conflict on workgroup outcomes. *International Journal of Conflict Management*, 8(4), 287–305.

Judge, T. A., & Piccolo, R. F., (2004). Transformational and transactional leadership: A meta-analytic test of their relative validity. *Journal of Applied Psychology, 89*(5), 755.

Kerckhoff, A. C., (1995). Institutional arrangements and stratification processes in industrial societies. *Annual Review of Sociology, 21*(1), 323–347.

King, E. B., Dawson, J. F., West, M. A., Gilrane, V. L., Peddie, C. I., & Bastin, L., (2011). Why organizational and community diversity matter: Representativeness and the emergence of incivility and organizational performance. *Academy of Management Journal, 54*(6), 1103–1118.

Kirkman, B. L., Tesluk, P. E., & Rosen, B., (2004). The impact of demographic heterogeneity and team leader-team member demographic fit on team empowerment and effectiveness. *Group and Organization Management, 29*(3), 334–368.

Konrad, A. M., (2003). Special issue introduction: Defining the domain of workplace diversity scholarship. *Group and Organization Management, 28*(1), 4–17.

Kosnik, R. D., (1990). Effects of board demography and directors' incentives on corporate greenmail decisions. *Academy of Management Journal, 33*(1), 129–150.

Leonard, M., Graham, S., & Bonacum, D., (2004). The human factor: The critical importance of effective teamwork and communication in providing safe care. *BMJ Quality and Safety, 13*(1), i85–i90.

Lopez, D., & Alegado, S., (2018). *Terms of Service Violation.* Retrieved from: https://www.bloombergquint.com/onweb/call-center-nation-fights-to-keep-automation-from-stealing-jobs (accessed on 24 February 2020).

Madan, T. N., (1992). *Religion in India.* Oxford University Press, USA.

Martins, L. L., Milliken, F. J., Wiesenfeld, B. M., & Salgado, S. R., (2003). Racioethnic diversity and group members' experiences: The role of the racioethnic diversity of the organizational context. *Group and Organization Management, 28*(1), 75–106.

McLeod, P. L., Lobel, S. A., & Cox, Jr. T. H., (1996). Ethnic diversity and creativity in small groups. *Small Group Research, 27*(2), 248–264.

Mees, H., (2016). *China as the World's Factory: In the Chinese Birdcage* (pp. 21–32). Palgrave Macmillan, New York.

Milliken, F. J., & Martins, L. L., (1996). Searching for common threads: Understanding the multiple effects of diversity in organizational groups. *Academy of Management Review, 21*(2), 402–433.

Mintzberg, H., (1983). The case for corporate social responsibility. *Journal of Business Strategy, 4*(2), 3–15.

Mitroff, I. I., (1978). *Methodological Approaches to Social Science.* Jossey-Bass Incorporated Pub.

Morrison, A. M., (1992). *The New Leaders: Guidelines on Leadership Diversity in America.* Jossey-Bass Management Series. Jossey-Bass, Inc., Publishers, 350Sansome Street, San Francisco, CA 94104.

Newcomb, T. M., (1961). *The Acquaintance Process.* New York Holt, Rinehart & Winston.

Parekh, B. C., (2011). *Rethinking Multiculturalism: Cultural Diversity and Political Theory.* Basingstoke: Palgrave Macmillan.

Pickett, C. L., & Brewer, M. B., (2001). Assimilation and differentiation needs as motivational determinants of perceived in-group and out-group homogeneity. *Journal of Experimental Social Psychology, 37*(4), 341–348.

Powell, W. W., (1991). Expanding the scope of institutional analysis. In: Powell, W. W., & DiMaggio, P. J., (eds.), *The New Institutionalism in Organizational Analysis: 183 203.* Chicago: University of Chicago Press.

Priem, R. L., Harrison, D. A., & Muir, N. K., (1995). Structured conflict and consensus outcomes in group decision making. *Journal of Management*, *21*(4), 691–710.

Reporting Requirements, (2018). Retrieved from: https://www.eeoc.gov/employers/reporting.cfm.

Richard, O. C., & Johnson, N. B., (2001). Understanding the impact of human resource diversity practices on firm performance. *Journal of Managerial Issues*, 177–195.

Richard, O. C., (2000). Racial diversity, business strategy, and firm performance: A resource-based view. *Academy of Management Journal*, *43*(2), 164–177.

Richard, O. C., Barnett, T., Dwyer, S., & Chadwick, K., (2004). Cultural diversity in management, firm performance, and the moderating role of entrepreneurial orientation dimensions. *Academy of Management Journal*, *47*(2), 255–266.

Richard, O. C., Stewart, M. M., McKay, P. F., & Sackett, T. W., (2017). The impact of store-unit-community racial diversity congruence on store-unit sales performance. *Journal of Management*, *43*(7), 2386–2403.

Richard, O., (2000). Racial diversity, business strategy, and firm performance: A resource-based view. *Academy of Management Journal*, *43*, 164–177.

Riordan, C. M., (2000). Relational demography within groups: Past developments, contradictions, and new directions. In: *Research in Personnel and Human Resources Management* (pp. 131–173). Emerald Group Publishing Limited.

Rosenbaum, M. E., (1986). Comment on a Proposed Two-Stage Theory of Relationship Formation: First, Repulsion; Then, Attraction. *Journal of Personality and Social Psychology*, *51*(6), 1171–1172.

Schneider, B., (1987). The people make the place. *Personnel Psychology*, *40*(3), 437–453.

Schuler, R. S., & Jackson, S. E., (1987–1989). *Linking Competitive Strategies with Human Resource Management Practices* (pp. 207–219). The Academy of Management Executive (1987–1989).

Shi, X., (2013). The glocalization of English: A Chinese case study. *Journal of Developing Societies*, *29*(2), 89–122.

Snyder, C. R., & Fromkin, H. L., (2012). *Uniqueness: The Human Pursuit of Difference*. Springer Science and Business Media.

Struch, N., & Schwartz, S. H., (1989). Intergroup aggression: Its predictors and distinctness from in-group bias. *Journal of Personality and Social Psychology*, *56*(3), 364.

Transparency International: India, (2018). Retrieved from: https://www.transparency.org/country/IND (accessed on 24 February 2020).

Tsui, A. S., & Gutek, B. A., (1999). *Demographic Differences in Organizations: Current Research and Future Directions*. Lexington Books.

Turner, J. C., (1999). Some current issues in research on social identity and self-categorization theories. *Social Identity: Context, Commitment, Content*, *3*(1), 6–34.

US Census Bureau. Race. n.d. Retrieved from: https://www.census.gov/topics/population/race.html (accessed on 24 February 2020).

Zhang, K. H., (2006). *China as the World Factory*. Routledge.

FURTHER READING

Abbas, M. S., (2019). *The Promise of Political Blackness?* Contesting blackness, challenging whiteness and the silencing of racism: A review article. Ethnicities. 1468796819834046.

Abji, S., Korteweg, A. C., & Williams, L. H., (2019). Culture talk and the politics of the new right: Navigating gendered racism in attempts to address violence against women in immigrant communities. *Signs, 44*(3), 797–822.

Acharya, A., & Gupta, M., (2016a). An application of brand personality to green consumers: A thematic analysis. *Qualitative Report, 21*(8), 1531–1545.

Acharya, A., & Gupta, M., (2016b). Self-image enhancement through branded accessories among youths: A phenomenological study in India. *Qualitative Report, 21*(7), 1203–1215.

Ackermann, R. R., (2019). Reflections on the history and legacy of scientific racism in South African paleoanthropology and beyond. *Journal of Human Evolution, 126*, 106–111.

Alang, S. M., (2019). Mental health care among blacks in America: Confronting racism and constructing solutions. *Health Services Research*.

Anderson, B., Narum, A., & Wolf, J. L., (2019). Expanding the understanding of the categories of dysconscious racism. *Educational Forum, 83*(1), 4–12.

Auld, G., (2018). Is there a case for mandatory reporting of racism in schools? *Australian Journal of Indigenous Education, 47*(2), 146–157.

Baak, M., (2019). Racism and othering for South Sudanese heritage students in Australian schools: Is inclusion possible? *International Journal of Inclusive Education, 23*(2), 125–141.

Bahler, B., (2018). How levinas can (and cannot) help us with political apology in the context of systemic racism. *Religions, 9*(11).

Beck, A., (2019). Understanding black and minority ethnic service user's experience of racism as part of the assessment, formulation, and treatment of mental health problems in cognitive behavior therapy. *Cognitive Behavior Therapist, 12*.

Bell, C. N., Kerr, J., & Young, J. L., (2019). Associations between obesity, obesogenic environments, and structural racism vary by county-level racial composition. International Journal of Environmental Research and Public Health, 16(5), 861.

Benz, T. A., (2019). Toxic cities: Neoliberalism and environmental racism in flint and Detroit Michigan. *Critical Sociology, 45*(1), 49–62.

Bernasconi, R., (2019). A most dangerous error: The Boasian myth of a knock-down argument against racism. *Angelaki-Journal of the Theoretical Humanities, 24*(2), 92–103.

Bernhard, P., (2019). The great divide? Notions of racism in Fascist Italy and Nazi Germany: New answers to an old problem. *Journal of Modern Italian Studies, 24*(1), 97–114.

Bhambhani, Y., Flynn, M. K., Kellum, K. K., & Wilson, K. G., (2019). Examining sexual racism and body dissatisfaction among men of color that have sex with men: The moderating role of body image inflexibility. *Body Image, 28*, 142–148.

Bonam, C. M., Nair, D. V., Coleman, B. R., & Salter, P., (2019). Ignoring history, denying racism: Mounting evidence for the Marley hypothesis and epistemologies of ignorance. *Social Psychological and Personality Science, 10*(2), 257–265.

Bonilla-Silva, E., (2019). "Racists," "class anxieties," hegemonic racism, and democracy in trump's America. *Social Currents, 6*(1), 14–31.

Botma, G., (2018). Spy fly and confessional: The fight against racism in the post-apartheid South African media [Loerbroer en biegbank: Die uitwysing van rassisme deur die post apartheid Suid Afrikaanse media]. *Tydskrif vir Geesteswetenskappe, 58*(4), 736–751.

Brewster, L. M., (2019). Race and racism in a medical context in the Netherlands [Ras en racisme in een medische context]. *Nederlands Tijdschrift Voor Geneeskunde, 163*.

Britton, J., (2019). Challenging the racialization of child sexual exploitation: Muslim men, racism, and belonging in Rotherham. *Ethnic and Racial Studies, 42*(5), 688–706.

Brooks, J. S., (2018). The unbearable whiteness of educational leadership: A historical perspective on racism in the American principal's office. *Whiteucation: Privilege, Power, and Prejudice in School and Society* (pp. 35–51).

Caldwell, L. D., & Bledsoe, K. L., (2019). Can social justice live in a house of structural racism? A question for the field of evaluation. *American Journal of Evaluation, 40*(1), 6–18.

Castle, B., Wendel, M., Kerr, J., Brooms, D., & Rollins, A., (2019). Public health's approach to systemic racism: A Systematic literature review. *Journal of Racial and Ethnic Health Disparities, 6*(1), 27–36.

Charania, G. R., (2019). Revolutionary love and states of pain: The politics of remembering and almost forgetting racism. *Women's Studies International Forum, 73*, 8–15.

Daniel, B. J., (2019). Teaching while Black: Racial dynamics, evaluations, and the role of white females in the Canadian academy in carrying the racism torch. *Race Ethnicity and Education, 22*(1), 21–37.

De Noronha, L., (2019). Deportation, racism, and multi-status Britain: Immigration control and the production of race in the present. *Ethnic and Racial Studies*.

Dennis, S. N., Gold, R. S., & Wen, F. K., (2019). Learner reactions to activities exploring racism as a social determinant of health. *Family Medicine, 51*(1), 41–47.

Dhillon-Jamerson, K. K., (2018). Euro-Americans favoring people of color: Covert racism and economies of white colorism. *American Behavioral Scientist, 62*(14), 2087–2100.

Doebler, S., McAreavey, R., & Shortall, S., (2018). Is racism the new sectarianism? Negativity towards immigrants and ethnic minorities in Northern Ireland from 2004 to 2015. *Ethnic and Racial Studies, 41*(14), 2426–2444.

Douxami, C., (2019). Brazilian black theatre: A political theatre against racism. *TDR-The Drama Review: A Journal of Performance Studies, 63*(1), 32–51.

Dyett, J., & Thomas, C., (2019). Overpopulation discourse: Patriarchy, racism, and the specter of ecofascism. *Perspectives on Global Development and Technology, 18*(43467), 205–224.

Eastwood, J., (2019). Reading Abdul Fattah Al-Sharif, reading Elor Azaria: Anti-Mizrahi racism in the moral economy of Zionist settler colonial violence. *Settler Colonial Studies, 9*(1), 59–77.

Elliott-Cooper, A., (2018). The struggle that cannot be named: Violence, space and the re-articulation of anti-racism in post-Duggan Britain. *Ethnic and Racial Studies, 41*(14), 2445–2463.

Fanning, B., & Michael, L., (2018). Racism and anti-racism in the two Irelands. *Ethnic and Racial Studies, 41*(15), 2656–2672.

Faulkner, N., & Bliuc, A. M., (2018). Breaking down the language of online racism: A comparison of the psychological dimensions of communication in racist, anti-racist, and non-activist groups. *Analyses of Social Issues and Public Policy, 18*(1), 307–322.

Flintoff, A., & Dowling, F., (2019). 'I just treat them all the same, really': Teachers, whiteness and (anti) racism in physical education. *Sport, Education and Society, 24*(2), 121–133.

Fox, J. E., & Mogilnicka, M., (2019). Pathological integration or, how East Europeans use racism to become British. *British Journal of Sociology, 70*(1), 5–23.

Franco, M., (2019). Let the racism tell you who your friends are: The effects of racism on social connections and life-satisfaction for multiracial people. *International Journal of Intercultural Relations, 69*, 54–65.

Gee, G. C., Hing, A., Mohammed, S., Tabor, D. C., & Williams, D. R., (2019). Racism and the life course: Taking time seriously. *American Journal of Public Health, 109*(S1), S43–S47.

Gillam, C., & Charles, A., (2019). Community wellbeing: The impacts of inequality, racism, and environment on a Brazilian coastal slum. *World Development Perspectives, 13*, 18–24.

Gilroy, P., Sandset, T., Bangstad, S., & Høibjerg, G. R., (2019). A diagnosis of contemporary forms of racism, race, and nationalism: A conversation with Professor Paul Gilroy. *Cultural Studies, 33*(2), 173–197.

Gorski, P. C., & Erakat, N., (2019). Racism, whiteness, and burnout in antiracism movements: How white racial justice activists elevate burnout in racial justice activists of color in the United States. *Ethnicities*.

Gorski, P. C., (2019). Fighting racism, battling burnout: Causes of activist burnout in US racial justice activists. *Ethnic and Racial Studies, 42*(5), 667–687.

Grant, J., & Guerin, P. B., (2018). Mixed and misunderstandings: An exploration of the meaning of racism with maternal, child, and family health nurses in South Australia. *Journal of Advanced Nursing, 74*(12), 2831–2839.

Griswold, M. K., Crawford, S. L., Perry, D. J., Person, S. D., Rosenberg, L., Cozier, Y. C., & Palmer, J. R., (2018). Experiences of racism and breastfeeding initiation and duration among first-time mothers of the black women's health study. *Journal of Racial and Ethnic Health Disparities, 5*(6), 1180–1191.

Gupta, M., & Kumar, Y., (2015). Justice and employee engagement: Examining the mediating role of trust in Indian B-schools. *Asia-Pacific Journal of Business Administration, 7*(1), 89–103.

Gupta, M., & Pandey, J., (2018). Impact of student engagement on affective learning: Evidence from a large Indian University. *Current Psychology, 37*(1), 414–421.

Gupta, M., & Ravindranath, S., (2018). Managing physically challenged workers at Microsign. *South Asian Journal of Business and Management Cases, 7*(1), 34–40.

Gupta, M., & Sayeed, O., (2016). Social responsibility and commitment in management institutes: Mediation by engagement. *Business: Theory and Practice, 17*(3), 280–287.

Gupta, M., & Shaheen, M., (2017a). Impact of work engagement on turnover intention: Moderation by psychological capital in India. *Business: Theory and Practice, 18*, 136–143.

Gupta, M., & Shaheen, M., (2017b). The relationship between psychological capital and turnover intention: Work engagement as mediator and work experience as moderator. *Journal Pengurusan, 49*, 117–126.

Gupta, M., & Shaheen, M., (2018). Does work engagement enhance general well-being and control at work? Mediating role of psychological capital. *Evidence-Based HRM, 6*(3), 272–286.

Gupta, M., & Shukla, K., (2018). An empirical clarification on the assessment of engagement at work. *Advances in Developing Human Resources, 20*(1), 44–57.

Gupta, M., (2017). Corporate social responsibility, employee-company identification, and organizational commitment: Mediation by employee engagement. *Current Psychology, 36*(1), 101–109.

Gupta, M., (2018). Engaging employees at work: Insights from India. *Advances in Developing Human Resources, 20*(1), 3–10.

Gupta, M., Acharya, A., & Gupta, R., (2015). Impact of work engagement on performance in Indian higher education system. *Review of European Studies, 7*(3), 192–201.

Gupta, M., Ganguli, S., & Ponnam, A., (2015). Factors affecting employee engagement in India: A study on off shoring of financial services. *Qualitative Report, 20*(4), 498–515.

Gupta, M., Ravindranath, S., & Kumar, Y. L. N., (2018). Voicing concerns for greater engagement: Does a supervisor's job insecurity and organizational culture matter? *Evidence-Based HRM, 6*(1), 54–65.

Gupta, M., Shaheen, M., & Das, M., (2019). Engaging employees for quality of life: Mediation by psychological capital. *The Service Industries Journal, 39*(5/6), 403–419.

Gupta, M., Shaheen, M., & Reddy, P. K., (2017). Impact of psychological capital on organizational citizenship behavior: Mediation by work engagement. *Journal of Management Development, 36*(7), 973–983.

Hammer, P. J., (2019). The flint water crisis, the karegnondi water authority and strategic-structural racism. *Critical Sociology, 45*(1), 103–119.

Harris, R. B., Cormack, D. M., & Stanley, J., (2019). Experience of racism and associations with unmet need and healthcare satisfaction: The 2011/12 adult new Zealand health survey. *Australian and New Zealand Journal of Public Health, 43*(1), 75–80.

Hart, C. L., & Hart, M. Z., (2019). Opioid crisis: Another mechanism used to perpetuate American racism. *Cultural Diversity and Ethnic Minority Psychology, 25*(1), 6–11.

Hirsch, S., (2019). Racism, 'second generation' refugees, and the asylum system. *Identities, 26*(1), 88–106.

Howell, A., & Richter-Montpetit, M., (2018). Racism in Foucauldian security studies: Biopolitics, liberal war, and the whitewashing of colonial and racial violence. *International Political Sociology, 13*(1), 2–19.

Hunt, W., (2019). Negotiating new racism: 'It's not racist or sexist. It's just the way it is.' *Media, Culture and Society, 41*(1), 86–103.

Ikuenobe, P., (2018). The practical and experiential reality of racism: Carter's and Corlett's realism about race and racism. *Journal of African American Studies, 22*(4), 373–392.

Ilten-Gee, R., (2019). Complicating moral messages through multimodal composition: Wrestling with revenge and racism. *Ethnography and Education, 14*(1), 84–100.

Ince, A., (2019). Fragments of an anti-fascist geography: Interrogating racism, nationalism, and state power. *Geography Compass, 13*(3).

Jaiswal, J., Singer, S. N., Siegel, K., & Lekas, H. M., (2019). HIV-related 'conspiracy beliefs': Lived experiences of racism and socio-economic exclusion among people living with HIV in New York City. *Culture, Health and Sexuality, 21*(4), 373–386.

Jamieson, E., (2018). Systemic racism as a living text: Implications of Uncle Tom's Cabin as a fictionalized narrative of present and past black bodies. *Journal of African American Studies, 22*(4), 329–344.

Jarvis, S., (2019). Standing up to racism. *Veterinary Record, 184*(3), 73.

Jeyasingham, D., & Morton, J., (2019). How 'racism' is understood in literature about black and minority ethnic social work students in Britain? A conceptual review. *Social Work Education*.

Kilvington, D., & Price, J., (2019). Tackling social media abuse? Critically assessing English football's response to online racism. *Communication and Sport, 7*(1), 64–79.

Kroon, Å., (2019). Recontextualizing racism and segregation by ways of "cozification" in a TV sports broadcast. *Social Semiotics, 29*(1), 112–128.

Kühlbrandt, C., (2019). Confronting racism in family planning: A critical ethnography of Roma health mediation. *Sexual and Reproductive Health Matters, 27*(1).

Kuznetsova, I., & Round, J., (2019). Postcolonial migrations in Russia: The racism, informality and discrimination nexus. *International Journal of Sociology and Social Policy, 39*(43467), 52–67.

Leonardo, Z., & Dixon-Román, E., (2018). Post-colorblindness, or, racialized speech after symbolic racism. *Educational Philosophy and Theory, 50*(14), 1386-.

Limb, M., (2019). 'It's time to call out racism in the profession.' *Veterinary Record, 184*(3), 81–84.

Lotem, I., (2018). Beyond memory wars: The indigènes de la république's grass-roots anti-racism between the memory of colonialism and antisemitism. *French History, 32*(4), 573–593.

Lykes, M. B., Lloyd, C. R., & Nicholson, K. M., (2018). Participatory and action research within and beyond the academy: Contesting racism through decolonial praxis and teaching "against the grain." *American Journal of Community Psychology, 62*(43528), 406–418.

Markwick, A., Ansari, Z., Clinch, D., & McNeil, J., (2019a). Experiences of racism among Aboriginal and Torres Strait islander adults living in the Australian state of Victoria: A cross-sectional population-based study. *BMC Public Health, 19*(1).

Markwick, A., Ansari, Z., Clinch, D., & McNeil, J., (2019b). Perceived racism may partially explain the gap in health between Aboriginal and non-Aboriginal Victorians: A cross-sectional population based study. *SSM-Population Health, 7*.

Marom, L., (2019). Under the cloak of professionalism: Covert racism in teacher education. *Race Ethnicity and Education, 22*(3), 319–337.

Matsuzaka, S., & Knapp, M., (2019). Anti-racism and substance use treatment: Addiction does not discriminate, but do we? *Journal of Ethnicity in Substance Abuse*.

Miller, A. K., (2019). "Should have known better than to fraternize with a black man": Structural racism intersects rape culture to intensify attributions of acquaintance rape victim culpability. *Sex Roles*.

Moxley, R. C., (2019). Liberal bias: The new "reverse racism" in the trump era. *American Anthropologist, 121*(1), 172–176.

Neblett, E. W. Jr., (2019). Racism and health: Challenges and future directions in behavioral and psychological research. *Cultural Diversity and Ethnic Minority Psychology, 25*(1), 12–20.

Opperman, R., (2019). A permanent struggle against an omnipresent death: Revisiting environmental racism with Frantz fanon. *Critical Philosophy of Race, 7*(1), 57–80.

Ortiz, S. M., (2019). "You can say i got desensitized to it": How men of color cope with everyday racism in online gaming. *Sociological Perspectives*.

Pandey, J., Gupta, M., & Naqvi, F., (2016). Developing decision making measure a mixed method approach to operationalize Sankhya philosophy. *European Journal of Science and Theology, 12*(2), 177–189.

Patel, T. G., & Connelly, L., (2019). 'Post-race' racisms in the narratives of 'Brexit' voters. *Sociological Review*.

Pearce, S., (2019). 'It was the small things': Using the concept of racial micro aggressions as a tool for talking to new teachers about racism. *Teaching and Teacher Education, 79*, 83–92.

Piracha, A., Sharples, R., Forrest, J., & Dunn, K., (2019). Racism in the sharing economy: Regulatory challenges in a neo-liberal cyber world. *Geoforum, 98*, 144–152.

Preston, C. J., (2018). Hissing, bidding, and lynching: Participation in Brandon Jacobs-Jenkins's an octoroon and the melodramatics of American Racism. *TDR-The Drama Review: A Journal of Performance Studies, 62*(4), 64–80.

Pulido, L., Bruno, T., Faiver-Serna, C., & Galentine, C., (2019). Environmental deregulation, spectacular racism, and white nationalism in the trump era. *Annals of the American Association of Geographers, 109*(2), 520–532.

Ratner, D., (2019). Rap, racism, and visibility: Black music as a mediator of young Israeli-Ethiopians' experience of being 'black' in a 'white' society. *African and Black Diaspora, 12*(1), 94–108.

Rhee, T. G., Marottoli, R. A., Van Ness, P. H., & Levy, B. R., (2019). Impact of perceived racism on healthcare access among older minority adults. *American Journal of Preventive Medicine, 56*(4), 580–585.

Russell, J. G., (2018). Darkies never dream: Race, racism, and the black imagination in science fiction. *New Centennial Review, 18*(3), 255–277.

Rzepnikowska, A., (2019). Racism and xenophobia experienced by Polish migrants in the UK before and after Brexit vote. *Journal of Ethnic and Migration Studies, 45*(1), 61–77.

Sambaraju, R., & Minescu, A., (2019). 'I have not witnessed it personally myself, but': Epistemics in managing talk on racism against immigrants in Ireland. *European Journal of Social Psychology, 49*(2), 398–412.

Sayan, P., (2019). Enforcement of the anti-racism legislation of the European Union against antigypsyism. *Ethnic and Racial Studies, 42*(5), 763–781.

Seet, A. Z., (2019). Racialised self-marketization: The importance of accounting for neoliberal rationality within manifestations of internalized racism. *Journal of Intercultural Studies, 40*(2), 155–171.

Seider, S., Clark, S., Graves, D., Kelly, L. L., Soutter, M., El-Amin, A., & Jennett, P., (2019). Black and Latinx adolescents' developing beliefs about poverty and associations with their awareness of Racism. *Developmental Psychology, 55*(3), 509–524.

Seikkula, M., (2019). Adapting to post-racialism? Definitions of racism in non-governmental organization advocacy that mainstreams anti-racism. *European Journal of Cultural Studies, 22*(1), 95–109.

Sheth, M. J., (2019). Grappling with racism as foundational practice of science teaching. *Science Education, 103*(1), 37–60.

Singh, V., (2018). Myths of meritocracy: Caste, karma and the new racism, a comparative study. *Ethnic and Racial Studies, 41*(15), 2693–2710.

Smedley, B. D., (2019). Multilevel interventions to undo the health consequences of racism: The need for comprehensive approaches. *Cultural Diversity and Ethnic Minority Psychology, 25*(1), 123–125.

Solic, K., & Riley, K., (2019). Teacher candidates' experiences taking up issues of race and racism in an urban education fellowship program. *Action in Teacher Education.*

Souto-Manning, M., (2019). Toward praxically-just transformations: Interrupting racism in teacher education. *Journal of Education for Teaching, 45*(1), 97–113.

Sriprakash, A., Tikly, L., & Walker, S., (2019). The erasures of racism in education and international development: Re-reading the 'global learning crisis.' *Compare.*

Temple, J. B., Kelaher, M., & Paradies, Y., (2019). Prevalence and context of racism experienced by older aboriginal and Torres Strait islanders. *Australasian Journal on Ageing, 38*(1), 39–46.

Thomas, V. G., Madison, A., Rockcliffe, F., DeLaine, K., & Lowe, S. M., (2018). Racism, social programming, and evaluation: Where do we go from here? *American Journal of Evaluation, 39*(4), 514–526.

Torevell, D., (2019). Racism, anger and the move towards reconciliation: A modest proposal about returning to a stable base. *Journal of Beliefs and Values, 13*(1).

Trieu, M. M., (2019). Understanding the use of "Twinkie," "banana," and "FOB": Identifying the origin, role, and consequences of internalized racism within Asian America. *Sociology Compass.*

Urquidez, A. G., (2018). What accounts of 'racism' do. *Journal of Value Inquiry, 52*(4), 437–455.

Uzogara, E. E., (2019). Gendered racism biases: Associations of phenotypes with discrimination and internalized oppression among Latinx American women and men. *Race and Social Problems, 11*(1), 80–92.

Van Sterkenburg, J., Peeters, R., & Van Amsterdam, N., (2019). Everyday racism and constructions of racial/ethnic difference in and through football talk. *European Journal of Cultural Studies*, 1367549418823057.

Venkatesan, S., (2019). Violence and violation are at the heart of racism: The 2017 debate of the group for debates in anthropological theory, Manchester. *Critique of Anthropology*, *39*(1), 12–51.

Wills, J. S., (2019). "Daniel was racist": Individualizing racism when teaching about the civil rights movement. *Theory and Research in Social Education*, pp. 1–30.

Wilson, S. A., (2019). Racism is real. Racism is complicated. Racism is real complicated. *Family Medicine, 51*(1), 8–10.

Winder, T. J. A., & Lea, C. H. III., (2019). "Blocking" and "filtering": A commentary on mobile technology, racism, and the sexual networks of young black MSM (YBMSM). *Journal of Racial and Ethnic Health Disparities, 6*(2), 231–236.

Wiseman, A. M., Vehabovic, N., & Jones, J. S., (2019). Intersections of race and bullying in children's literature: Transitions, racism, and counter narratives. *Early Childhood Education Journal*, 1–10.

Zapolski, T. C. B., Faidley, M. T., & Beutlich, M. R., (2019). The experience of racism on behavioral health outcomes: The moderating impact of mindfulness. *Mindfulness, 10*(1), 168–178.

CHAPTER 4

Education Divide at Work

NEHA GANGWAR

Doctoral Fellow, National Institute of Industrial Engineering (NITIE), Mumbai; Room No 411, Gilbreth Hall, NITIE, Powai, Mumbai, Tel: 9871623987/9415230212, E-mail: gangwar.n12@gmail.com

ABSTRACT

The concept of diversity management is so wide and deep that no single chapter can explain all its the edges. This chapter is prepared by keeping in view the discrimination with concern to diversely educated employees. The discussion is started with different categories of diverse workforces, then moves towards the major theme of educational diversification within the organization. This is the newest in the category of diversity management. The educational divide depicts the situations in which employees were treated differently based on their educational institutional background. The concept is related to an employee's educational institutional value rather than their qualification or abilities. The chapter contains cases for a better understanding of the concept in the organizational practices.

4.1 INTRODUCTION

Educational qualification is the base for employment decisions in any organization but when employers believe that the institutional value is better than employees' skill then it leads to the selection discrimination. Different institutes have different quality status in India and abroad but this can't be taken as the only parameter for judging candidates knowledge and skills, it has been observed in many cases that organizations want to take an employee only from certain institutes which is not the essential requirement for the selection procedure. This chapter will talk about the issues based on educational qualification discrimination. The trend is increasing at a very fast pace as every organization wants to mention the tags in their company profiles and at some corner, it is becoming very difficult for the candidates to fight this scenario.

Based on Cambridge Dictionaries Online (2013), "In human social affairs discrimination is treatment or consideration of, or making a distinction in favor of or against, a person based on the group, class, or category to which the person is perceived to belong rather than on individual attributes. Includes treatment of an individual or group, based on their actual or perceived membership in a certain group or social category, in a way that is worse than the way people are usually treated." According to Norton and Company Inc (2009), "it involves the group's initial reaction or interaction going on to influence the individual's actual behavior towards the group leader or the group, restricting members of one group from opportunities or privileges that are available to another group, leading to the exclusion of the individual or entities based on logical or irrational decision making. In addition to this discrimination develops into a source of oppression. Thompson, Neil (2016) mentioned, "it is similar to the action of recognizing someone as 'different' so much that they are treated degraded."

During the past decade, the term "diversity" has been widely used to refer to the demographic composition of a team. In empirical studies, team diversity is usually measured using the compositional approach (Tsui and Gutek, 2000), which focuses on the distribution of demographic attributes—e.g., age, ethnicity, gender—within teams. Miles and Snow (1986) described a futuristic network organization characterized by constantly evolving inter-team linkages that allow organizations to quickly respond to technological and market changes, and thereby improve their chances of survival. Lazear (1999) has emphasized that the gains from diversity are greatest when the individuals have separate, but complementary information sets and the information can be learned at low cost. In studies of team diversity and organizational demography, numerous attributes have proved to be of interest, including age, gender, and ethnicity, length of tenure in the organization, functional specialization, educational background, cultural values, and personality. We refer to these attributes as the content of diversity (following Jackson, May, and Whitney, 1995). Although substantive differences in perspective may actually be beneficial to the team's performance on some types of tasks (see Jackson et al., 1991; Jackson, May, and Whitney, 1995), educational diversity is also likely to stimulate conflict and reduce cooperation.

When the women and the minority groups get isolated from the social and instrumental exchanges in organizations (Ibarra, 1992, 1995), their lack of access to social capital acts as a barrier to advancement (Friedman and Krackhardt, 1997; Ragins and Sundstrom, 1990). Increasing the representation of women and minorities throughout an organization—increasing the diversity of the organization—is one way to improve their access to social capital (Morrison and Von Glinow, 1990). Another way that employers can increase the access to social capital of women and minority employees is by supporting identity network groups (Friedman, 1996). Furthermore, men and women not only hold different job titles in organizations, but also occupy different hierarchical positions, with men dominating the managerial and supervisory ranks and women concentrated in the lower organizational ranks (Wolf and Fligstein, 1979a, b). The few studies that have included organizational variables have treated them as given and labeled as discrimination only that portion of the observed wage gap between men and women not due to their different individual characteristics and organizational positions. For instance, if men and women with equal education, experience, and other relevant individual characteristics hold different jobs within an organization, and as a result receive unequal wages, such studies would conclude that the wage differential between the two groups is the result not of discrimination but rather of their having different job titles. Thus, the wage differences between people holding different organizational positions are linked to the legitimate portion of the wage differential between men and women (Haberfeld Yitchak, 1992).

Organizational location, or position in a unit such as a division or a department, may affect a worker's exposure to organizational rewards and career opportunities. Those who work close to a company's headquarters, for example, may enjoy better working conditions than those employed in peripheral divisions simply because the former have more information regarding organizational policies and better access to rewards and opportunities. Likewise, access to high-level positions is determined by promotion processes, which are probably the major avenue through which organizations affect earnings inequality, since they create formally stratified social structures (Scott, 1987; Simon, 1957).

4.1.1 TYPES OF DISCRIMINATION

Since the turn of the century, the history of discrimination within the workplace has been developing by adding new clauses and understandings. Many businesses have been accused of workplace discrimination over the past century or so. As such, the government has tried to regulate discrimination in order to protect employees' rights. State and federal agencies are in charge of overseeing that workplace discrimination and ensuring that it does not occur of the best of their abilities (Nicole, 2016).

- **Employment Discrimination:** Denying someone employment, or disallowing one from applying for a job, is often recognized as employment discrimination when the grounds for such an exclusion is not related to the requirements of the position, and protected characteristics may include age, disability, ethnicity, gender, gender identity, height, nationality, political affiliation, religion, sexual orientation, skin color, and weight.
- **Age:** Ageism or age discrimination is discrimination and stereotyping based on the grounds of someone's age. It is a set of beliefs, norms, and values which used to justify discrimination or subordination based on a person's age.
- **Caste:** Discrimination based on caste, as perceived by UNICEF, is mainly prevalent in parts of Asia, (India, Sri Lanka, Bangladesh, China, Pakistan, Nepal, Japan), Africa, and others.
- **Disability:** Discrimination against people with disabilities in favor of people who are not is called ableism or disablism. Disability discrimination, which treats non-disabled individuals as the standard of 'normal living,' results in public and private places and services, education, and social work that are built to serve 'standard' people, thereby excluding those with various disabilities.
- **Language:** Discrimination exists if there is a prejudicial treatment against a person or a group of people who speak a particular language or dialect.
- **Name:** Research has further shown that real-world recruiters spend an average of just six seconds reviewing each résumé before making their initial "fit/no fit" screen-out decision and that a person's name is one of the six things they focus on most.
- **Nationality:** Discrimination on the basis of nationality may show as a "level of acceptance" in a sport or work team regarding new team members and employees who differ from the nationality of the majority of team members.
- **Race or Ethnicity:** Racial and ethnic discrimination differentiates individuals on the basis of real and perceived racial and ethnic differences and leads to various forms of the ethnic penalty.
- **Region:** Regional or geographic discrimination is discrimination based on the region in which a person lives or was born. It differs from

national discrimination in that it may not be based on national borders or the country the victim lives in, but is instead based on prejudices against a specific region of one or more countries.
- **Religious Beliefs:** Religious discrimination is valuing or treating a person or group differently because of what they do or do not believe or because of their feelings towards a given religion.
- **Sex, Sex Characteristics, Gender, and Gender Identity:** Though gender discrimination and sexism refer to beliefs and attitudes in relation to the gender of a person, such beliefs and attitudes are of a social nature and do not, normally, carry any legal consequences. Sexual discrimination can arise in different contexts. For instance, an employee may be discriminated against by being asked discriminatory questions during a job interview, or by an employer not hiring or promoting, unequally paying, or wrongfully terminating, an employee based on their gender.
- **Other Forms of Discrimination:**
    1. Drug use;
    2. Othering;
    3. Redlining;
    4. Reverse discrimination.

Above are the some studied topics under discrimination at work, different researcher were talking about many aspects of discrimination but there are very few studies which are talking about the discrimination at work because of any employee's education qualification. The present chapter will talk about this issue.

4.2 NEED FOR DISCUSSING THIS THEME

Managing the diverse workforce in the most important issue for HR, and this diversification may lead to discrimination for the employees and it leads to dissatisfaction among the employees. Discrimination in the workplace is a major concern in today's business community. The increase in cultural and gender diversity in the workplace has obligated employees from different ethnicities and backgrounds to work together to meet the goals of the company. Unfortunately, differences between people have a tendency to lead to misunderstandings, and result in conflict and discrimination. Employers have a responsibility to their workers to protect them from discrimination and unfair treatment in the workplace. Bridging the gap between employee and employer views will require substantial changes from HR. The gap should be managed and reduced.

4.3 REVIEW OF THIS THEME

Discrimination is quite old as the concept but educational discrimination is the most recent form of it, there are many forms of discrimination described by different laws in different countries but so far there is no specific law or policy is present for education discrimination. Following are some of the evidences for discrimination and polices in relation to them:

In the U.S., effective federal legislation banning employment-related discrimination did not exist until the 1960s, when Congress passed Title VII of the Civil Rights Act (1964). In the years since several other important federal laws have been passed. In addition to the myriad federal laws banning discrimination on the basis of race, color, sex, religion, national origin, age, disability, and veteran's status, almost all states have anti-discrimination laws affecting the workplace. Some state laws also attempt to prevent discrimination against individuals and groups that are not included in federal law.

Title VII is probably the most valuable tool that employees have for remedying workplace discrimination because it covers the greatest number of protected classifications. Title VII of the Civil Rights Act of 1964 has had an enormous impact on the human resource management (HRM) practices of many companies, by forcing them to take a close look at the way they recruit, hire, promote, award pay raises, and discipline their employees. As a result of this self-scrutiny, many firms have changed their practices, making them more systematic and objective. A number of Supreme Court decisions in the mid-to-late 1980s made discrimination claims under Title VII more difficult for employees to substantiate. To put more teeth into the law, Congress amended it by enacting the Civil Rights Act of 1991. Moreover, the CRA of 1991 adds additional bite to the 1964 law by providing a more detailed description of the evidence needed to prove a discrimination claim, making such claims easier to prove.

- The Equal Pay Act of 1963 prohibits discrimination in pay on the basis of sex when jobs within the same company are substantially the same. The company is allowed to pay workers doing the same job differently if the differences are based on merit, seniority, or any other reasonable basis other than the workers' gender.

- The Age Discrimination in Employment Act (ADEA) of 1967 bans employment discrimination on the basis of age by protecting applicants and employees who are 40 or older.
- The Pregnancy Discrimination Act of 1978 amended the CRA of 1964 by broadening the interpretation of sex discrimination to include pregnancy, childbirth, or related medical conditions. It prohibits discrimination against pregnant job applicants or against women who are of child-bearing age.
- The immigration Reform and control act of 1986 prohibited discrimination based on national origin and citizenship.
- The Americans with Disabilities Act (ADA) of 1990 was designed to eliminate discrimination against individuals with disabilities.

Navon (2009) measured knowledge diversity by a Herfindahl index that accounted for both the number of skilled workers and their disciplines and found positive productivity effects with Israeli data. She also found that hiring workers who possess diversified specific knowledge (a university degree) is beneficial for the plants' productivity. In Grund and Westergård-Nielsen (2008) the standard deviation of education was negatively related to labor productivity in fixed effects estimation (although positively related in OLS), whereas in Ilmakunnas et al. (2004) it was positively related to total factor productivity. Diversity of education is also relevant because teams working on complex cognitive tasks in organizations are typically com prized of people with different educational backgrounds (cf. Bantel and Jackson, 1989) representing distinct "thought worlds" (Dougherty, 1992).

4.3.1 DESCRIPTION OF THE TOPIC

After the introduction of so many concepts and issues about diversity in the workforce, organizations are becoming more aware about the discrimination among the employees. As time is increasing, we are facing new challenges, in the Indian context, we can justify the diversity which is based on the employee's educational institutes. In India, we have so many layers of educational institutes that fulfill the workforce's requirements of the different organizations. The educational institutes are dived into different categories such as Premier institutes (IITs, IIMs, IISERs, NITs, AIIMS, and Top Ranked Institutes, etc.), Central, and State Universities, Private Universities, University Affiliated and Autonomous Institutes. The discriminations start when an employee is differentiated based on his/her higher educational institute, the employee is not getting due recognition based on the ability and skills. Now in recent, it is being noticed that organizations are preferring employees from the premier institutes which in any case cannot fulfill the complete organizational workforce requirement and this preference in leading the organization with blank posts. On the other hand students of other institutes has to compromise for getting job, they are being paid less for the same work and duties, allocation of extra working hours and less recognition are some of the common issues.

Employees are diverse in the knowledge and their education background, which gives employers a freedom to choose from the variety of knowledge pool. But unfortunately, this freedom becomes nightmare for the candidates because of the brand conscious nature of job market. This consciousness leads in the discrimination from employer's side. Discrimination exists at almost everywhere and it can be based on many factors in general we categories the organizational discrimination based on the issues of person's or group's race, color, sex, religion, national origin, age, educational background, disability, or veteran's status. Any employee who comes in the organization has own set of skills and knowledge, but most often we connect their skills and knowledge with their education background. Many employers require you to have a specific type and level of education to qualify for certain jobs. Workplace Educational divide happens when an employer requires a specific institutional degree of education that isn't necessary for the job. Educational background is still important in determining a person's position. Considering degrees only the base of learning is the discrimination with the people and it is not the only way to learn and acquire knowledge. We may acquire knowledge and skills by learning and on job experience. The employers have the perception towards ordinary educated employees that they are not as good as the premier educated employees, which in reality is not the case. The issue which employees face in the organization is that their skills are being judged on the bases of their degrees and they will not get equal chances of showing their skills. For dealing with this issue the management and organizations have to become open, and be adoptive for the skills rather than degrees. Educational divide is also within the organizations, employees who are already working in the organization are struggling for the fair recognition and responsibilities. There are so many cases when deserving employees are suggested to get a degree for his/her promotion. Even now, organizations are supporting their employees to pursue the degree for getting the recognition in the organization; they feel that promoting an employee purely based on his/her knowledge is not justifiable.

Organizational policies of recruitment, selection, and promotions often perpetuate segregation based on gender or race (Ely, 1995; Nkomo, 1992; Wharton, 1992). While upper management levels in organizations may be predominantly White or male, minorities and women are often confined to entry levels. These characteristics of organizational demography reinforce identification on the basis of gender and race (Ely, 1995; Nkomo, 1992; Wharton, 1992) as well as the formation of segregated social networks within an organization (Ibarra and Smith-Lovin, 1997).

Jehn et al. (1997) found that when team members differed in terms of educational background they perceived greater conflict in the group. In a study of a household goods moving firm, John, and her colleagues found that greater informational diversity (which could be created by educational differences) in teams was associated with more task conflict (Jehn, Northcraft, and Neale, 1999). In their study of top management teams, Knight et al. (1999) found that educational diversity was associated with lower levels of strategic consensus.

The challenge is for the HR to understand the core requirement of skills rather than running for the background of the employees. As there are many talented employees which an organization can get but losing because of this one aspect. The chapter is trying to focus on this issue for the better understanding and minimizing the discrimination and making organization more diversified.

4.4 IMPLICATIONS

To manage diversification based on the educational institutes of the employees as a big concern for the managers. The concept of educational background is very much important for the managers to understand employee's skills set and it will help them to understand the employees more clearly and to help them learn and grow in the organization. The concept is about understanding the difference among the two and uses of their skills and degrees as per the requirement of the organization. It's not only the degree that is required for any organization to grow that's why this becomes outmost importance for the companies to understand that the discrimination should not be based on the education institutions. The following will be the probable issues which can be resolved by understanding the education diversification.

- Help the managers to understand the employee's skills rather than focusing the degrees.

- Premier education institutes can serve only a small need of the workforce, so in those cases, employers can look for the employee's skills rather than Institutes.
- Not all the positions can be filled only by the ordinary or premier educated employees; a perfect mix is required for organizational growth.
- Clarifies the need requirement for the role or job in the organization.
- Helps the organization to draw policies for diversified educated employees.
- The candidates from non-premier institutes have a high future expectation and urge to prove themselves, companies can use this for their own growth.

4.5 CONCLUSION

The chapter Education Divide at Work is focusing on discrimination based on someone's educational institutional background. It can be further understood with two aspects, first is when in the organization two employees having same experience were differentiated on the basis of their educational institute's reputation and another is when for hiring the employees the preference is given on the educational institute's reputation rather than the knowledge or experience. This was the broad theme of this chapter. For explaining it further, the chapter includes different types of discrimination, the beginning of educational discrimination, its managerial implications, and why it is important to understand. Cases are also included for understanding the concept more clearly and understanding the practical aspect of the theme. In the subsequent section, a case covering the above concept has been provided (refer Appendix 4.1 and 4.2 for plausible case discussion).

4.6 CASE 1: NG PHARMACEUTICALS: HIRING ESCAPE

In the summer of 2018, Rajat Singh is facing a tough time in getting the suitable profiles for the urgent vacant post of Marketing Head in the NG Pharmaceuticals. Rajat is HR executive in NG pharmaceuticals and he is having an urgent requirement for filling up the vacancy, as marketing unit is one of the most important functions and the current head is leaving the organization in a month. He starts posting the vacancy in all the leading job portals as well as all the social professional platforms. The days are increasing in numbers and no match found till now. Rajat decides to

Education Divide at Work 117

open employee referral for the vacancy and starts preparing the database for resumes received so far. As the organization follows a set standard for qualification and experience for hiring any employee, it has become very complicated doing initial screening.

4.6.1 COMPANY BACKGROUND

NG Pharmaceuticals in the leading company in the Pharmaceutical industry and has a big brand value in the market. They offer very high incentives and good perks to their employees, Mr. Bhalla current marketing head is going to start his own business and thus leaving the organization in the short notice of one month though he is ready to provide full support to the new join as he spent 15 years in the organization, and he is one of the building pillars of the organization. The organization is known for its marketing strategies in the industry and this is reason that organization is very much concerned about this vacancy. Management don't want to take any risk for filling up this position that's why team of HR is doing a very rigorous job in making the job specification. The hr team comes up with the idea of hiring the head from one of the premier institutes of the country. This will make this situation somewhat better. Management is happy with this idea and they start the advertisements. After passing 10 days of advertisement the HR is not able to find the desired person for the organization. The HR is getting very good profile but none of them having degree from the premier management institute. After the second meeting with the management, HR team has decided to go with the currently available resumes and start the interview process so that they will find the best candidate before Mr. Bhalla left the organization.

4.6.2 RAJAT'S HIRING RACE

Rajat from HR team is responsible for doing the screening and interview scheduling, he has all the instructions from the management still, he feels that with the criteria of taking only the premier management institute pass outs is going to be very difficult in finding the right candidate. He came across with many good profiles with good experience, so he decides to inform the seniors about the situation and waiting their response. He received the positive response and fellows the guidelines and updates the team about the status of application, after getting the permission for making the screening list and scheduling the interviews he is happy because now he feels that many good candidates will get a chance to prove them. And he prepares a list for 10 best suitable candidates and after the initial telephonic round he lined up the interview to be held on Tuesday 10, July, 2018.

4.6.3 THE INTERVIEW DAY

In the morning of 10 July, 2018 the interview room is all set for welcoming the best candidates shortlisted after initial scanning, the day is started at 10:15 with the first candidate and the first round ends for lunch after shortlisting of Mr. Alex and Mr. Vijay. The lunch is over and management is all set to meet the candidates and discuss on different roles and duties expected from them and a small discussion over their expectation from the company. After the deep discussion with both the candidates management found that Mr. Vijay, a 38-year-old energetic man with full of market knowledge and capabilities to move the business to its heights is the best suitable candidate for them. Rajat is said to make the offer letter and inform Mr. Vijay unofficially that we will be requiring an urgent joining from him. In 2 days of time, he will receive the final letter for selection. The day ends well and Rajat is very happy in closing this hiring.

4.6.4 HIRING ESCAPE

Mr. Vijay is very much happy and excited in the next morning after the interview and planning to resign in the current organization. As it was communicated to him that he has to join on urgent bases and his organization has a month notice period so he is planning to submit his resignation today only so that he can serve the maximum of notice period duration. He is waiting for the call from Rajat and its five in the evening he has not received any call. On the other side in NG, pharmaceutical Rajat is busy in scheduling the interview process on an urgent basis after receiving the resume of Ms. Madhuri an alumnus of one of the premier institutes of the country. The management is very much excited after getting the news of Ms. Madhuri's interview and they are hoping for filling the vacancy at the earliest. The interview with Ms. Madhuri finished only in one round and she got the final offer on the same day for the job.

The main question is: What is your opinion about the incidents happening in NG pharmaceuticals? The specific questions are:

- Analyze the situation of Mr. Vijay, what he will understand about NG pharmaceuticals if he won't get any call from Rajat.
- Present your opinion from the perspective of HR on the selection procedure followed by NG pharmaceuticals.

4.7 CASE 2: RIDHI'S PROMOTION SWAG

Ridhi Sales manager of NCPL electronics is waiting to meet Mr. Umashanker Chief Sales Executive at NCPL. It has been three years now since Ridhi joined the organization. NCPL is the milestone company for her career after graduating in Business Management. She has a very aggressive approach for marketing. In graduation her batch mates were surprised as very few girls were opting for marketing in her institute; on the other hand, Ridhi is very much excited about the thought of going into the marketing domain.

Summer of 2012 in PRV Institute of Management placement was started and Ridhi is waiting for her dream placement as every other candidate, she is hoping for the best package and profile. She have heard from her seniors that very less no of students got good salary package so she should consider profile at this point of time. Ridhi was the topper of the batch till her third semester and expecting the same result in the last semester too. Many campus drives were happen but Ridhi is not getting her dream profile and the semester exams were about to start and her family wants her to concentrate on exams at this point of time. After completing her exams she started the search for good job, many of the big firms were not shortlisting her for managerial profiles then she decided for going any initial profile she would get in good company and by her hard work, and she will grow in the organizational hierarchy.

4.7.1 THE WALK-IN-INTERVIEW

Priya one of the MBA classmates called Ridhi on Friday morning and while talking to each other they discuss about one email received a day before for walk-In the NCPL electronic a well-known firm in the electrical producers. The firm is looking for different positions in the marketing domain and the job profiles will be offered based on the candidate's performance in interview. They both decided to go for interview on Tuesday 27 July, 2012.

The walk-In start time is 9:00 am, Ridhi, and Priya both reached to the location at 8:45 am they were considering that they must be the first candidate for the day but after reaching the venue they found that almost 20 candidates were already reported to the location and the HR executives were giving them forms for doing the initial formalities. And the count is increasing by passing of every single minute and by 10 o'clock the number reached to approx, 50 candidates and the HR have announced that they are close for today's walk-In intake.

The situation in NCPL making Priya very nervous and she suggested to Ridhi that they should leave as they will not compete with all the candidates present there. Ridhi convinced Priya that at least they should try and give their best and getting the job is another point. While waiting for their turn, they discuss with some of the candidates and they came to know that many of them are, also fresher's and some are in their initial career. The status of other candidates gives them a little strength and they eagerly starting to wait for their turn. Priya got the interview call before Ridhi, this is the first round taken by the HR as screening of the candidate is done at this phase and later on, they have to face some other rounds for the final selection. Ridhi and Priya both get threw from the initial two rounds and the lunch break is announced. At this point of time, they are 15 in number selected for three vacancies at different levels. The post-lunch interviews were taken by separate panels for the different positions, Ridhi is shortlisted for the "Sales Representative" designation, which is the entry-level job in NCPL and Priya is shortlisted for office front desk representative. They were separated into other rooms specially allotted as per their shortlisting. They wished luck to each other and moved to their respective rooms, Ridhi clears one more round and now they are four in number left, she is very much happy that the made it till last round. She is expecting a good salary offer from the company. While discussing with the other three candidates she understood that salary will not be as per her expectation. On the other hand, Priya is not able to qualify the last round and she is out of the selection process and waiting for Ridhi to come with good news. Ridhi is in the final round in front of the Chief Sales Executive Mr. Umashanker and HR Head Ms. Misha Dutt. They offer a job with a salary of 2.25 Lac PA with the variable of 5% which is target based and all the job-related expectations were cleared to her.

The offer is far less from her expectations and it will be very disappointed after the struggle of entire day she is getting this offer, Mr. Umashanker is very much impressed with her performance in the interview and thus he suggests Ridhi to take this job and prove herself and she can receive salary hike in the six months of time. And he also suggests to her that NCPL is the company which knows to value their talented staff. Words of Mr. Umashanker calm her down and she accepted the job offer.

4.7.2 TWO YEARS LATER

Ridhi is with Mr. Umashanker for the same discussion but the fact is that it's been two years in the organization and she learnt a lot and NCPL got so many benefits from her. This is the yearly time of appraisal and Ridhi is thinking of her all the achievements of past two years, starting as a fresher in the organization achieving so much in the two years of span is really an achievement for her, still her ambitions are high and her dream is far away. After joining NCPL in the initial 6 months she worked day and night to learn the work and prove herself as committed at the time of interview and her efforts paid her very well, she got her first promotion and salary hike in just six months of time. She conquers many important deals for the organizations and she helped in building the image of organization in the market. In the first yearly review, she got the position of "Asst. Sales Manager" with an almost a double of initial salary. She is on cloud nine because her plan is working out very well, at the time of her graduation completion she is hoping for this career path only. The first year in the organization went very well and now she is all set to fight for her dreams and her plan is to get "Sales Manager" designation this year, her journey is going very good, and she is very much into her work, trying very hard to get all the major deals for the organization. One of the very important clients is handed over to her and did wonders have a deal which was almost impossible to think for a new employee like her. She is very much hopeful for her review this year. The annual employee award function in near the corner, and Ridhi got the best employee of the year and best achiever award. Ridhi is every happy and everyone in the organization praising her and expecting her to be a manager in the annual appraisal.

Mr. Umashanker offered her seat and they started the discussion about her journey from a fresher to the best achiever of the organization. Ridhi is very much happy that she is able to impress the person who hired her as a fresher and put so much of trust in her. Ridhi is eagerly waiting for the discussion about her promotion. Mr. Umashanker started the discussion about her performance in the last two years and how happy he is having her in the organization. He recommends that she should undergo some executive program from the reputed institutes, this will make her career even more fast and bright, and he also mentioned that in NCPL they have a policy to appoint somebody on managerial designations only when they have degrees from reputed institutes that's why she could not be promoted this year, it will only be the designation which is not going to be changed otherwise they are increasing salary and perks same as the managerial designation, as it is the policy matter so they can't change anything. Ridhi is so much disheartened after hearing this, she was thinking that her efforts and hard is the only thing which the organization is requiring from her but all of sudden this condition is making her depressed. Mr. Umashanker encouraged her that it is not like that company is not valuing her efforts but it's just a matter of degree, there are so many institutes which offers part-time and executive programs for the working professionals even the management wants her to help in her education she just take the admission and organization will provide her required leaves and they will also sponsor her education fee. He wishes her luck and reminds that it should be a well-known and reputed institute. Ridhi left the room with deep thoughts and confusion.

4.7.3 THE ASSESSMENT DAY

One year has passed since her last meeting with Mr. Umashanker, this year passed with so many remarkable changes in Ridhi's life. She again concentrated on the studies, with the job she prepared for the common aptitude test for the National Management Institutes which is considered as one of the difficult tests in the country. She has an extra burden on her for qualifying this test and going for the studies again, she is also not very sure that what qualities she will gain after completing the degree as she already dings her work very well. She is also confused about her monetary aspects as it is not the right time for her to invest that much of amount for her qualification. Having all these questions in mind she worked hard for the entrance exam and qualified the national entrance exam for top institutes. After qualifying the exam she has to appear the next round for Personal Interview which is again a hurdle for her but somehow she managed that as well and she got the admission in the institute and now she is all set for the admission and going for the EPGDP (executive post-graduate diploma program) from National Management Institute (leading institute of management in the country). She shared this news with Mr. Umashanker and going to meet him personally for the discussion about her education, promotion, education expenses, job security and many more issues which are there in her mind right now.

This meeting is going to be a milestone in her career if things went well then her all the efforts were in the right direction on the other hand if this will not work well then she has to choose either to go with this program or find another job which is again a task which is not going to be easy for her with an immense puzzlement and pressure she is ready to meet the authorities.

The panel along with Mr. Umashanker is in front of her; the management has decided that in what manner they can deal with the case and what are the things they should make her clear. The conversation is started with Ridhi and Mr. Umashanker shared with Ridhi that they are very happy for her decision and excited to welcome her back after completion of the degree. The NCPL has an offer to share her half education expenses, only if she accepts to comeback in the organization and commits to staying with them for a fixed period of two years without any further promotion. She will be joining back with the designation of Sales Manager with the annual package of 15 Lakhs after her completion of the degree, this one year will be treated as her education break and once she joins back her services considered to be in continuation. They gave her time for thinking about the offer; Ridhi is happy and excited but puzzled as well because she is able to estimate the payback from this deal. She has lots of dreams for the coming years and for her bright future, and expects a lot many things from this deal. The main question is: What is your opinion about the degree requirement for an achiever like Ridhi? Specific questions that need to be answered are as follows:

- Discuss about the pros and cons of having such promotion policies in the organization?
- Discuss about the issues mentioned in the case and identified the diversity concept mentioned.
- Ridhi's promotion is justified without the National level recognized degree or not, discuss about Ridhi's state of mind after knowing her promotion condition.

APPENDIX 4.1: CASE DISCUSSION FOR CASE 1

Synopsis

Rajat Singh, an HR executive at NG pharmaceuticals is facing a tough time in getting the suitable profiles for marketing head as the current head Mr. Bhalla is going to leave the organization in a month, NG pharmaceutical is a leading company in the industry and it is known for its marketing strategies in the industry. Rajat has to fill the vacant position before Mr. Bhalla will leave the organization, he post the vacancy in all the portals and when Rajat realized that the criteria of selecting only from premier institutes are leading nowhere, he decides to inform top management about some good profiles so that the process will start for now. He received a positive response from management and the recruitment process gets started. The interview is scheduled to be held on 10/July/2018; Mr. Alex and Mr. Vijay are the shortlisted candidates. After a discussion with both of them, management found Mr. Vijay the best-suited candidate for the job. Mr. Vijay is informed about his section and immediate joining need but in the informal way, and then he has been assured that he will receive final offer letter in a day or two. Next day after Mr. Vijay's section Rajat received a resume of the dream candidate, Ms. Madhuri alumnus from premier institute of country the job is offered to her after interview and discussion on the other side Mr. Vijay is waiting for his formal letter of appointment.

Main Question

What is your opinion about the incidents happening in NG pharmaceuticals?

Probing Questions

- Analyze the situation of Mr. Vijay, what he will understand about NG pharmaceuticals if he won't get any call from Rajat.
- Present your opinion from the perspective of HR on the selection procedure followed by NG pharmaceuticals.

Intended Audience

The audience for this case will be MBA students and it will help them to understand the Educational divide happening in the organizations. This will help them to broaden their perceptive regarding the hiring process and requirements of the companies.

Teaching Experience of the Case

This case was used in the MBA class, the students find it interesting and they were asking many questions in relation to the recruitment process, organizational norms of selecting, legal aspects, impact on the reputation of the organization, actual scenario of organizational discrimination. The students have a different perspective for the premier educational institute's selection as well as a debate happens for the pros and cons of fixing the criteria.

APPENDIX 4.2: CASE DISCUSSION FOR CASE 2

Synopses

The case is talking about Ridhi, a young management graduate and her career issues. The case started with Ridhi's job search after degree completion and it also explained how she got the job in NCPL electronics after facing a long procedure. Then it talks about her hard work and promotions in the two years of time in the organization, it was also mentioned that she got the best employee of the year and best achiever award. The last section of the case is talking about companies' promotion policies and Ridhi's confusion for the suggestion given by Mr. Umashanker. Mr. Umashanker suggests that she should join a good degree from any premier institute for further promotion and growth, then Ridhi's struggle for admission and job continuation is mentioned and it was also mentioned that Ridhi is confused with the further study and job change choice. At the last Ridhi is ready to join the EPGDP (executive post-graduate diploma program) from National Management Institute and then the company is her an offer for dream promotion with a very good salary but with certain conditions which are again a point of judgment for Ridhi.

Main Question: What is your opinion about the degree requirement for an achiever like Ridhi?

Probing Questions:

- Discuss about the issues mentioned in the case and identified the diversity concept mentioned.
- Ridhi's promotion is justified without the National level recognized degree or not, discuss about Ridhi's state of mind after knowing her promotion condition.
- Discuss about the pros and cons of having such promotion policies in the organization?

Intended Audience: The audience for this case will be *undergraduate* and *postgraduate* students and it will help them to understand the Educational divide happening in the organizations. This will help them to broaden their perceptive regarding career growth and the requirements of the companies.

Teaching Experience of the Case: This case was used in the MBA class, entire class was divided into groups of two and they were different perspective one from Ridhi's and other from NCPL. The students find it interesting and they were having a good discussion for such polices which the organization's frame without taking common logic into consideration. The discussion for the requirement of executive education went in detail and students identified many benefits of encouraging employees for going higher education on the other hand necessity of having a branded degree is not accepted by the students. The students have a different perspective for the premier educational institute's value in the mid-career growth as well as a debate happens for the pros and cons of having such policies.

KEYWORDS

- **discrimination at work**
- **diversity management**
- **educational discrimination**
- **educational diversity**
- **importance of an educational degree**
- **institutes role in skills**
- **institutional discrimination**
- **knowledge divide**
- **the relation between degree and skills**
- **selection discrimination**
- **workplace discrimination**

REFERENCES

Cambridge Dictionaries Online. Discrimination definition, *Cambridge University.*

Carmichael, F., & Woods, R., (2000). Ethnic penalties in unemployment and occupational attainment: Evidence for Britain. *International Review of Applied Economics, 14,* 71–98. [Online]. 10.1080/026921700101498.

Christiane, S., & Glunk, U., (2008). Mechanisms underlying nationality-based discrimination in teams. Quasi-experiment testing predictions from social psychology and microeconomics. *Small Group Research, 39*(6), 643–672.

Discrimination in Education. https://en.wikipedia.org/wiki/Discrimination_in_education (accessed on 24 February 2020).

Dougherty, D., (1992). Interpretive barriers to successful product innovation in large firms. *Organization Science, 3*, 179–202.
Fisher, C. (2019). *Educational Discrimination in the Workplace*. http://woman.thenest.com/educational-discrimination-workplace-22669.html (accessed on 24 February 2020).
Free Advice. https://forum.freeadvice.com/hiring-firing-wrongful-termination-5/discrimination-against-my-educational-background-176601.html (accessed on 24 February 2020).
Friedman, R. A., & Krackhardt, D., (1997). Social capital and career mobility: A structural theory of lower returns to education for Asian employees. *Journal of Applied Behavioral Science, 33*, 316–334.
Friedman, R., (1996). Defining the scope and logic of minority and female network groups: Can separation enhance integration? In: Ferris, G., (ed.), *Research in Personnel and Human Resource Management* (Vol. 14, pp. 307–349). Greenwich, Conn.: JAI.
Giddens, A., Duneier, M., & Appelbaum, P. R., & Carr, D., (2008). *Introduction to Sociology* (7th edn., p. 334). W. W. Norton and Company Inc. New York.
Grund, C., & Westergård-Nielsen, N., (2008). Age structure of the workforce and firm performance. *International Journal of Manpower, 29*, 410–422.
Haberfeld, Y., (1992). Employment discrimination: An organizational model. *Academy of Management Journal, 35*(1), 161–180.
Hankyoreh. *Discrimination*. http://www.hani.co.kr/arti/english_edition/e_national/583972.html (accessed on 24 February 2020).
Ibarra, H., & Smith-Lovin, L., (1997). New directions in social network research on gender and organizational careers. In: Cooper, C. L., & Jackson, S. E., (eds.), *Creating Tomorrow's Organizations: A Handbook for Future Research in Organizational Behavior* (pp. 359–384). New York: John Wiley & Sons.
Ibarra, H., (1992). Homophily and differential returns: Sex differences in network structure and access in an advertising firm. *Administrative Science Quarterly, 37*, 422–447.
Ibarra, H., (1995). Race, opportunity, and diversity of social circles in managerial networks. *Academy of Management Journal, 38*, 673–703.
Ilmakunnas, P., Maliranta, M., & Vainiomäki, J., (2004). The roles of employer and employee characteristics for plant productivity. *Journal of Productivity Analysis, 21*, 249–276.
Jackson, S. E., May, K. E., & Whitney, K., (1995). Understanding the dynamics of diversity in decision-making teams. In: Guzzo, R. A., & Salas, E., (eds.), *Team Effectiveness and Decision Making in Organizations* (pp. 204–261). San Francisco: Jossey-Bass.
Jehn, K. A., Chadwick, C., & Thatcher, S., (1997). To agree or not to agree: Diversity, conflict, and group outcomes. *International Journal of Conflict Management, 8*, 287–306.
Jehn, K. A., Northcraft, G. B., & Neale, M. A., (1999). Why differences make a difference: A field study in diversity, conflict, and performance in workgroups. *Administrative Science Quarterly, 44*, 741–763.
Kirkpatrick, G. R., Katsiaficas, G. N., Kirkpatrick, R. G., & Mary, L. E., (1987). Introduction to critical sociology. *Ardent Media, 261*. ISBN: 978-0-8290-1595-9.
Kislev, E., (2016). Deciphering the 'ethnic penalty' of immigrants in western Europe: A cross-classified multilevel analysis." *Social Indicators Research*. 10.1007/s11205-016-1451-x.
Kleiman, S. L. (2018). *Discrimination*. http://www.referenceforbusiness.com (accessed on 4 December 2018).
Kleiman, S. L. (2019). *Revised by Barnett Tim Discrimination*. http://www.referenceforbusiness.com/management/De-Ele/Discrimination.html (accessed on 24 February 2020).

Knight, D., Pearce, C. L., Smith, K. G., Olian, J. D., Sims, H. P., Smith, K. A., & Flood, P., (1999). Top management team diversity, group process, and strategic consensus. *Strategic Management Journal, 20*, 445–465.
Labor Law Center. *Pre-Employment and Hiring.* www.laborlawcenter.com (accessed on 24 February 2020).
Lazear, E. P., (1999). Globalization and the market for team-mates. *Economic Journal, 109*, C15–C40.
Miles, R., & Snow, C., (1986). Network organizations: New concepts and new forms. *California Management Review, 28*, 62–73.
Morrison, A. M., & Von Glinow, M. A., (1990). Women and minorities in management. *American Psychologist, 45*, 200–208.
Navon, G., (2009). Human capital spillovers in the workplace: Labor diversity and productivity. *Bank of Israel, Research Department.* Discussion paper no. 5.
Nedumala, J. (2010). *Bridging the Educational Divide in India.* Youth spring. https://www.iyfnet.org/sites/default/files/library/YOUth_Spring13_Nedumala.pdf (accessed on 24 February 2020).
Nicol (2016). *History of Discrimination Within the Workplace.* https://www.laborlawcenter.com/education-center/history-of-discrimination-within-the-workplace (accessed on 24 February 2020).
Nkomo, S., (1992). The emperor has no clothes: Rewriting "race in organizations." *Academy of Management Review, 17*, 487–513.
Oxford Dictionaries. Definition of Ageism, *Oxford University Press.*
Ragins, B., & Sundstrom, E., (1990). Gender and power in organizations: A longitudinal perspective. *Psychological Bulletin, 105*, 51–88.
Relocate Magazine. www.relocatemagazine.com (accessed on 24 February 2020).
Scott, S. (2019). *Discrimination at the Workplace.* http://smallbusiness.chron.com/discrimination-workplace-2855.html (accessed on 24 February 2020).
Scott, W. R., (1987). *Organizational Rational, Natural, and Open Systems.* Englewood Gliffs, NJ: Prentice-Hall.
Simon, H. A., (1957). *Administrative Behavior.* New York: Macmillan.
Smith, J., (2014). *Here's What Recruiters Look at in the 6 Seconds They Spend on Your Resume."* Business Insider, Nov.
The Power in Demography: Women's Social Constructions of Gender Identity at Work, (1995). *Academy of Management Journal, 38*, 589–634.
Thompson, N., (2016). *Anti-Discriminatory Practice: Equality, Diversity, and Social Justice.* Palgrave Macmillan, 2016, ISBN 978113758667.
Tsui, A., & Gutek, B., (2000). *Demographic Differences in Organizations.* New York: Lexington Books.
Wharton, A., (1992). The social construction of gender and race in organizations: A social identity and group mobilization perspective. In: Tolbert, P., & Bacharach, S., (eds.), *Research in the Sociology of Organizations* (Vol. 10, pp. 55–84.). Greenwich, Conn.: JAI Press.
Wilkinson, J., & Ferraro, K., (2002). Thirty years of ageism research. In: Nelson, T., (ed.), *Ageism: Stereotyping and Prejudice Against Older Persons.* Massachusetts Institute of Technology.
Wolf, W. G., & Fligstein, N. D., (1979a). Sex and authority in the workplace: The causes of sexual inequality. *American Sociological Review, 44*, 235–252.
Wolf, W. G., & Fligstein, N. D., (1979b). Sexual stratification: Differences in power in the work setting. *Social Forces, 58*, 94–107.
Youthrights.org, Young and Oppressed, Retrieved on 11 April 2012, Archived July 28, 2011, at the Way Back Machine.

FURTHER READING

Abou Al-Shamat, H., (2009). Educational divide across religious groups in nineteenth-century Lebanon: Institutional effects on the demand for curricular modernization. *Journal of Islamic Studies*, 20(3), 317–351.

Acharya, A., & Gupta, M., (2016a). An application of brand personality to green consumers: A thematic analysis. *Qualitative Report*, 21(8), 1531–1545.

Acharya, A., & Gupta, M., (2016b). Self-image enhancement through branded accessories among youths: A phenomenological study in India. *Qualitative Report*, 21(7), 1203–1215.

Andrews, J. E., Carnine, D. W., Coutinho, M. J., Edgar, E. B., Forness, S. R., Fuchs, L. S., et al., (2000). Bridging the special education divide. *Remedial and Special Education*, 21(5), 258–260.

Conley, M. W.., (2016). Collaborating across educational divides: Why it is important for literacy and what it takes. *Journal of Adolescent and Adult Literacy*, 59(5), 497–502.

Edwards, K., (2014). Education, technology and the disruptive innovations challenging the formal/informal education divide. *Informal Education, Childhood and Youth: Geographies, Histories, Practices*, pp. 140–162.

Fernández, J. J., & Eigmüller, M., (2018). Societal education and the education divide in European identity, 1992–2015. *European Sociological Review*, 34(6), 612–628.

Green, B., (2017). Engaging curriculum: Bridging the curriculum theory and English education divide. *Engaging Curriculum: Bridging the Curriculum Theory and English Education Divide* pp. 1–303.

Gupta, M., & Kumar, Y., (2015). Justice and employee engagement: Examining the mediating role of trust in Indian B-schools. *Asia-Pacific Journal of Business Administration*, 7(1), 89–103.

Gupta, M., & Pandey, J., (2018). Impact of student engagement on affective learning: Evidence from a large Indian university. *Current Psychology*, 37(1), 414–421.

Gupta, M., & Ravindranath, S., (2018). Managing physically challenged workers at micro sign. *South Asian Journal of Business and Management Cases*, 7(1), 34–40.

Gupta, M., & Sayeed, O., (2016). Social responsibility and commitment in management institutes: Mediation by engagement. *Business: Theory and Practice*, 17(3), 280–287.

Gupta, M., & Shaheen, M., (2017a). Impact of work engagement on turnover intention: Moderation by psychological capital in India. *Business: Theory and Practice*, 18, 136–143.

Gupta, M., & Shaheen, M., (2017b). The relationship between psychological capital and turnover intention: Work engagement as mediator and work experience as moderator. *Journal Pengurusan*, 49, 117–126.

Gupta, M., & Shaheen, M., (2018). Does work engagement enhance general well-being and control at work? Mediating role of psychological capital. *Evidence-Based HRM*, 6(3), 272–286.

Gupta, M., & Shukla, K., (2018). An empirical clarification on the assessment of engagement at work. *Advances in Developing Human Resources*, 20(1), 44–57.

Gupta, M., (2017). Corporate social responsibility, employee-company identification, and organizational commitment: Mediation by employee engagement. *Current Psychology*, 36(1), 101–109.

Gupta, M., (2018). Engaging employees at work: Insights from India. *Advances in Developing Human Resources*, 20(1), 3–10.

Gupta, M., Acharya, A., & Gupta, R., (2015). Impact of work engagement on performance in Indian higher education system. *Review of European Studies, 7*(3), 192–201.

Gupta, M., Ganguli, S., & Ponnam, A., (2015). Factors affecting employee engagement in India: A study on off shoring of financial services. *Qualitative Report, 20*(4), 498–515.

Gupta, M., Ravindranath, S., & Kumar, Y. L. N., (2018). Voicing concerns for greater engagement: Does a supervisor's job insecurity and organizational culture matter? *Evidence-Based HRM, 6*(1), 54–65.

Gupta, M., Shaheen, M., & Das, M., (2019). Engaging employees for quality of life: Mediation by psychological capital. *The Service Industries Journal, 39*(5/6), 403–419.

Gupta, M., Shaheen, M., & Reddy, P. K., (2017). Impact of psychological capital on organizational citizenship behavior: Mediation by work engagement. *Journal of Management Development, 36*(7), 973–983.

Hsieh, M. Y., (2017). An empirical study of education divide diminishment through online learning courses. *Eurasia Journal of Mathematics, Science, and Technology Education, 13*(7), 3189–3208.

Kickbusch, I. S., (2001). Health literacy: Addressing the health and education divide. *Health Promotion International, 16*(3), 289–297.

Konietzka, D., & Kreyenfeld, M., (2010). The growing educational divide in mothers' employment: An investigation based on the German micro-censuses 1976–2004. *Work, Employment and Society, 24*(2), 260–278.

Krishnan, S., (2016). The great education divide. *Economic and Political Weekly, 51*(18).

Kumar, P., (2013). Bridging east and west educational divides in Singapore. *Comparative Education, 49*(1), 72–87.

Muttarak, R., & Chankrajang, T., (2015). Who is concerned about and takes action on climate change? Gender and education divides among Thais. *Vienna Yearbook of Population Research, 13*(1), 193–220.

November, N., (2015). Using social media to enhance critical thinking: Crossing socio-educational divides. *The Palgrave Handbook of Critical Thinking in Higher Education*, pp. 509–523.

Pandey, J., Gupta, M., & Naqvi, F., (2016). Developing decision making measure a mixed method approach to operationalize Sankhya philosophy. *European Journal of Science and Theology, 12*(2), 177–189.

Rao, M., Singh, P. V., Katyal, A., Samarth, A., Bergkvist, S., Renton, A., & Netuveli, G., (2016). Has the Rajiv Aarogyasri community health insurance scheme of Andhra Pradesh addressed the educational divide in accessing health care? *PLoS One, 11*(1).

Yusof, N. S. M., Hashim, R., & Rashim, S. N. A., (2012). Assessment of educational divide and computer literacy among primary school students in Selangor Malaysia. *ISBEIA 2012-IEEE Symposium on Business, Engineering and Industrial Applications*, pp. 437–442.

CHAPTER 5

Gen-Z, the Future Workforce: Confrontation of Expectations, Efforts, and Engagement

SHEEMA TARAB

Department of Commerce, Aligarh Muslim University, Aligarh, Uttar Pradesh, India, Tel.: (+91) 8791537797, E-mail: sheematarab@gmail.com

ABSTRACT

Over the decade's scholarly literature in organization studies has witnessed an increasing interest in work engagement and diversity inclusion aspects. In an era of disruption, equipping younger people in a team requires a reflection of the future workforce concerning their attitude at work in order to predict their engagement levels. This chapter fills in the gap of describing Gen-Z's behavior, qualities, and expectations to work but a considerable light is also pondered upon the amount of efforts they are willing to render. This comprehensive preview is expected to enable organizations to understand the new generation and adjust appropriately in a multi-generational workplace context.

5.1 INTRODUCTION

This chapter discusses the youngest workforce in organizational reference as an attempt to understand the opportunities and challenges related to their work engagement. Engaged employees go beyond the call of duty to perform their roles in distinction. The conceptualization of engagement is not limited to physical presence, but it also includes the cognitive and emotional involvement at work (Kahn, 1990). However, engaging employees is a matter of sheer concern for managers, especially in the present scenario where the prospects are open to numerous opportunities and the strategists are hooking talents. At the moment, the changing expectations of workers have completely influenced the labor market situations. As we are moving with time, new philosophies are touching upon and hitting the workspaces that come in with the entry of new human resources (HR).

Organizations are now designing their policies and structures in accordance with the new work expectations. In the past, diversity has been confined to race, gender, sexual orientation, cultural background, and/or ability. But now race and gender take a back seat and the present modern workplaces are unfolding fresh dimensions of diversity with respect to work and working philosophies. Employee well-being has been a serious HR issue in business organizations because it is closely related to work outcomes such as organizational commitment and work satisfaction (Bakker et al., 2008). Among the factors associated with employee well-being, work engagement has garnered enormous attention both from academia and industry, since it represents affirmative work experience (Sonnentag, 2003).

Committed employees are the backbone of organizational success. They are bliss for both the company and the colleagues. There is so much that an engaged employee can disseminate at work by not just performing, but also by interacting and exhibiting positive energy. The main aim of recruiters is not confined only to recruit the best talent, but also to bring the best out of their skills and capabilities which is best nurtured in a constructive work environment, supported by job resources and self-efficacy beliefs that lead to organizational prosperity (Alessandri et al., 2018). The realization of the fact that the future of organizations is with the young minds, the problems related to their engagement at work become crucial hence cannot be overlooked.

5.2 NEED FOR DISCUSSING THIS THEME

Majorly, workforce comprises of people belonging to the different age group that works together under one roof, this dynamics of generational heterogeneity is always challenging for the administrators. Gender equality and work autonomy have become a forerunner in explicating the multi-generational organizations. And ultimately, all organizations are influenced by the values and preferences of their next generation. Failure on the part of managers to understand the demands of the new generation and adjust appropriately to generational differences who tend to enter the workplace can result in misunderstandings, miscommunications, and mixed signals (Fyock, 1990). Further, it may affect employee productivity, innovation, and corporate citizenship eventually resulting in problems related to work engagement and organizational commitment (Kupperschmidt, 2000).

Drawing upon the traits, the upcoming generation as the future workforce is throwing a big challenge to organizations and thereby employers are making efforts in advance to understand these newbies. Interfacing competitive advantage and global exposure, managers, and decision-makers acknowledge and accept the worth of these changing facets at workplaces and are considerate enough to look beyond a perspective to become proactive and ready to offer what is being expected next from them. Enormous research and theory could be traced on employee and work engagement, but there is a dearth of literature explicating Generation Z and their prospective work behavior. This chapter fills in the gap of describing the future workforce Gen-Z's behavior, qualities, and expectations to work but a considerable light is also pondered upon the amount of efforts they are willing to render.

5.3 REVIEW OF THIS THEME

Over the years, employee well being, commitment, and work engagement are often becoming major agendas on the table. Almost with the retirement of baby boomers in present times, and joining of younger generations in formal work teams reflects the shifting culture of organization settings, a wave of balancing their diversity has become a focal center of managerial discussions and decisions. Though several studies and literature provide support to discuss this issue, the increasing data related to disengagement of employees at work insist attention. The younger generation is too forward and fast in terms of thinking and doing, their presence within an organization in different roles requires a thought process. Noting the evolving nature of a working hub, the organizations are now keen to formulate policies and foster an environment that can help in engaging this young cohort of workers. A practical exposure of the expectations and problems of the youngest generation is required and of much significance in comprehending this segment of the future workforce. This will help in understanding them better and may assist in handling their engagement issues proactively.

5.3.1 WHO IS GEN-Z?

Generational difference is a diversity that exists among the people born in a different period of time and happens to work together, so the values and kind of preferences they carry with themselves at work becomes noteworthy to mull over. Speaking about the Traditionalists (1925–1945) to almost a decline; contemporary organizations are facing challenges in managing the three generations viz.; Baby Boomers (1946–1964); Xers or Gen-X (1965–1980); and Generation Y or Millennials (1981–1996). But soon the fifth generation widely acknowledged as the Generation Z (1997–2010) is ready to take off and join the work platforms. Parallel to this, due to their entrepreneurial orientation, few of them have already joined in the shape of startups and opened a segment for the policymakers to reflect upon. This chapter will focus on Gen Z workforce. They are the digital natives capable of moving to the next level in a blink of an eye. Gen-Z is the upcoming subdivision of workforce, very demanding, impatient, devotee, of virtual reality and technology, self-reliant in nature.

It is imperative for organizations to become upbeat in handling this fragment so as to harness their potential for a better outcome. In order to make better decisions, it is crucial to figure out the values and aspirations across generations, but also to keep in mind that the national and social context also does influence this very notion. According to a study across 19 countries, all generations were alarmed about whether their personalities fit where they work. Danish, Japanese, and Indians were found to be the most concerned as per the survey (Bresman and Rao, 2017). Another survey by Ernst and Young reflects upon the belief of trust in the employer as a key determinant of engagement as per Gen Z (Twaronite, 2016). So, it is important to emphasize the traits of Gen Z to know them better: They are the ones grown up in a global, volatile economy and the technologically driven world. The following features highlight the prominent habits and behavior observed in this young generation as a prospective concept:

- **Lack of Patience:** Being born in a global techno-oriented world, where every minute detail is shared with personalized and/or general groups, Zers tend to have little or less amount of patience, as they are quick in reacting they also expect rather demand a quick understanding, quick processing, and a quick redressal of the issues and concerns.
- **Miniscule Span of Attention:** As stated above, that Gen Z is quick in reacting, they tend to be less attentive, owing to which their interest is not fully drawn towards the concerned issue. Moreover, this generation is too much occupied with social networking, gaming, and operating multiple applications simultaneously while communicating that they tend to lose concentration and miss out a few details of the subject.

- **Early Career Focus:** Being connected globally via technology Zers tend to be well-informed and much aware of the existing opportunities, magnitude of prevailing competition and worth of becoming self-dependent. This has caused career consciousness among them so they attempt to map it properly well within the time in the early age of their studies. Along with the basic education, this generation also looks for different interests and credentials.
- **Speed Lovers:** Manual speed and mechanic speed is so obviously different. Since Gen-Z is totally technologically driven, their learning, practice, the search is all based upon a click that facilitates information in their palms. They ought to become an admirer of speed and short of patience. Generally, those tasks and activities that take a needlessly long time to annoy them.
- **Cause-Focused Passion and Socially More Responsible:** Generation Z is more careful and concerned about their responsibility towards society and future generations. They yearn for associating with such activities and tasks that can help them perform their social duty. Organizations may take a note of this interest and utilize this as an avenue for engaging Gen-Z. Zers are not only concerned about their own selves but are also sensitive about the depleting natural resources, non-human species, future generations and the planet as a whole. They prefer workplaces which are ethical and active in making charities and performing their social responsibilities. In this way, they foresee the possibility of volunteering that can enable them to contribute their time, energy, or resources towards the betterment of society. As compared to Gen X and Y, Gen Z is more active and positive in spending and working for the activities that are associated with some noble cause or charity.
- **Expressive and Creative:** This generation style of work is unique and bold. As Generation Z is exposed to worldwide networking with reference to social and professional interactions, this generation is found to be more creative in their approach and delivery. Digital Natives strongly advocates raising voices through trends and filing petitions. For example, to discard a political action or unjust policy they might use the social networking platform and create a poll for it, or they may generate support by adding a hashtag to the event and trending it for the sake of dealing with it. Zers express their happiness, anger, support, or opinion on several problems, stigmas, and taboos, etc., over different social platforms.

- **Idealistic and Forward Thinking:** Zers are receptive to diversity and open to accept differences, they are more empathetic to situations and people. Having awareness and resources within their capability and reach, Zers aim very high for their career and life. Their assumptions are optimistic and they believe to revolutionize the system by joining hands, protests, and striking. Regardless of their naive attitude, this generation is confident, stresses on forward-thinking, and creative work.
- **Honest and Confident:** This lot of young generation is honest, straightforward, and confident enough to present their ideas and discuss them directly with the bosses, they dislike any indirect channel in between. One reason that this generation is likely to become honest that they are born in an era of transparent work cultures and accountability (Staff, 2018). They are much keen to get their due credit and quick appreciation. This is why they are motivated by rewards.
- **Receptive to Faith, Creed, and Orientation Diversities:** Work with no boundaries may be geographic, social, or political is the mantra of Zers. Generation-Z is open to accept and respect the difference among humans belonging to a different creed, color, race, disabilities, and variety of sexual orientations. However, receptivity to sexual diversity is subjective owing to the values and customs practiced in different regions, or nations but broadly; Gen-Z is adaptive to diversity.
- **Stumpy Work Ethics:** Drawing upon the impatience observed in Zers and the dynamic fabrication of nature and values explicating their work behavior are challenging the organizations. It is crucial to understand that their work behavior is stumpy. Zers have a propensity to lose interest in prolonging tasks. Their passion is intense; they become excited with novelty, but how far their efforts move in that direction is a question to focus upon.
- **Crave for Feedback:** They prefer no regulations, yet they look up to guidance, right directions, and constant feedback support. If they receive the right atmosphere, their commitment to work becomes unmatchable, this generation is much keen to celebrate the victory, even the smallest of achievements and share it with the world.
- **Comfy Zone Generation:** Post-millennials are ultra-modern, obsessed with great work autonomy, appreciation, success, fame, and comfort at the same time. This generation is hugely influenced by media as it can be witnessed with the increasing number of you tubers, gamers, and freelancers. This generation's utmost priority is comfort starting from their dressing to communicating, working, and negotiating. They prefer

to talk on their terms and do not wish to follow any code of conduct and rules. They demand flexibility, transparency, and honesty at work.
- **Urban Language:** Another distinctive highlight of Zers is that they have their own metropolitan world of communication. Their language comprises abbreviations, signs, and emoticons. It is an era of urban language, and other people (not the Zers) often found exploring different search engines, consulting urban dictionaries and distinctive apps to showcase their up-to-date and moving- with-time philosophies. To ensure that Zers will remain intact with their work and teams, it is essential for fellow members (from other generations) adapted to such changes, as communication is the key to better and productive work.

Since traditionalist and baby, boomers are almost out of the work systems due to their age, retirements, and mortality. The inter-generational heterogeneity which should be taken care of directs the attention to focus Gen X, Y, and now Z working altogether. Let us look at these three prominent generations at a glance with respect to their values, work preference and habits in general (refer Table 5.1).

TABLE 5.1 Three Generations Overview

Basis	Gen-X/Xers (1965–1980)	Gen-Y/Millennials (1981–1996)	Gen-Z/Post Millennials (1997–2010)
Values	Spiritual	Moral and Self-Esteem	Social
Attitude	Submissive	Impulsive	Inquisitive
Interest	Job Security	Work-Life Balance	Intrapreneurship/ Autonomy
Nature	Dependent/Security seekers	Multi-tasking/Confident	Entrepreneurial/ Self- reliant
Approach	Pragmatic	Realist	Idealist
Focus	Initiative	Innovative	Creative
Thoughts	Skepticism	Optimism	Proactive
Commitment	Loyal to Organization	Loyal to Work	Loyal to employer
Work-life concerns	Job satisfaction/ Quality of Work Life	Work-life balance and Gender Parity	Technologically driven/ challenging
Magnitude of expectations	Low Expectations	High Expectations	Demanding

The outline of inter-generational value differences may assist the organizations and policymakers to realize that tackling this upcoming workforce along with existing previous two generations is not an effortless task. With such a demanding cadre, the challenge to make them work in an engaged and involved manner is a serious phenomenon.

5.3.2　ECONOMIC AND SOCIAL ENVIRONMENT DURING THE UPBRINGING OF ZERS

The eldest Gen-Z is approximately now of age 21, likely to move in as a part of the workforce landscape and in some regions already occupying a role in form of start-ups or part-time internships. The big challenge for the organizations is to adopt the budding heterogeneity for smooth functioning and enhanced performance. Before delving into the suggestions to deal, let us peep into the genesis of this recent generational era about how these said values probably have evolved and are continually blooming. Generation Z is the one born in the late 90s and brought up in the new millennium, hence they are also known as *'millennium babies'* and *'digital natives.'* This millennium era is completely driven by technology; Zers have the world of information in their palms right from the very beginning of their growth years. Drawing from the globalization factor, any event or mishap in one part of the planet may impact all over the other parts of the region and countries, so boundaries goes blur when we mention about an era of volatility on global economic fronts, inclusive diversity to a boom in context of social and demographic shifts, administrative, and regulatory reforms and what not. They are the generation most of which is born to working parents, owing to which in their initial years of growth they were being left to play and deal on their own (with the caretaker/crèche) that owe greatly to their habit of freedom and less/no control. They are also considered as the conscious generation who comprehend the significance of aptitude and competitiveness admitting the fact of cutting-edge competition which apparently has been sensed by them within their homes (being born to working parents), and also at the very first external world exposure like playschools or pre-nursery where they undergo the procedures of entrances/exam criterion, etc.

Another dominant characteristic observed among them is their tech-savvy attitude, as majorly they learned rhymes on gadgets and are habitual of playing with talking apps in their toddler age, which sometimes is also an escape for the parents, who are already equipped with work stress. Apparently, they are been considered the ones with poor writing skills though. But Gen-Z is also a well informed and a vigilant generation, effect of which could be seen in organizational practices illuminating less gender disparity, the glass ceiling and lower sexual abuse moans in the recent past. However, a dilemma is observed in their dependency for guidance and observance along with their need of self-defined rules at working space, this will create a great deal of concern pertaining to their inclusion in organizations as a team and also about their independent performances. Being raised in a digital society, where privacy is social and profession is more like a trend, Zers lack in taking initiative for taking up challenges and ought to be lazy. Being the speed lovers, they are smart enough in delegating a task to humans or shifting the workload to machines. Being exposed to information at a global extent, their expectations rise, and it is easily reflected in their career choices and education. In a nutshell, Zers are raised in a closed, pampered, and overprotective environment, full of facilities, resources, and information which apparently has made them demanding but also dependent, conscious but also lazy, creative but not dedicated, smart but not intelligent. It has also been observed that Gen Z also lacks leadership skills; although they are tending to be good, learners (refer Table 5.2).

TABLE 5.2 Antecedents and Outcomes of Gen-Z Behavior

Antecedents	Outcome Behavior
• Technology	• Tech-Savvy
	• Speed Lovers
	• Fine Learners
	• Career-Focused
• Intelligent Systems	• Creative
	• Lazy
	• Dependent
• Virtual Reality	• Well-informed
	• Inquisitiveness
	• Receptive to diversity
	• Individualism
• Working Parents	• Demanding
	• Inpatient
• Social Media	• Trends
	• Information and Sharing
	• Idealistic
	• Socially Responsible
• Transparent Work Culture	• Honest
	• Ethical
	• Straightforward

The distinctive characteristics which are observed in Zers are challenging the competitive landscapes and triggering a need to ponder upon the redefinitions of work philosophies. Drawing upon their traits managers will face issues if their capabilities do not find a suitable atmosphere to prosper. Hence, it is an imperative consideration.

5.3.3 EXPECTATIONS OF ZERS

Organizations look up to the employees as hands that strengthen the system with all their heart and mind, but practically issues related to low energy at work, dissatisfaction end up in de-motivated and unhappy workers. Problems like absenteeism, attrition, and job-hopping are not new anymore rather they are becoming a new normal with the changing work trends. A big concern of managers is not to acquire but to retain the talent they have. These and several related concerns affect the working of employees and it has created awareness among the administrators to undertake feasible and possible best actions that can serve the purpose. One way is to understand what employees want at their job. The author interviewed about 50 undergraduate and postgraduate students (age group 20–23 years) about their future work expectations. By resorting to free elicitation technique in order to spot the expectations of Zers regarding work and workplace, students were asked to speak about the elements that top their priority at work and are essential to keep their work engagement intact.

A list of factors have been reported, the most significant element that this generation is seeking at work is respect. Almost 80% of respondents have put respect in first place followed by positive relationships with peer groups and bosses along with good pay plus monetary benefits. Zers are pertinent about the respect quotient that in case fair treatment will not be done to them, they are more likely to leave, but as long as their dignity is well-taken they prone to stay with their organization in the long run. Their loyalty to the organization can also be traced with their laziness, making extra efforts for the sake of higher benefits, or imbalance in personal life seems not so logical to this generation. Rather, they are more comfortable working in the friendly environment where they receive decent reverence. Drawing upon the responses, it is palpable that the young generation does not prefer any line of a gap between the bosses and workers. They believe in an informal environment, where there is the flexibility of rules and comfort at work with respect to information and resources.

5.4 IMPLICATIONS

Apparently, there seems a conflict between the expectations and efforts in Zers, but the real management challenge is to strike a balance and tackle this loophole. The future of work depends upon the evolving philosophies. Since globally, organizations are witnessing the highest variety of diversity these days, therefore, recognizing, and accommodating Gen Z along with former generations is quite a task. Certain recommendations are suggested here for managers to handle this confrontation.

- As compared to millennials, Gen Z prefers to have a direct dialogue and instant messaging rather than emails and formal meetings. There should be minimum levels between the employee and the authority, as Zers prefer having direct communications.
- The organization should adopt the policy of monitoring the performances by providing instant feedbacks, preferably using technology.
- Owing to the lethargy in Gen Z the details of the task should be very clear, also, the task should be designed to get accomplished in short span. Division of task can be adopted, instead of expecting them to perform the whole task single-handedly; the small task can be designated to them.
- Gen Z is likely to stay in their comfort zone, they resist formal environment. Managers should consider developing a friendly environment at work wherein productivity remains intact.
- Digital natives get boosted up by appreciation, fame, and rewards. The akin fragmenting task, responsibilities or rewards can also be planned in a phased but consistent manner. Small designation promotions, awards, or letter of appreciation which may act as a proof of their achievement can make them happy.
- Gen Z demands equal respect irrespective of any seniority or contribution. If they find that respect is no longer served, they are likely to lose interest.
- Their interest in societal development can be well taken by the organizations. They can associate a cause with the task to improve their engagement at work.
- Gen Z becomes impatient with rigid rules and works ethics, they believe in autonomy. Generally, too much freedom may hamper the work progress, organizations can think of making teams in which there is a combination of three generations since task will be allocated as per the interest, the team is more likely to synergize.

- Drawing from their idealistic attitude, post-millennials have high expectations, the organizations should approach a subtle way to educate them about the organization vision, and work standards. It is observed, that if the organization or the team becomes successful in attaining the confidence of Zers, they work with their best of efforts.
- Zer's approach to work is fresh and novel. They believe in working on ideas, not following old traditional routines. The organization may adopt urban vision to succeed.

5.5 CONCLUSION

Work engagement is a topic of much debate in both academic and practitioner circles. As the organization recruit young people on board, the issues of different opinion, work habits and life preferences start to matter. In a team dynamics, it is crucial to appreciate and channelize the group's heterogeneity in the best interest of the organization. It is the need of the hour for organizations to become proactive and responsive to the rampant work trends. Younger employees have a different way to see things in comparison of existing work patterns and philosophies, it is imperative to understand their mindset and offer a congenial environment at work via inclusive and adaptive approach. In this chapter, we see a range of characteristics of Gen Z that can guide the task allocation and their involvement at work to some extent.

In the subsequent section, a case covering the above concept has been provided (refer Appendix 5.1 for plausible case discussion).

5.6 CASE: IS GEN Z SEEKING CHALLENGE OR NOT?

Kabeer is a young 32 years old, dynamic faculty who teaches Marketing and HRM to UG and PG students in a reputed Indian Business Studies Institution. As a student, Kabeer was hardworking, focused, and responsible who proved his talent in academics by attaining meritorious positions and now with his zeal making a notable mark at his workplace as well. Kabeer believes in doing things differently, enjoying challenging tasks, and making a constant effort to learn and grow. He is a keen explorer and enthusiast about comprehending new concepts and theories. In the present scenario of competition and talent overload, he is completely aware of the vitality of new and innovative teaching practices which is not just the need of the hour but also something that excites him. He always strives to infuse the same level of enthusiasm and spirit in his students also. New generation students being born in digital age are smarter as compared to the older students, in this era of virtual classrooms, the expectations are far more than just a chalk and talk lecture, Kabeer comprehend the relevance of engaging students in classrooms with content and practice as a big challenge, especially with an informed and impatient group of students.

5.6.1 BUNDLE OF EXPECTATIONS

It was the first day of the fresh semester at the campus. Kabeer entered into the PG class and sensed the high spirit; students who belong to the age group of 21–24 years were excited and joyful. He became happy to see the enthusiasm in the batch. With a couple of minutes, and short personal introductions, Kabeer initiated the theme of the subject and related talk, and the class got over on a positive note. As per academic routine, he prepared and uploaded the session plan and deliberated his course of action. In the second session, he addressed the class about the recent trends in marketing communications but noticed that students are disinterested in lecture and more engrossed in using mobile phones, or chatting as and when they manage to do it during the lecture. To break this, Kabeer announced few authors and important books concerning a topic, to bring the concentration at unity which actually helped, and continued with modulated pace, involving interactions and facts into his session.

After a few lectures, he observed that students are interestingly coming to his classes and tend to become more regular. This motivated him to share more pertinent and constructive information, case studies, videos, etc. to engage students in the classes. Gradually, post-lecture also few students approached him not only seeking guidance regarding the concerned topic but also about other subjects and related competitive exams. This led to his observation that students are looking for right direction and are willing to be evolved. Academic evaluations are important parameters to check the progress in learning and are an excellent opportunity to monitor the performance and career mapping. The usual practice in the institution regarding internal assessment was a combination of assignment and written exam, but it has been observed that as a trend, students do not pay much attention to assignments and submit copy-paste material from the internet. After having discussed with the students about this problem and to sort out this issue he received few responses like, *"Sir we want to do something by which we can learn and that we have never done before."* Another student said, *"We want to do some task more creative, where the lectures and concepts could become applicable"* another one said, *"Sir, we are bored with writing assignments on some concept or theory, or giving presentations on a selected topic…so something else,"* etc., and some more similar responses. Kabeer discovered that students are seeking some creativity in work that can push them to make effort and where they can learn something of practical relevance.

5.6.2 THE CHALLENGE BEGINS

The positive response of students charged up Kabeer's emotions and he decided to allot a challenging assignment ensuring that students will have an avenue of learning and suffice opportunity to showcase their creativity. Next day Kabeer announced that students will work in a group and will design a marketing campaign. The products or service that they are choosing can either be hypothetical or real, but the students must avoid plagiarism and replicating the original. After briefing the exercise, Kabeer gave students the freedom to choose their group members among themselves and ask them to finalize and inform him within two days. Though he expected the prompt response, but surprisingly he didn't received any rejoinder in the next two days also.

Kabeer followed up with the class on which they requested Kabeer to allow forming asymmetrical groups, for example, one group may have 5 members, while another will have 8, etc., as all close friends want to work as one team. Kabeer was fine with this proposal but he guided students to allocate the task equally among the groups, so that each individual may have an opportunity to perform. The students agreed happily and promised to submit the team names with their topics next week. Kabeer neither received any query nor any detail of any of the team in the mean time before the next scheduled class. However, he believed to receive it surely in the next class.

Next week, Kabeer seeks the details of final teams with their topics. To his surprise, students fail to finalize their teams and when asked the reason they responded that they could not get time to communicate with each other. Kabeer thought deeply for some time and ask students if they have any confusion in understanding the task? *Let me explain it you again, he said*. He asks the class to think of any product or service hypothetically and design a marketing campaign referring to the past lectures and discussions made in this regard. He used an example of a hypothetical product and then elaborated about the major decisions required to be considered like specific objective of the campaign, product lifecycle, a budget required for the campaign, etc. He also shared a template of marketing plan as a road map to follow. He asked now finally to submit team names after the class, but the students after a while requested him to assign the groups. Listening to this Kabeer wondered about the students but then he quickly announced the sequence of teams with 5 members in the order of their roll numbers. Class resisted a bit, but later agreed when Kabeer told them that they had enough time to choose, but they themselves wasted the opportunity, and since they all are batch mates, so working as a team will be a good chance to know each other well.

To retain the enthusiasm of students, Kabeer further add that out of all the presentations the best eight groups (50% of the total) will be given an opportunity to present their idea in front of Dean and other senior faculties of the department. Suddenly, Kabeer can sense a positive vibe in the class, one of the students asked: *"Sir, when is the deadline?"* Kabeer was thinking that already a week is gone in deciding upon teams, he didn't want an unwanted extension, still he choose not to be autocratic in his approach, so he inquired from the students *"by when will you all, be able to submit it?"* Students requested a time of two weeks, as the festival of Holi was around. Two weeks was a good long time but since it was a festive time so Kabeer agreed to the submission on a date after 15 days without any further extension. In the later days, few students approached Kabeer to seek further guidance and help in the project work.

5.6.3 AN ALTERCATION OF EXPECTATIONS AND EFFORTS

Two weeks later, on the day of submission, Kabeer was excited to see the ideas, varied perspectives and to meet the even more curious lot of students who will be eager to receive the feedback. But to his utter shock, students were sitting quietly and showed no response to his question of project submission. When he again enquired, few students responded slowly, someone said: *"Sir, we could not find matter related to it, give us some more time we will do it."* With this response, Kabeer got a bit disappointed that the students were looking for some readymade material. He further asked, *is there any group who had done it?* But there was silence. He was thwarted but extended the submission date for another week without any further delay. This incident made him realize that too much liberty is not feasible in some situations. Nevertheless, next week again he received complete silence and no project. This time he became a bit angry, but he composed him and patiently asked: *What is the actual problem, can someone tell me? Don't you all understand that this is a task and you are expected to do it?* He again received no response. He rephrased his question, *"Did you people want me to explain the activity again?"*

Kabeer was making all possible efforts to understand the problem. He asked for the last time *"is there any group who has completed the assignment?"* Suddenly, someone asked, *Sir; can we prepare an assignment on some topic and submit?*

Kabeer was speechless at this moment, after a little while he announced to submit individual assignments within two days.

In the next two days, all assignments were submitted without even being asked for (as there is a component of evaluation and students were concerned about the grades). General topics have been chosen by students and almost picked up directly from the syllabus that gave an impression of an effortless submission.

5.6.4 THE DILEMMA

Drawing upon this example, it could be said that the Zers have high expectations but when it comes to performing, they look for easy options. Can it be concluded that they are not seeking a challenge? It will be unjust to make such strong judgment, as it is an incident quoted from one college setting, though various other studies conducted on Zers report the same confrontation of efforts and expectations, still, this cannot be generalized, as context also plays a crucial role in the discovery of facts. For instance, despite being connected globally, people grew in a specific environment and culture which has a deeper impact on the personalities and thought processes. In the Indian context, the author conducted a survey of 400 college students reporting the openness of Zers preference to work with an inclusive organization. Almost 90% of students were open to working in a team ignoring any discrimination referred to caste, creed, color, or any disabilities, but only 23% of students were open to working in LGBT organizations, rest 77% were reluctant to even apply for a job in such workplaces. So, the Indian traditions and culture being rigid are somewhere creating a dilemma among the younger generations to adapt in the evolving time with richer diversity. Results also reveal that this generation is high on emotions and traditional values; in fact, several habits of Zers are found to be similar to Gen-X. The new generation is also concerned with their job security and constant feedback.

Critical Questions:

- What are the attributes that differentiate millennials and post-millennials?
- Why are engaging employees crucial for management in recent times?
- *"One key cannot unlock different locks."* Elaborate.
- What factors affect the commitment levels of Gen Z?
- What are the work values of digital natives?

APPENDIX 5.1: CASE DISCUSSION

Case Synopsis

This case revolves around the confrontation of expectations and efforts of Generation Z students. Kabeer, the protagonist, is an instructor who acknowledges the ruthless global competition, owing to which the challenge of student engagement at study is alarming, as a solution, he abides to take up creative and innovative methods and practices in delivering his job proactively. As an educator, he is more open to interactive teaching and student's involvement in decision making, to ensure that a comfort level may develop in the class for smooth learning on both sides. In the present case, the incident of a classroom has been described regarding the allocation of a task for the student's internal assessment. In the past, he had realized that students do not perform well when it comes to writing exams or assignments so he decided to avoid the act of announcing unnecessary task that bores them. Kabeer enquires from the students about their interest and expectations from an activity related to the subject which they believe is of utility to them and they will perform willingly. Appreciating the enthusiasm of students and drawing upon the responses of students about their desire to learn through some practical industry requirement activity, Kabeer suggested a group task in which there is enough scope to showcase the creativity by the students in a different manner as per their choice and selection of information. But the prolonging attitude in students subsequent to allocation has surprised Kabeer and put him in a situation to figure out the discrepancy in the demand and delivery of efforts in Gen Z.

Pedagogical Objectives

The target audience for this case is ideally undergraduates and also Postgraduate students. This case can be taught for manifold topics like Student

Engagement, team dynamics and also intrinsic motivation. As it is a short length case, the students will be able to remain intact with the case and may avoid boredom. This case can be discussed in a class of 60 or 75 minutes easily. Being short, students may center on the learning outcomes without much distraction and the time can be utilized in dialogue and deliberations. Varied perspectives could be brought forward as the case involves students themselves, so they would be in a better situation to relate and discuss.

Broad Learning Plan

The objective in the case is to ponder upon the following:
- Engagement in Generation Z is a current issue which is the focal discussion point of industry and academia. One can understand this as Student engagement as well. It is significant to understand that young students are well-informed so the old practices of teaching will not work. To develop their interest in the subject and to avoid distractions in the class, some creative approach needs to be adopted for involving them in the lecture. This scenario can be related to an organization and employee engagement theme could be explained.
- It may also assist in discussing the situational leadership and transformational leadership concepts through this case. How far do involvement and participative management can help, and despite it, do the members make efforts in the direction of performance? The importance of understanding a situation and changing leadership role accordingly can become a topic of debate in this case.
- Positive environment, open communication among supervisor and team; support and guidance may sometimes also fail, as the intrinsic motivation is somewhere missing. This aspect can also be pinpointed through this case.
- There is also something very interesting that can be observed. Drawing from the Self- determination theory, the element of '*amotivation*' (Ryan and Deci, 2000) can also be discussed and the dialogue can actually create a platform of understanding the very phenomena of confrontation in a better way. Students may use this case referring to the SDT theory for making specific recommendations.

This small case is carrying the capacity of highlighting some crucial aspects of the behavioral field and important management theories can be reflected upon. To cover different topics, the case can be used and students will be able to connect easily as it revolves around their routine life.

Discussion Plan

1. **Gen-Z Expectations:** Since the organizations are looking up to the newcomers, and pondering upon their involvement, engagement, and performance issues in a stiff competitive environment. The Instructor can lead the discussion with inputs on the traits of Zers. Dealing with the students also is a challenge, and academia is constantly working on offering better, to receive the best. In continuation to the basic concept, the following questions can also be raised:
    i. As a student, what are your expectations for your teachers and classes?
    ii. What kind of task you think can help you to learn better and grow?

    This discussion than can be linked to the responses by the students. It is expected that there will be high proponents of practical learning, smart teaching abiding less of lecturing and short span assignments.

    At this point, the instructor may highlight the confrontation and introduce the concept of situational leadership.

2. **Situational Leadership:** In this case, there is a need to understand that in a class of diverse students, all students cannot be treated as same; actually Kabeer could have handled the students in different ways. But the problem is no reaction that came jointly from the class. This sudden shift in student's behavior has caused rage in Kabeer. The instructor may ask the students here either any of the following questions or if the discussion goes, smooth, and not distracted then maybe in respective order:
    i. Do you think that Kabeer is becoming lenient in this case, and offering too much liberty to students to decide?
    ii. What would have been your line of action in this situation of silent resentment regarding non-performance for an activity which is self-chosen?
    iii. Can you suggest a better way out to decide, other than open participative discussion?
    iv. Can you point out the flaws in the leadership of Kabeer, if any?

    During the discussion of discovering the expectations in the light of the efforts, the disparity can be addressed by the issues of engagement.

3. **Gen-Z Engagement:** Although the engagement phenomena are ongoing since the inception of class, the concept can be explained at this stage of discussion and how disengagement can severely affect the planning and create confusions while making policies and decision making. Parallel to individual interest, the task allocated is a group task. It is imperative to understand here, that students were reluctant to write individual assignments, but being told to work in teams has failed; despite given the discretion to choose their own teams. The students did not revert and ended up by submitting the same old individual assignment. This is dubious. The instructor may ask the following question:

   i. Can you identify the reason for the clash between what has been said and what has been done?

   In this stage of discussion, there is a possibility of a contrary opinion, which paves the way for bringing intrinsic motivation at the center of discussion.

4. **Intrinsic Motivation:** When the point is to keep people engaged in work, the guiding force is motivation. More than extrinsic, the intrinsic motivation is moderating the situation. Self- determination theory also involves a concept of amotivation. Here, the basic tenet of the theory can be explained by the instructor and further questions can be asked:

   i. Given the autonomy to decide upon a task for themselves by the students has failed in this case. According to you what is the probable reason for this failure? Do you agree that intrinsic motivation outshines extrinsic motivation?

   ii. Based on SDT try to recommend the conclusion for this case.

   This short case will help in student's brainstorming and enable fruitful discussions to keep a track of the learning outcomes.

Timeline

Topic	Time (in Minutes)
High Expectations and traits of Gen-Z	15 minutes
Dichotomy and Situational leadership	10 minutes
Gen-Z Engagement	15 minutes
Intrinsic Motivation	20 minutes
SDT/Amotivation	10 minutes

KEYWORDS

- amotivation
- autonomy
- demographic shifts
- dilemma
- gen Z
- intelligent systems
- millennials
- Zers

REFERENCES

Alessandri, G., Consiglio, C., Luthans, F., & Borgogni, L., (2018). Testing a dynamic model of the impact of psychological capital on work engagement and job performance. *Career Development International*.

Bakker, A. B., Schaufeli, W. B., Leiter, M. P., & Tarris, T. W., (2008). Work engagement: An emerging concept in occupational health psychology. *Work and Stress, 22*, 187–200.

Bresman, H., & Rao, V. D., (2017). *A Survey of 19 Countries Shows How Generations X, Y, and Z are-and Aren't Different*. Harvard Business Review.

Fyock, C., (1990). *America's Work Force is Coming of Age*. Toronto: Lexington Books.

Guillot, C., & Soulez, S. S., (2014). On the heterogeneity of Generation Y job preferences. *Employee Relations, 36*, 319–332.

Kahn, W. A., (1990). The psychological conditions of personal engagement and disengagement at work. *Academy of Management Journal, 33*, 692–724.

Kahn, W. A., (1992). To be fully there: Psychological presence at work. *Human Relations, 45*, 321–349.

Kupperschmidt, B., (2000). Multi-generation employees: Strategies for effective management. *The Health Care Manager, 19*, 65–76.

Ryan, R. M., & Deci, E. L., (2000). Intrinsic and extrinsic motivations: Classic definitions and new directions. *Contemporary Educational Psychology* (pp. 54–67).

Sonnentag, S., (2003). Recovery, work engagement, and proactive behavior: A new look at the interface between nonwork and work. *Journal of Applied Psychology*, 518–528.

Staff, R., (2018). *The Modern Manager's Guide to Communicating with Gen Z Employees*. Blog. [online] https://risepeople.com>blog>gen-z-co (accessed on 24 February 2020).

Twaronite, K., (2016). *A Global Survey on the Ambiguous State of Employee Trust*. Harvard Business Review.

FURTHER READING

Acharya, A., & Gupta, M., (2016a). An application of brand personality to green consumers: A thematic analysis. *Qualitative Report, 21*(8), 1531–1545.

Acharya, A., & Gupta, M., (2016b). Self-image enhancement through branded accessories among youths: A phenomenological study in India. *Qualitative Report, 21*(7), 1203–1215.

Adamson, M. A., Chen, H., Kackley, R., & Micheal, A., (2018). For the love of the game: Game- versus lecture-based learning with generation Z patients. *Journal of Psychosocial Nursing and Mental Health Services, 56*, 29–36.

Alaeddin, O., & Altounjy, R.., (2018). Trust, technology awareness, and satisfaction effect into the intention to use cryptocurrency among Generation Z in Malaysia. *International Journal of Engineering and Technology (UAE), 7*(4.29 Special Issue 29), 8–10.

Alvi, S., (2018). Marrying digital and analog with generation Z: Confronting the moral panic of digital learning in late modern society. *Studies in Health Technology and Informatics, 256*, 444–453.

Arkhipova, M. V., Belova, E. E., Gavrikova, Y. A., Pleskanyuk, T. N., & Arkhipov, A. N., (2019). Reaching generation Z. attitude toward technology among the newest generation of school students. *Advances in Intelligent Systems and Computing, 726*, 1026–1032.

Bencsik, A., Molnar, P., Juhasz, T., & Machova, R., (2018). Relationship between knowledge sharing willingness and life goals of generation Z. *Proceedings of the European Conference on Knowledge Management, ECKM, 1*, 84–94.

Biernacki, M., Zarzycka, E., & Krasodomska, J., (2019). Job preferences among Generation Z students of accounting. *Proceedings of the 32nd International Business Information Management Association Conference, IBIMA 2018–Vision 2020: Sustainable Economic Development and Application of Innovation Management from Regional expansion to Global Growth*, 1439–1443.

Binns, J., (2017). For ALDO Group, Global Growth is All About gen, Z. *Apparel, 59*(6), 10–12.

Chen, M. H., Chen, B. H., & Chi, C. G. Q., (2019). Socially responsible investment by generation Z: A cross-cultural study of Taiwanese and American investors. *Journal of Hospitality Marketing and Management, 28*(3), 334–350.

Chicca, J., & Shellenbarger, T., (2018). Connecting with generation Z: Approaches in nursing education. *Teaching and Learning in Nursing, 13*(3), 180–184.

Chicca, J., & Shellenbarger, T., (2018). Generation Z: Approaches and teaching-learning practices for nursing professional development practitioners. *Journal for Nurses in Professional Development, 34*(5), 250–256.

Chillakuri, B., & Mahanandia, R., (2018). Generation Z entering the workforce: The need for sustainable strategies in maximizing their talent. *Human Resource Management International Digest, 26*(4), 34–38.

Cho, M., Bonn, M. A., & Han, S. J., (2018). Generation Z's sustainable volunteering: Motivations, attitudes and job performance. *Sustainability (Switzerland), 10*(5).

Çobanoğlu, E. O., Tağrikulu, P., & Gül, A. C., (2018). Games from generation X to generation Z. *Universal Journal of Educational Research, 6*(11), 2604–2623.

Cobe, P., (2014). Feeding Gen Z: Kids' menus cater to fussier customers. *Restaurant Business, 113*(5), 49–50.

Dirga, R. N., & Wijayati, P. H., (2018). How can teachers assess reading skills of generation Z learners in German language class? *IOP Conference Series: Materials Science and Engineering, 296*(1).

Dwidienawati, D., & Gandasari, D., (2018). Understanding Indonesia's generation Z. *International Journal of Engineering and Technology (UAE), 7*(3), 250–252.

Frunzaru, V., & Cismaru, D. M., (2018). *The Impact of Individual Entrepreneurial Orientation and Education on Generation Z's Intention Towards Entrepreneurship*. Kybernetes.

Gentina, E., Tang, T. L. P., & Dancoine, P. F., (2018). Does gen Z's emotional intelligence promote iCheating (cheating with iPhone) yet curb iCheating through reduced nomophobia? *Computers and Education, 126,* 231–247.

Ghani, N., Mansor, M., & Zakariya, K., (2018). Gen Z's activities and needs for urban recreational parks. *Planning Malaysia, 16*(2), 141–152.

Goh, E., & Jie, F., (2019). To waste or not to waste: Exploring motivational factors of generation Z hospitality employees towards food wastage in the hospitality industry. *International Journal of Hospitality Management, 80,* 126–135.

Goh, E., & Lee, C., (2018). A workforce to be reckoned with: The emerging pivotal generation Z hospitality workforce. *International Journal of Hospitality Management, 73,* 20–28.

Grow, J. M., & Yang, S., (2018). Generation-Z enters the advertising workplace: Expectations through a gendered lens. *Journal of Advertising Education, 22*(1), 7–22.

Gupta, M., & Kumar, Y., (2015). Justice and employee engagement: Examining the mediating role of trust in Indian B-schools. *Asia-Pacific Journal of Business Administration, 7*(1), 89–103.

Gupta, M., & Pandey, J., (2018). Impact of student engagement on affective learning: Evidence from a large Indian university. *Current Psychology, 37*(1), 414–421.

Gupta, M., & Ravindranath, S., (2018). Managing physically challenged workers at micro sign. *South Asian Journal of Business and Management Cases, 7*(1), 34–40.

Gupta, M., & Sayeed, O., (2016). Social responsibility and commitment in management institutes: Mediation by engagement. *Business: Theory and Practice, 17*(3), 280–287.

Gupta, M., & Shaheen, M., (2017a). Impact of work engagement on turnover intention: Moderation by psychological capital in India. *Business: Theory and Practice, 18,* 136–143.

Gupta, M., & Shaheen, M., (2017b). The relationship between psychological capital and turnover intention: Work engagement as mediator and work experience as moderator. *Journal Pengurusan, 49,* 117–126.

Gupta, M., & Shaheen, M., (2018). Does work engagement enhance general well-being and control at work? Mediating role of psychological capital. *Evidence-Based HRM, 6*(3), 272–286.

Gupta, M., & Shukla, K., (2018). An empirical clarification on the assessment of engagement at work. *Advances in Developing Human Resources, 20*(1), 44–57.

Gupta, M., (2017). Corporate social responsibility, employee-company identification, and organizational commitment: Mediation by employee engagement. *Current Psychology, 36*(1), 101–109.

Gupta, M., (2018). Engaging employees at work: Insights from India. *Advances in Developing Human Resources, 20*(1), 3–10.

Gupta, M., Acharya, A., & Gupta, R., (2015). Impact of work engagement on performance in Indian higher education system. *Review of European Studies, 7*(3), 192–201.

Gupta, M., Ganguli, S., & Ponnam, A., (2015). Factors affecting employee engagement in India: A study on offshoring of financial services. *Qualitative Report, 20*(4), 498–515.

Gupta, M., Ravindranath, S., & Kumar, Y. L. N., (2018). Voicing concerns for greater engagement: Does a supervisor's job insecurity and organizational culture matter? *Evidence-Based HRM, 6*(1), 54–65.

Gupta, M., Shaheen, M., & Das, M., (2019). Engaging employees for quality of life: Mediation by psychological capital. *The Service Industries Journal, 39*(5/6), 403–419.

Gupta, M., Shaheen, M., & Reddy, P. K., (2017). Impact of psychological capital on organizational citizenship behavior: Mediation by work engagement. *Journal of Management Development, 36*(7), 973–983.

Gupta, T., (2018). Changing the face of instructional practice with Twitter: Generation-z perspectives. *ACS Symposium Series, 1274*, 151–172.

Haddouche, H., & Salomone, C., (2018). Generation Z and the tourist experience: Tourist stories and use of social networks. *Journal of Tourism Futures, 4*(1), 69–79.

Healy, K., (2019). Using an escape-room-themed curriculum to engage and educate generation Z students about entomology. *American Entomologist, 65*(1), 24–28.

Iványi, T., & Bíró-Szigeti, S., (2019). Smart city: Studying smartphone application functions with city marketing goals based on consumer behavior of generation Z in Hungary. *Periodica Polytechnica Social and Management Sciences, 27*(1), 48–58.

Julisar, (2017). Textbook versus E-book: Media for learning process in generation Z. *Proceedings of 2017 International Conference on Information Management and Technology, ICIMTech 2017*, pp. 139–143.

Kamenidou, I. C., Mamalis, S. A., Pavlidis, S., & Bara, E. Z. G., (2019). Segmenting the generation Z cohort university students based on sustainable food consumption behavior: A preliminary study. *Sustainability (Switzerland), 11*(3).

Kapusy, K., & Logo, E., (2018). Values derived from virtual reality shopping experience among generation Z. *8th IEEE International Conference on Cognitive Info Communications, Cog Info Com 2017-Proceedings*.

Kirchmayer, Z., & Fratričová, J., (2018). What motivates generation Z at work? Insights into motivation drivers of business students in Slovakia. *Proceedings of the 31st International Business Information Management Association Conference, IBIMA 2018: Innovation Management and Education Excellence through Vision 2020*, pp. 6019–6030.

Kishore, K. M., & Priyadarshini, R. G., (2018). Important factors of self-efficacy and its relationship with life satisfaction and self-esteem with reference to gen y and gen z individuals. *IOP Conference Series: Materials Science and Engineering, 390*(1).

Lemy, D. M., (2016). The effect of green hotel practices on service quality: The gen Z perspective. Heritage, culture and society: Research agenda and best practices in the hospitality and tourism industry. *Proceedings of the 3rd International Hospitality and Tourism Conference, IHTC 2016 and 2nd International Seminar on Tourism, ISOT 2016*, pp. 9–14.

Liberato, P., Aires, C., Liberato, D., & Rocha, Á., (2019). The destination choice by generation Z influenced by the technology: Porto case study. *Advances in Intelligent Systems and Computing, 918*, 32–44.

Lichy, J., & Kachour, M., (2017). Insights into the culture of young internet users: Emerging trends- move over gen Y, here comes gen Z! *Research Paradigms and Contemporary Perspectives on Human-Technology Interaction*, pp. 84–112.

Meret, C., Fioravanti, S., Iannotta, M., & Gatti, M., (2018). The digital employee experience: Discovering generation Z. *Lecture Notes in Information Systems and Organization, 23*, 241–256.

Miroššay, A., & Mirossay, L., (2007). MDR1/PGPG gene in epigenetics, pharmacogenetics and its impact in pharmacotherapy of multidrug resistance phenomenon [MDR1/PGP gén z pohl'aduepigenetiky, farmakogenetiky a jehoúlohavofarmakoterapii v kontextemnohopoče tnejliekovejrezistencie]. *Klinicka Onkologie, 20*(1), 13–17.

Mueller, J. T., & Mullenbach, L. E., (2018). Looking for a white male effect in generation Z: Race, gender, and political effects on environmental concern and ambivalence. *Society and Natural Resources*, *31*(8), 925–941.

Opriş, I., & Cenuşă, V. E., (2017). Subject-spotting experimental method for gen Z. *TEM Journal*, *6*(4), 683–692.

Pandey, J., Gupta, M., & Naqvi, F., (2016). Developing decision making measure a mixed method approach to operationalize Sankhya philosophy. *European Journal of Science and Theology*, *12*(2), 177–189.

Peres, P., & Mesquita, A., (2018). Characteristics and learning needs of generation Z. *Proceedings of the European Conference on e-Learning, ECEL*, pp. 464–473.

Persada, S. F., Miraja, B. A., & Nadlifatin, R., (2019). Understanding the generation z behavior on D- learning: A unified theory of acceptance and use of technology (UTAUT) approach. *International Journal of Emerging Technologies in Learning*, *14*(5), 20–33.

Pousson, J. M., & Myers, K. A., (2018). Ignatian pedagogy as a frame for universal design in college: Meeting learning needs of generation Z. *Education Sciences*, *8*(4).

Roberts, A., (2018). Targeting millennials and gen Z with packaging. *Package Printing*, *65*(4), 26–31. Rodriguez, M., Boyer, S., Fleming, D., & Cohen, S., (2019). Managing the next generation of sales, gen Z/millennial cusp: An exploration of grit, entrepreneurship, and loyalty. *Journal of Business-to-Business Marketing*, *26*(1), 43–55.

Roseberry-McKibbin, C., (2017). Generation Z rising: A professor offers some hints on engaging members of gen Z, who are taking college campuses by storm. *ASHA Leader*, *22*(12), 36–38.

Ruangkanjanases, A., & Wongprasopchai, S., (2017). Adoption of mobile banking services: An empirical examination between Gen Y and Gen Z in Thailand. *Journal of Telecommunication, Electronic and Computer Engineering*, *9*(3–5 Special Issue), 197–202.

Salubi, O. G., Ondari-Okemwa, E., & Nekhwevha, F., (2018). Utilization of library information resources among generation Z students: Facts and fiction. *Publications*, *6*(2).

Samuel, J. C., Magesh, K. R., & Grace, I., (2018). Relative importance of predictor variable on digital technology use for payments: An empirical study among generation Z in South India. *International Journal of Civil Engineering and Technology*, *9*(8), 763–770.

Sayavaranont, P., Piriyasurawong, P., & Jeerungsuwan, N., (2018). Enhancing Thai generation z's creative thinking with scratch through the spiral model. *International Journal of Learning Technology*, *13*(3), 181–202.

Shiny, K. G., & Karthikeyan, J., (2016). Review on the role of anxiety and attitude in second language learning among gen-X and gen-Z students. *Man in India*, *96*(4), 1247–1256.

Skinner, H., Sarpong, D., & White, G. R. T., (2018). Meeting the needs of the Millennials and Generation Z: Gamification in tourism through geocaching. *Journal of Tourism Futures*, *4*(1), 93–104.

Soysal, F., Çalli, B. A., & Coşkun, E., (2019). Intra and intergenerational digital divide through ICT literacy, information acquisition skills, and internet utilization purposes: An analysis of gen Z. *TEM Journal*, *8*(1), 264–274.

Steckl, M., Simshäuser, U., & Niederberger, M., (2019). Attraction of generation Z: A quantitative study on the importance of health-related dimensions at work [Arbeitgeberattraktivitätaus Sicht der Generation Z: Eine quantitative Befragungzur Bedeutunggesundheitsrelevanter Dimensionenim Betrieb]. *Prävention und Gesundheitsförderung*.

Stergiou, D. P., Airey, D., & Apostolakis, A., (2018). The winery experience from the perspective of generation Z. *International Journal of Wine Business Research*, *30*(2), 169–184.

Tavares, J. M., Sawant, M., & Ban, O., (2018). A study of the travel preferences of generation Z located in Belo Horizonte (Minas Gerais-Brazil). *E-Review of Tourism Research*, *15* (43499), 223–241.

Venter, C., & Myburgh, I., (2018). Adapting a dw/bi module for gen-z students: An action design research study. *Proceedings of the European Conference on e-Learning, ECEL,* pp. 577–584.

Vogelsang, M., Rockenbauch, K., Wrigge, H., Heinke, W., & Hempel, G., (2018). Medical education for "generation Z": Everything online?: An analysis of Internet-based media use by teachers in medicine. *GMS Journal for Medical Education*, *35*(2).

Wahab, A. N. A., Ang, M. C., Jenal, R., Mukhtar, M., Elias, N. F., Arshad, H., Sahari, N. A., & Shukor, S. A., (2018). Mooc implementation in addressing the needs of generation z towards discrete mathematics learning. *Journal of Theoretical and Applied Information Technology*, *96*(21), 7030–7040.

Wells, T., Fishman, E. K., Horton, K. M., & Rowe, S. P., (2018). Meet generation Z: Top 10 trends of 2018. *Journal of the American College of Radiology*, *15*(12), 1791–1793.

Williams, C. A., (2019). *Nurse Educators Meet Your New Students: Generation Z.* Nurse Educator, *44*(2), 59–60.

CHAPTER 6

Cross-Generational Engagement Strategies

MUSARRAT SHAHEEN[1] and FARRAH ZEBA[2]

[1]*Assistant Professor, Department of Human Resource, ICFAI Business School (IBS), Hyderabad, a constituent of IFHE (Deemed to-be-university), India. Tel.: (+91) 8978219231, E-mail: shaheen.musarrat@gmail.com (M. Shaheen)*

[2]*Farrah Zeba, Assistant Professor, Department of Marketing and Strategy, ICFAI Business School (IBS), Hyderabad, a constituent of IFHE (Deemed to-be-university), India, Tel.: (+91) 8367574051, E-mail: drfarrahzeba@gmail.com (F. Zeba)*

ABSTRACT

The modern workplace is found grappling with the challenges of increased generational diversity. In today's workplace, the five generational cohorts are working side by side. These different age group workers have varied work expectations and value systems which make it difficult for the organizations to manage. This calls for an effective framework of workplace strategies that can be used by the organizations to involve and engage the cohorts at the workplace. In this view, the chapter draws attention towards the increasing need for managing the generational diversity. The chapter begins with the discussion of the evolution of generations and then outlines the work expectations and value systems of the five generational cohorts. The concluding part of the chapter delineates the strategies to engage and involve the cohorts group in their work. The organizations can implement these strategies to manage and handle the challenges associated with the generation cohorts.

6.1 INTRODUCTION

A workforce is considered as one of the sources of competitive advantage. The human, social, and psychological capital of the workforce is considered as one of the intellectual assets of the organization. But, organizations can manage and reap benefits from the workforce only when it learns the art of managing and handling the dynamics that exists due to the diverse nature of the workforce. Managing diversity at the workplace is considered as one of the major strategic moves in today's volatile environment (Gerhardt and Hebbalalu, 2018). Recent trends have witnessed that organization is formulating policies and strategies to manage the dynamics of workforce diversity (Joshi et al., 2010). Workplace diversity refers to the differences between the employees in an organization. The differences could be for various factors and encompass race, gender, age, cognitive style, education, personality, socioeconomic status, religion, cultural value systems, and physical abilities and disabilities, education level, and generation cohorts (Galston, 2017). To have different ideas and capabilities, the organization employs a diverse workforce. The prime objective is to have a 'larger pool of applicants.' Diverse workforce makes the organization agile and engenders diverse capabilities and expertise. Diverse work groups bring with them different capabilities, expertise, perspectives, and value systems (Galston, 2017).

Maintaining diversity at the workplace has a host of potential benefits to the organization, such as expanded talent pool, improved decision making, more ideas and innovation, happier customers (diverse workgroup catering to the diverse customers), and improved firm reputation (Parry, 2014a, 2014b). But, along with benefits, workplace diversity also leads to several challenges to the organization (Kaminska and Borzillo, 2018). The diverse workgroup comes with diverse needs and has different expectations from the organization. These needs and expectations are to be met to keep them motivated and engaged in their work. To do this, organizations adopt different strategies to keep them engaged and motivated towards their work role (Smola and Sutton, 2002).

A diverse workforce, when connected, and motivated at the workplace, can provide varying viewpoints and a larger pool of ideas and innovative solutions. Ethno Connect, a consulting company specializing in workforce diversity, suggested that managing workforce diversity is important as employees who come from diverse backgrounds bring in a variety of solutions on means and ways to achieve a common goal (Johnson, 2018). Employees who are having a direct interface with customers can be paired up with customers from their specific demographic. It will ease and make the customer feel comfortable with the employees and with the company (Johnson, 2018).

6.2 NEED FOR DISCUSSING THIS THEME

Out of several factors which outline workforce diversity, generational diversity has recently drawn attention from the scholars and researchers (Carter, 2018). Generational diversity is comprised of different generational cohorts who are born in different time periods. People born in similar years have similarities in life and career experiences as they have experienced similar histories and environmental challenges (Kaminska and Borzillo, 2018). On the basis of the findings from the previous studies, Carter suggested that the organization should start focusing on generation cohorts, as it leads to diversity at the workplace (2018). The author stated that the economic, political, and social environments of most organizations operating in different countries will become more diverse in the next 25 years which leads to a significant impact on the composition of the workforce in terms of beliefs, value system, attitudes, and behaviors. Similarly, from a survey of 18,000 professionals and students across the three generations (X, Y, and Z) spread over 19 countries, it was found that the generation holds different aspirations and value systems (Bresman and Rao, 2017). The differences in needs, expectations, and values raise a concern that whether organizations should remap its strategies to keep these generations engaged and motivated in work?

Findings from different surveys and studies (Anonymous, 2017; Gerhardt and Hebbalalu, 2018) have confirmed that the workforce composition is varying and organizations cannot afford to ignore such variation. For instance, Lynch conducted a survey and suggested that by the year 2014, around 36% and by the year 2020, close to half (that is 46%) of the U.S. workforce will be millennial (Wright, 2013). Millennial is the generation Y cohorts who are born in the time period of 1981–1997. Another survey states that almost 20% of the world's population will be in the age bracket of 15 to 24 years, and surprisingly the global workforce will comprise of three-quarters of generation Y employees by the year 2025 (Catalyst, 2018). A similar opinion is given by Richard Fry, a senior researcher at Pew Research Center analysis of U.S. Census Bureau data, who points out that 'more than one-in-three American labor force participants (35%) are millennial, making them the largest generation in the U.S. labor force' (Fry, 2018). Similar is the case with other countries. Such as in India by the year 2020, more than 50% of the population would be millennial having age less than 25 years (Ray, 2017).

The growing participation of millennial (Gen Y) in the workforce across the world has led to a concern to understand their needs and the factors which will influence their work motivation and work engagement level. The belief that one size fits all will hold little credence to these growing generations.

6.3 REVIEW OF THIS THEME

The generational cohort refers to an "identifiable group that shares birth years, age location, and significant life events at critical developmental stages" (cf. Macky et al., 2008). There are currently four generational cohorts available in today's workplace—'Veterans,' 'Baby Boomers,' 'Generation X' and 'Generation Y' (Macky et al., 2008; Gerhardt and Hebbalalu, 2018). For the time period and please refer to Figure 6.1.

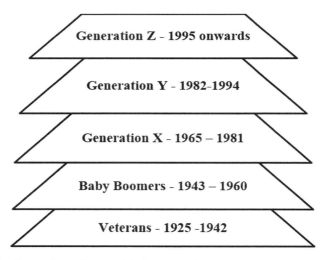

FIGURE 6.1 Generation cohorts and their age brackets.

These generation cohorts are called by many names. For instance generation Y is called 'Nexters,' 'Millennials,' Generation 'Why?' and the 'Internet Gen.' The notion 'generation' is widespread to understand differences between age groups. The members of a generation share common life experiences which tie them together into what is termed as 'cohorts.' A cohort apart from the summation of the individual experience histories also has distinct composition and character which explains the generational personality. A review of the past studies has confirmed that there exist four generational cohorts in today's workplace, viz., 'Veterans,' 'Baby Boomers,' 'Generation X' and 'Generation Y.' Those who are born after 1995 are considered as 'Z' (Clausing et al., 2014).

6.3.1 VETERANS

Veterans are the oldest generation in the workforce who are born between 1925 and 1942. And most of them have already retired. Popularly they are also referred from 'the Silent Generation' to 'Matures' or 'Traditional Generation' as well.

Veterans' life journey is characteristically intertwined with the connotations of a tough life owing to the aftermath of economic times. All through their life, they were raised in a social set-up that emphasized on morality, obligations, social norms, tradition, loyalty, self-denial, and hard work as part of their value system in their everyday life. It was common for them to grow up between two world wars coupled with the depression, and consequently, scarcity, and learning to go without was also accepted by them.

In the workplace, the traditional executive decision-making command model of management is welcomed by the Veterans who believe in its effectiveness, and thus they respect authority. They preferred to have a paternalistic employment relationship with the organization and expected the organization will provide benefits like safe working conditions as well as job security. In return, they did their job well and built their work ethic on commitment, responsibility, and conformity. And this not only helped them to give back to the organization but they also derived satisfaction out of their job.

Thus, veterans believe in lifetime employment, loyalty to the organization, and living up to one's expectation so as to gain respect, power, and status, and conclusively to escalate in the corporate ladder. However, with the new generation induction the workforce, they are challenged to cope with co-workers with diverse values, lifestyles, and demands, cutting edge technology, and non-conventional managers.

6.3.2 BABY BOOMERS

Baby boomers are the next oldest and also the largest generational cohorts in today's workforce. They are born between 1943 and 1964. The world witnessed the post-war baby boom not only in New Zealand but across the world that created the most positive and loved generation to date. Boomers flourished in an era of unprecedented national wealth and expansion. This equipped them to claim the world by right of inheritance and the belief that every other generational cohort should follow their example.

Boomers are highly competitive as they had to fight for most of the things owing to the widespread peers' rivalry amongst them. When they transitioned as young adults, they witnessed a transformation of gender roles and family constellations and an upheaval in the social set-up. Boomers' growing years were characteristically impacted by the dramatic changes in the society which include the women's empowerment, a workforce powered on technological advance and service orientated workforce and working on the global canvas. Striving for self-realization is of top priority for the self-absorbed soul searching baby boomers. Boomers became the change agents by forming or the initiation of self-help movements to facilitate implementations of every fad management program that they believed in with the hope that it would be the quick fix that they were striving for. Boomers' attitude towards their work was that of self-immersion, an insatiable desire for self-satisfaction and a feeble sense of community. As they were inclined to work more from their emotion and intuition than from the objective reason.

In the workplace, boomers are identified as the employees who are workaholic with a strong will who are concerned with not only the work content but also material gain. There was no line of separation between their work and/or personal lives, this was to the extent that their work became their personal identities. The slogan 'live to work' drives them to the extra mile at their workplace. In their attempt to go an extra mile, Boomers were expected to arrive early and often left late as for them the key to success was their visibility at the workplace. However, in return Boomers also run high expectations from their organizations ranging from promotions, coveted titles, exclusive corner offices, and to premium reserved car parking spaces. Boomers were good in building consensus, mentoring, and implementing change as they were backed by their excellent interpersonal and communication skills. They employed their keen appreciation for democracy and teamwork to create workforces to accomplish organization goals aligned with their individual goals.

6.3.3 GENERATION X

After Baby Boomers, Generation X is the largest generational cohort in today's workplace. They are born between 1965 and 1981 (Dimock, 2019).

Author Douglas Coupland wrote gave them the title 'Generation X' to the late baby-boomers. But the title 'Generation X' came to the forefront from ubiquitous usage of the title by the media tycoons who popularized the phrase during the mid-part of the year 1990. Xers grew up in a family of dual-income where parents were immersed in consumerism. This made them the 'latch-key' kids—a kid who comes to empty home from school as both the parents are out on work i.e., they grow up with little parental supervision. Therefore, Xers are self-learners who grew up teaching themselves what worked and what did not. This made them very independent generation. Being adversely affected by their parents' ever-increasing divorce rate and their dis-balance work-life, Xers got fixated and the notion that they will never to make the same mistakes as their parents. Hence, Xers idolize quality of life, strive for the balance and create boundaries on the work so as their work does not infringe on their personal lives. Consequently, they changed the slogan from 'live to work' to 'work to live.'

The information revolution in which Xers are being bought shaped the way they learn, think, and communicate. Comfortable with the new technology, Xers have easily mastered the art of generating and analyzing the huge amounts of facts and figures required in today's workplace. As a result, Xers have learned to value diversity: diverse nationalities, diverse family constellations, and diverse technology. Xers learned early on that loyalty was not a two-way street, and that the 'cradle to grave' job security of previous generations was a thing of the past. In return, Xers provide 'just in time loyalty' by performing a good job and by meeting the job demands. To maintain career security and enhance their employability Xers expect high challenging jobs in which they learn and grow simultaneously. In order to do this, Xers seek alignment with organizations that value their competencies, reward productivity, rather than longevity, and create a sense of community.

Xers are pragmatic, hardworking, ambitious, selfish, and determined to succeed financially. It is being observed that Xers collectively say 'no' to traditional management approaches in the workplace. They also expect being trusted and given freedom and flexibility to design their own working hours. Xers are tech-savvy generation, as a result, they demand for a high technological work environment, competent, and credible managers and colleagues, and leaders who coach and mentor them, rather than commanding and micromanaging. The generation, Xers are also found having high determination and independent. They believe in their own entrepreneurial spirit to add value to current work. To motivate and keep the Xer employee's performing; employers should offer job variety, high stimulating jobs with a constant change to maintain their interest. To inspire Xers motivation managers need to reward innovation, make public displays of success, support personal growth, create opportunities for satisfying teamwork and personal responsibility and create a culture of fun!!

6.3.4 GENERATION Y

Generation Y are popularly referred to as 'Nexters,' 'Millennials,' Generation 'Why?' and also the 'Internet Gen.' They are born between 1982 and 1994. They are the youngest generation in the workforce till today. The most are yet to enter the workforce. Due to recent economic expansion, Generation Y is more optimistic and positive who are shifting back to virtue and value. They are very close to their parents. And they also show more concern for religion and community. Consequently, they are generally more comfortable and believe in their abilities than their previous generations (Clausing et al., 2014).

When Generation Y was growing they had to deal with strict parental supervision—full parental attention, organized structure, chaperones, and also various after-school activities. This gave them very little or no unplanned free time. Consequently, their expectation from the organization is to furnish structure in the workplace and this entails sometimes lack of spontaneity. Like Xers, Yers are increasingly educated and technologically superior. They perceive work as a learning experience that leads them to something better. So that their work can help them avoid any dead-end if they ever encounter in their work life. As a 'Why' generation, Yers are fearless who are not afraid to voice their concerns and sportingly share their opinions and adamantly question the authority. They have a keen sense of fairness and fair play at their workplace. So they believe rules are meant to be rules and hold the notion that superiors should enforce them and not try to bend them. They are comfortable with authority but will listen to authority only when seeing that the said authority is competent and display integrity.

Yers maintain a remarkable balance between their personal and professional lives. They value their family and friends above all else. In the workplace, for them, the team is of paramount importance. They are effective as a productive member of organized teams. Consequently, they can get the things done and encouragingly expect to be evaluated as a unit. Thus, they are motivated by the remuneration based on group performance.

6.3.5 GENERATION Z

Demographic cohorts after the millennials are the Generation Z. They also popularly called post-millennial, i-Generation, founders, plurals, or the homeland generation. The difference between the Millennial and the Generation Z is strategically important for the business so that they can prepare their business, in the case when they want to repositions, align their leadership, and last but not the least is to fine-tune their recruiting efforts to prepare for the future. This is because Gen Z lives in a VUCA world (volatile, uncertain, complex, and ambiguity), a world of continuous updates. What is interesting to note that Gen Z is depts. In processing information faster and effectively than any other generational cohorts thanks to the social interaction apps like SnapChat and Vine. But the other side of this is that their attention spans might be significantly lower than Millennials. Overall, Gen Z is agile as they can quickly and efficiently shift between both work and pleasure even if they encounter multiple distractions going on in their background. Multitasking at the same is the key for them. More precisely to say is that they accomplish multi-multi-tasking. For the recruiters, this opens the avenues for them to deliberate about how this kind of flow might reshape the workplace behavior. Independence and knowledge are what is truly valued by Gen Z.

Generation Z takes any opportunity to learn be it learning something themselves or it can be through a non-traditional technique. Teens between the ages of 16 and 18 are expected to opt-out of the traditional education system to join the workforce straight away after finishing the school online. Deep Patel, Gen Z marketing strategist, explained that the newly developing world driven by technology and networked society has created any generation whose thought and action are rooted in an entrepreneurial approach to all problems. This entrepreneurial approach aspires them for a more independent work ecosystem. If we see it in quantifiable terms, almost 72% of teens aspire to start a business someday (Patel, 2017).

Marcie Merriman, executive director of growth strategy at Ernst and Young, voiced this same point and said that "When it doesn't get there that fast they think something's wrong," said "They expect businesses, brands, and retailers to be loyal to them. If they don't feel appreciated, they're going to move on. It's not about them being loyal to the business" (McCumber, 2017). Generation Z is increasingly displaying a more global thinking process and, social, and professional interactions. 58% of adults in the age bracket of thirty-five years across the world agree that "kids today have more in common with their global peers than they do with adults in their own country" (Jenkins, 2017). Diversity is of prime importance for Generation Z.

6.3.6 WORK AND PERSONAL VALUES OF GENERATION COHORTS

The current market scenario is really exciting, as the generational cohorts are redefining the workplace dynamics. Specifically, what is interesting to acknowledge is that for the very first time in this millennium, there are five generations with specific nuances who are working side-by-side in a team. It is left up to the leaders and top management alike to capitalize on this diversity and engage all the generational cohorts in a team together to accomplish a shared goal (Gerhardt and Hebbalalu, 2018). But, a key point to be deliberated here is that the shared goal as a team can only be achieved if the organization can tap on the knowledge about the specific engagement drivers for each generational needs and wants. The following section discussed the strategies related to job roles by which different generation cohorts can be engaged at their workplace. By making a relative change in the attributes of job organizations can keep the different generation cohorts motivated, committed, and engaged to their work (Van de Broeck et al., 2008; Hernaus and Pološki Vokic, 2014).

- **Veterans:** These generational cohorts are in the small proportions in the workforce today as they are born between 1928 and 1945. One does not see them now so often in the workplace. But, they still significantly comprise around 3% of the workforce. Most strikingly, these cohorts strongly hold onto the motto that an honest day's pay must be matched by an honest day's work. They are extreme loyalist. They seek pleasure from this fact that they are respected for being a loyalist. In alignment with this, they can be perceived as conformists, someone who values job titles and the accolades and money that comes with that. Organizations should reap their loyalty and should enrich their jobs to satisfy their urge of contributing to the work (Hernaus and Pološki Vokic, 2014). They should be provided an opportunity to identify their contribution in the job and can perceive the significance of their contribution. These attributes will keep them contented and dedicated to their work (Hallberg and Schaufeli, 2006).
- **Baby Boomers:** Born between 1946 and 1964, this group of baby boomers is curiously called the 'Me' generation. They are majorly in their 40s and 50s and consequently are in the commanding positions in their careers. Given this, they are in the positions of power and authority, i.e., in the decision making positions. Thus, they are quite likely to display varied behaviors from being ambitious to be loyal, to show a work-centric attitude, and lastly can be cynical too. They

prefer both monetary as well as non-monetary rewards. Non-monetary rewards that engage them can include retirement plan which is flexible and recognition from their peers. They do not prefer or seek or more precisely require constant feedback. They have a mindset of—all is well unless someone says something. Their goal-oriented approach at the work-place can be capitalized by engaging them through lucrative promotions, realistic professional development, and also valuing and acknowledging their acumen and expertise.

- Also, prestigious job positions, titles, and recognition coupled with the size of the office and basic parking spaces can also be significant engagement drivers for them. Apart from general engagement drivers mentioned here, they can also be engaged through higher levels of job responsibility, coveted perks, accolades, and challenging job profiles. It is extrapolated that as up to as 70 million of this generational cohort is going to retire by the year 2020. This makes them engaged to the opportunities pertaining to sabbaticals from their work and funding which can be retirement funding and/or something like 401(K) matching fund (Lyon et al., n.d.).
- Baby boomers are optimistic, self-cantered, goal-driven, and materialistic. Organizations can provide them with opportunities to earn rewards and incentives for them. Their job should provide challenges to them to keep them hooked with the work. They should be promoted on the basis of their performance and their goals should be designed in mutual discussion with them (Hernaus and Pološki Vokic, 2014). They are good team players and should be allowed to work in teams.

- **Generation X:** Generation Xers are born between 1965 and 1980. They are smaller than the veterans and succeeding Boomers. But Xers are often known for bringing work-life balance. This is because they were the first to witness how their overworked parents suffered from becoming so burnout.
  - Another important aspect is that the members of this generation are that they are in their 30s and 40s. Their childhood was spent alone with little parents' supervision. Due to this, they imbibe an entrepreneurial spirit within themselves. It is now becoming obvious as to why the Xers contribute as the founders to the success of several start-ups.
  - When in the case they are not creating their own ventures, then they are found to be inclined towards working independently with

least supervision from anyone. They place great value to avenues where they can grow and decide for themselves. Also, they crave for meaningful relationships with their mentors. They have a mindset that the promotions should be based on competence and not be judged by being the titles, age factors, or seniority levels (Lyon et al., n.d.). Generation Xers can be engaged by flexible schedules with ample benefits of telecommuting, recognition from the superiors, and lucrative bonuses, which can be in the form of stock, or gift cards and other monetary rewards. They are cynical and skeptical in nature, but they are comfortable with competition and value quality over quantity. Organizations can provide them such working conditions which allow them to work on a challenging task, have more variety of jobs, and should provide them with autonomy (Hernaus and Pološki Vokic, 2014).

- **Generation Y:** Millennials are born after 1980. The tech pro generation which is currently the largest age group in an emerging market likes India. As 20, something, and they have been started to join the workforce. Today, they are the fastest growing employee segment of the current workforce. Most of the Millennials are open to demonstrate their skills to the highest payer. From this, it can be not wrong to say that unlike Boomers this generation is not loyal. They have their rationality to justify why they have no problem in switching from one organization to another organization.
    - Acknowledging this, one cannot say that they cannot engage this generation. They can be engaged effectively by offering them training to enhance their skills, mentoring them in right career directions and giving them constructive feedback. Lastly, organizational culture is also significantly critical for the Millennial. They want to work in a collaborative environment where there is less of hierarchy and more team spirit. Anything forms flexible work schedules, to frequent, and timely time offs, and to technologically connect social units at the workplace are equally important engagement drivers for this generation. Millennial is also engaged when they work in an organized structure with stability, various learning avenues, and effective feedback. If they are not offered monetary rewards, they are open to stock options too.
    - Millennial are confident, socially aware and responsible. They have global and diversity consciousness, hence organizations should use them as team players and allow them to work in

diverse groups. They enjoy collaborative works as such they should be provided opportunity to lead groups and design their work accordingly (Hernaus and Pološki Vokic, 2014).
- **Generation Z:** This generation is the successor of the Millennial who joining the workplace at a very early age when they should have been the traditional educational system. What is more interesting is that in the developed countries like American they make-up one-quarter of the country's population. This makes this generation larger than both the previous generation—baby boomers or Millennial.
  - This generation can be engaged by rewards which are socially oriented, mentors stewarding their career ahead, and continuous and effective feedback. They also want to get engaged by doing meaningful tasks and their engagement is expected to be more if they are given broader responsibilities. Also, it is to be highlighted that like their previous generational cohorts, they also like engage at the workplace which facilitates more flexible schedules.
  - Other ways to engage this generation is through rewards which are not only social but are also experiential. For instance, awarding badges which one can earn in gaming and other opportunities for their personal growth. Apart from the innovative engagement starting, these cohorts also fervently seek structure at their workplace couple with clear directions as well as transparency. And also, it is quite intriguing to note about the Zers is that 53% of them prefer to communicate face-to-face.

Please refer to Table 6.1 for the workplace aspiration, personal values, and beliefs of different generation cohorts.

6.3.7 ENGAGING A WORKFORCE COMPRISING OF MULTI-GENERATIONS

It is noted that to engage the workforce across the generations the organization has to learn to be mindful of the generational diversity of each individual and consequently treat them is the same way. The point to be deliberated upon is that no matter what generation we are from, it is for us to stick to old ways of doing what we are doing now and so as to act like that each generation can be engaged by the same things that engaged them in the past (Rampton, 2017). Most of our professional management instincts propel us to acknowledge the fact that same engagement strategy does fit across all the

TABLE 6.1 Workplace Aspiration, Values, and Belief System of Generation Cohorts

Cohorts	Workplace Aspiration	Personal Value and Beliefs
Veterans	They prefer to have a paternalistic employment relationship with the organization and expected the organization will provide benefits like safe working conditions as well as job security. They value job security, an opportunity for advancement and public recognition, commitment to teamwork and collaboration.	They emphasize morality, obligations, social norms, tradition, loyalty, self-denial, and hard work as part of their value system. They are dedicated, consistent, committed, practical, and realistic.
Baby Boomers	Boomers' tare generally immersed and engrossed in their work which fulfills their insatiable desire of self-satisfaction. They were also found to have a feeble sense of community. They were inclined to work more from emotion and intuition than from objective reason. In fact, they are both process-and results-oriented and find identity in their work. They can be called as risk takers, and always on the hunt for meaningfulness, purposefulness in work. They strive for continuous growth (self-improvement).	The top priority for self-absorbed bommers was to understand themselves in terms of self-realization about their own desires and interests. They are highly competitive, goal-driven, critical, and materialistic.
Gen X	Xers seek alignment with organizations that value their competencies, reward productivity, rather than longevity, and create a sense of community. They desire to be trusted and being given the freedom and flexibility to design their own work schedules. They also want a tech-savvy work environment, credible, and competent supervisors, and leaders who can coach and mentor them rather than engagement into micromanagement.	Xers idolize quality of life, strive for the balance and create boundaries on the work so as their work does not infringe on their personal lives. Consequently, they changed the slogan from 'live to work' to 'work to live.' Scholars have found generation X as more practical, hardworking, ambitious, self-oriented and focused towards financial successes.
Gen Y (Millennial)	They believe rules are meant to be rules and hold the notion that superiors should enforce them and not try to bend them. They are motivated by the remuneration based on group performance.	They are more optimistic and positive and are very close to their parents. They show more concern for religion and community and have a strong belief in their own abilities than their previous generations.
Gen Z	They expect organizations and retailers to be loyal to them. They want independence and appreciation. If they don't feel appreciated, they're going to move on.	Gen Z is agile; they can quickly and efficiently shift between both work and pleasure even if they encounter multiple distractions going on in their background. Multitasking at the same is the key for them.

distinctly varied generational cohorts. But management professional rarely 'walk the talk.'

What is the need of the hour is their mindfulness in their actions so that they are actively tuned into creates custom-made engagement strategies encompassing all the generational diversities? That is they have to use everyone's abilities and goals for creating their engagement strategies. Particularly, it is their responsibility to make every employee, regardless of their generation, feel engaged. They also need to integrate their employees into their organizational culture so as to make them feel valued and cherished (Rampton, 2017). This may come across as if a herculean task but the organization can achieve this by the simple step i.e., making sure that they have hired the right person for the job. Also, the organization should make sure that the selected candidates are a good fit for the job profile embedded in their organizational culture. Also, the organization should ensure that the employees' purpose and meaning behind their work are clear and well defined. Creating and sharing a mission or vision should also equip the employees to be cognizant of why their job exists or the matter of fact created in the first place (Arellano, 2015). Most importantly, the organization engagement strategies should not overlook the work-life balance, should encompass benefits covering health and welfare of the recruits. In other words, work engagement can be achieved through such rewards that the employees would care about.

In this direction, the first step in crafting the organizational engagement strategy is to step back and remember the composition of their workforce which should be just an abstract generational category (Arellano, 2015). If the organization builds its engagement approach relying solely on insights from literature about millennial or boomers or Gen X, it will not have a strategy truly tailored to the needs of their diverse workforce. In this process, the organization will end instead with a strategy catering to someone else's idea of what they think the organization's workforce is like (Arellano, 2015). This change can only be triggered starts when the organizational leadership makes themselves available to employees, connect with them personally and foster meaningful professional relationships. Generational cohorts have different workplace needs and want. But before employees, they are human first. And for human, the need to belong is a universal phenomenon. So strong work relationships are the secret recipe for a strong engaged workforce. Different generation groups belong to different eras, and accordingly, they have undergone various changes be it economic or cultural or political or social (Gerhardt and Hebbalalu, 2018). As already mentioned it is challenging for the CEO of an organization to reach out with thousands of employees to connect with them on an individual basis.

Thus, in that case, managers are required to do their part and limit themselves to connect with their corresponding teams. The end goal is to design robust communication channels across every layer of the organization so that the information flows freely from top levels to bottom levels. For example, when the time comes to evaluate how and why an organization needs to change its engagement strategies, the middle-level managers need to be informed so that the needs of the lower-level employees can be accurately represented. In case there a breakdown in the communication channels then the managers will never receive the information that they need and real change in the engagement strategies can never occur. Apart from a strong thrust on communication channels and meaningful workplace relationships, the organizations are required to consider prioritizing the cross-generational engagement strategies (Anonymous, 2018) (Figure 6.2).

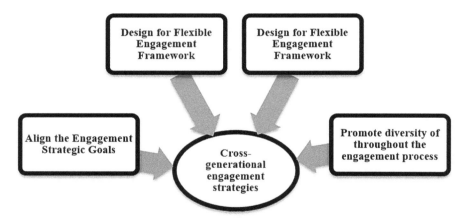

FIGURE 6.2 Four-pronged cross-generational engagement strategies.

6.3.8 ALIGN THE ENGAGEMENT STRATEGIC GOALS

From an engagement perspective, it is a prerequisite that the organization's engagement strategic goals are aligned with the specific goals of each of the generational cohorts. For this, the organization should be wired in to gauge what it wants to achieve before it can craft its engagement strategies. Regardless of the environment, an organization is striving to create; synergies must be at the core of their workplace engagement efforts. To summarize, it will not be wrong to attribute the success of the engagement strategies that an organization wants to achieve is hinged on whether or not everyone including executives, managers, even the bottom line employees are working towards the same endpoint.

6.3.9   DESIGN FOR FLEXIBLE ENGAGEMENT FRAMEWORK

The engagement strategists should be flexible in their thought process so as to systematically implement the engagement strategies. Some forty years ago, it was simple for leadership to design programs that inspired to a large workforce, given that work was quite homogeneous in that period. However, in this time period, the tradition engagement strategies are less effective especially if one consider the need and wants which are in so contrasts and differing between the generational cohorts. To overcome these engagement strategies should be investing in stable systems which allow the organization to adopt a flexible framework. For example, such engagement frameworks can be made more flexible when more options are furnished to the employees in terms of flexible work shifts and/or arrangements for office space i.e., benefits offerings which have a better chance of engaging a more diverse workforce.

6.3.10   AVOID STEREOTYPES AND BIASES

One of the pitfalls in designing engagement strategies is when the organization thinks in silos. That is when it views its employees as categories and not from a holistic point of view. Such a narrow silo perspective creates stereotypes and introduces uncalled for and misleading biases that end up in unfair and often inaccurate views of employees. The organization can actively overcome this syndrome and by combating this bias formation through the implementation of cross-generational teams wherever it is possible. Over categorizations of employees yields nothing good, especially if leads to misleading perception about the employees (Parry, 2014a). Thus, the organization should create opportunities for all the generational cohorts to interact and work together. So, they can work on their differences and engage in their work as a team. This is because not only do these generational varied coworkers learn more from each other than any formal training programs or cross-generational teams can ever provide the concrete layout for strong and meaningful workplace relationships.

6.3.11 PROMOTE DIVERSITY OF THROUGHOUT THE ENGAGEMENT PROCESS

Diversity should not just be initiated and curtailed with during organizational hiring practices only. Diversity needs due weightage throughout the engagement process of the organization with its workforce. For this, the organization has to cultivate an environment which backs diversity of thought through pervasive workplace engagement practices. Especially, this practice should just not focus on the most obvious top performers. What will make the practice more effective is that the organizations make sure to put the spotlight on all the categories of its workforce across every generational cohort? Recognition is a strong engagement tool, and such robust practices make everyone feel valued and cherished.

The way forward is that it is critical for the engagement strategists to be fully aware of this inevitable change. The bottom line is that it is the pre-requisite for any engagement strategies to encompass the varied needs and wants of generational cohorts of tomorrow as well as those of today. As, any engagement programs with too much inclination towards today's generations only, the engagement strategies will be ill-equipped to engage emerging generations of tomorrow. To conclude, it can be inferred that present systems can sustain and flourish only when they have a sharp focus on the needs of humans in a broader sense instead of the narrow perspective of categories.

6.4 IMPLICATIONS

Individual coming from the same generational cohort has certain distinct shared work characteristics, which has a huge impact on their work and life (Glass, 2007). Actually, these characteristics affect the way these individuals' views their work relationships, ethical work practices, and their actual engagement with the work (Glass, 2007; Pita, 2012). To meet the diverse needs and expectation, organizations should, therefore, design their HRM policies only through an understanding of the different generation of their employees (Domeyer, 2006).

Contemporary workers (Veterans and Baby Boomers) of yesterday were more compliant and were easier to manage. But, to retain and manage today's workforce organizations are facing several challenges (Kaminska and Borzillo, 2018). As today's workforce comprised of generation Y and Z people who are more demanding. Veterans and baby boomers generational cohort of yesterdays have their values rooted in the loyalty mantra towards their organization, but now the generation Y and Z cohorts have transformed themselves into being more being individualistic and achievement orientated who are striving for more authenticity and challenging tasks.

As a result, the traditional management policies and procedure are not adequate to engage the current workforce. Human resource (HR) professionals are required to reposition management and engagement strategies which have close alignment with the generational savvy strategies. It can be achieved when the diversity is valued. New work engagement strategies which recognize the current workforces' value and attitudes should be employed. A deep understanding of the values and attitudes of each of the generational cohorts will lead to a more effective talent management strategies which will motivate and retain today's diverse workforce.

6.5 CONCLUSION

Organizations are struggling to engage their workforce (Shaheen et al., 2018). It becomes more difficult when the workforce is diverse and belongs to different age groups. It is primarily because people who are born and brought up in different time periods have different work aspirations and value systems (Parry, 2014b). They are influenced by varied economic, cultural, and societal changes of their time periods. For instance, people who are born during industrialized revolution period will have different attitudes towards the work compare to the people who are born in great depression period a will have different work aspiration from those who are born in an industrial revolution world (Yang et al., 2018). An engaged workforce is beneficial and provides a competitive advantage to the organization, but for the organization should understand the different values of their employees to have a right fit with the job (Hurst and Good, 2009). The chapter discusses different work aspiration and value system of five generation cohorts will be useful to HR professionals in the different talent management strategies.

6.6 CASE: DILEMMA OF ENGAGING TEAM MEMBERS

Harish was having a tough time with his targets. He was recently hired as a business development manager in a multi-national insurance company. He was asked to lead a team having five members. He was given full autonomy to design the work and allocate targets for his team members. He was quite excited at the beginning, but in the last quarter, his team's performance was below the expectation. In a review meeting, he was not able to defend the poor performance of his team. The country head-Gunjan Chabbra, who was part of his recruitment's interview panel, was quite upset. In the meeting, he said, "We had high hope on you Harish!" His words were reverberating in his mind even after the meeting was over. He was not feeling hungry that night. His wife Meenu noticed everything and requested him to share his problems. She is working as a talent manager for a pharmaceutical company. After listening to the problems, she advised him to first understand the diverse characteristics of his team members and try to find out the reason why the team members are not working as a team and what keeps them disengaged.

When going through the profile of his team members, Harish found out that he is leading a team whose composition is quite diverse in nature. Ramana, the one who is handling the agency recruitment part was the 1940s born. Abdus who was handling the in-house customer-lead generation was of the age of 45 years. Shelly who was handling the over-seas client was of the 1990s born, and Surekha who is always busy with his tech devices is the millennial youth. Surekha was asked to handle the SMEs clients. He reviewed the work profiles and discussed the same with his wife. Meenu smiled and asked him to develop different strategies to handle the different age group people, as they may have different work aspirations and value system which if not having congruent with the current job will result in a lower level of commitment and engagement.

The next morning Harish called a review meeting with his team members and asked them to prepare an exhaustive list of different work activities of their team. After this, he asked them to provide suggestions to complete the listed activities in the best efficient way. As expected, he found that Surekha was able to provide more solutions for improving customer interactions proving data-driven and tech-based solutions. Abdus was more responsive to improving the relationship and congruency between team members. Ramana provided relevant points to improve the compiling and reporting structure of the in-house members. Later, in the evening Harish was thinking that he needs to revise the KRAs of his team members to achieve more congruency with the job and improve the engagement of the team members. Do you agree with his decision?

APPENDIX 6.1: CASE DISCUSSION

The case provides insight into how a mismatch with work and employees, with different age groups, can result into a lower level of performance. Attempts should be made to design the work in such a way that it matches the work aspirations and value system of the employees. As, Hernaus et al. (2014) rightly pointed out that job design is an important management tool which if used effectively can lead to highly motivated, committed, and engaged employees. Managers should align the job duties according to the personal capacities and interest areas. It will result in a right fit with the job and higher level of work performance.

As mentioned in the case, Harish was handling a team of five diverse members. He used traditional methods to manage the team such as allocating the job without understanding the diverse characteristics of its members. He has not taken any initiative to understand the workplace aspirations and personal values, and beliefs of his team members. Empirical evidences have been provided that different age groups (generation cohorts) have different aspirations from their job and they have different style of working (Parry, 2014a; Gerhardt and Hebbalalu, 2018). When Harish understands this and examined his team closely, he was quite surprised that he was managing a diverse age group member. When he redesigned the work by aligning it with the diversity of his team member, he found his team members not only engaged in their work but also performing.

Managing diversity is one of the major challenges of today's workplace. To reap the benefits from the workforce, one needs to understand the different expectations and value systems of the workforce. It will provide a good fit of the workforce and the job. Neglecting the differences will lead into poor fit between the workforce and the job leading to lower level of performance and involvement of the employees in the job. Poor fit lowers the moral and motivation of the employees and they feel disconnected with the work. They don't find meaning and significance in their work, as they have different working beliefs and styles. For instance, Surekha was a millennial and was more comfortable with technology, but she was not handling a profile where she can display her technical skills. Later on, when Harish provided her autonomy to design her work she feels connected as now she can use her tech knowledge to accomplish her tasks. It not only makes her comfortable with her job, but also somewhere down the line she finds the work manageable where she can perform and complete her targets.

KEYWORDS

- baby boomers
- gen X
- gen Y
- gen Z
- generation 'X' and generation 'Y'
- generation cohorts
- veterans
- work engagement strategies

REFERENCES

Anonymous. *How to Engage Employees from All 5 Generations*. https://bonfyreapp.com/blog/engage-employees-5-generations (accessed on 24 February 2020).

Anonymous. *Millennials in the Workplace: Top Stats Every CEO Should Know*. https://www.optimalnetworks.com/2017/11/30/millennials-workplace-stats/ (accessed on 24 February 2020).

Arellano, K. (2015). *The Generational Shift in the Workplace: Are WE Ready?* http://integral-leadershipreview.com/12937-47-the-generational-shift-in-the-workplace-are-we-ready/ (accessed on 24 February 2020).

Autry, A. (2019). *Millennial Employee Engagement and Loyalty Statistics: The Ultimate Collection*. https://blog.accessperks.com/millennial-employee-engagement-loyalty-statistics-the-ultimate-collection (accessed on 24 February 2020).

Bresman, H., & Rao, V. (2017). *A Survey of 19 Countries Shows How Generations X, Y, and Z are and aren't different*. https://hbr.org/2017/08/a-survey-of-19-countries-shows-how-generations-x-y-and-z-are-and-arent-different (accessed on 24 February 2020).

Carter, C. M. (2016). *The Complete Guide to Generation Alpha, the Children of Millennials*. https://www.forbes.com/sites/christinecarter/2016/12/21/the-complete-guide-to-generation-alpha-the-children-of-millennials/#773125203623 (accessed on 24 February 2020).

Catalyst, *Generations: Demographic Trends in Population and Workforce*. https://www.catalyst.org/knowledge/generations-demographic-trends-population-and-workforce (accessed on 24 February 2020).

Deloitte, (2018). *Millennial Survey 2018 | Deloitte | Social Impact, Innovation.* [Online] https://www2.deloitte.com/global/en/pages/about-deloitte/articles/millennialsurvey.html (accessed on 24 February 2020).

Dimock, M. (2019). Defining generations: Where Millennials end and Generation Z begins," retrieved on March 21, 2019 from https://www.pewresearch.org/fact-tank/2019/01/17/where-millennials-end-and-generation-z-begins/ (accessed on 22 March 2020).

Fry, R., (2018). *Millennials are the Largest Generation in the US Labor Force*. Pew Research Center.

Galston, W. A. (2017). *Millennials Will Soon be the Largest Voting Bloc in America*. https://www.brookings.edu/blog/fixgov/2017/07/31/millennials-will-soon-be-the-largest-voting-bloc-in-america/ (accessed on 24 February 2020).

Hallberg, U. E., & Schaufeli, W. B., (2006). "Same same" but different? Can work engagement be discriminated from job involvement and organizational commitment? *European Psychologist, 11*(2), 119–127.

Hernaus, T., & Pološki, V. N., (2014). Work design for different generational cohorts: Determining common and idiosyncratic job characteristics. *Journal of Organizational Change Management, 27*(4), 615–641.

Hurst, J. L., & Good, L. K., (2009). Generation Y and career choice: The impact of retail career perceptions, expectations and entitlement perceptions. *Career Development International, 14*(6), 570–593.

Jenkins, R. (2017). *4 Reasons Generation Z will be the Most Different Generation*. https://www.inc.com/ryan-jenkins/who-is-generation-z-4-big-ways-they-will-be-different.html (accessed on 24 February 2020).

Johnson, R. (2019). *What Are the Advantages of a Diverse Workforce?* https://smallbusiness.chron.com/advantages-diverse-workforce-18780.html (accessed on 24 February 2020).

Joshi, A., Dencker, J. C., Franz, G., & Martocchio, J. J., (2010). Unpacking generational identities in organizations. *Academy of Management Review, 35*(3), 392–414.

Lyon, K., Legg, S. & Toulson, P. (2005). Generational cohorts. *International Journal of Diversity in Organisations, Communities & Nations, 5*(1), 89–98 (accessed on 24 February 2020).

Macky, K., Gardner, D., & Forsyth, S., (2008). Generational differences at work: Introduction and overview. *Journal of Managerial Psychology, 23*(8), 857–861.

McCumber, E. (2017). *Five Things you Need to Know About Gen Z, Starting Yesterday*. https://medium.com/madison-ave-collective/five-things-you-need-to-know-about-gen-z-starting-yesterday-27ad98410c15 (accessed on 24 February 2020).

Patel, D. (2017). *5 Differences Between Marketing to Millennials vs. Gen Z*. https://www.forbes.com/sites/deeppatel/2017/11/27/5-differences-between-marketing-to-millennials-vs-gen-z/#1a2076282c9f (accessed on 24 February 2020).

Price Waterhouse Cooper (PWC). *Millennials at Work: Re-Shaping the Workplace*. https://www.pwc.de/de/prozessoptimierung/assets/millennials-at-work-2011.pdf (accessed Jan 22, 2019).

Rampton, J. (2017). *Different Motivations for Different Generations of Workers: Boomers, Gen X, Millennials, and Gen Z*. https://www.inc.com/john-rampton/different-motivations-for-different-generations-of-workers-boomers-gen-x-millennials-gen-z.html (accessed on 24 February 2020).

Ray, A. (2017). *PM Modi is Banking on India's Millennials for Growth but They aren't Sure They can Pull it off*. https://economictimes.indiatimes.com/jobs/demographic-dividend-indian-millennials-are-not-sure-if-they-can-pull-it-off/articleshow/59594185.cms (accessed on 24 February 2020).

Shaheen, M., Zeba, F., & Mohanty, P. K., (2018). Can engaged and positive employees delight customers? *Advances in Developing Human Resources, 20*(1), 103–122.

Van Den Broeck, A., Vansteenkiste, M., De Witte, H., & Lens, W., (2008). Explaining the relationships between job characteristics, burnout, and engagement: The role of basic psychological need satisfaction. *Work and Stress, 22*(3), 277–294.

Wey Smola, K., & Sutton, C. D., (2002). Generational differences: Revisiting generational work values for the new millennium. *Journal of Organizational Behavior, 23*(4), 363–382.

Wright, A. D. (2013). *Study: Millennials Expect to Use Latest Tech Tools at Work.* https://www.shrm.org/resourcesandtools/hr-topics/technology/pages/millennials-expect-to-use-latest-tech-tools-at-work.aspx (accessed on 24 February 2020).

Yang, J., Yu, C. S., & Wu, J., (2018). Work values across generations in China. *Chinese Management Studies, 12*(3), 486–505.

FURTHER READING

Acharya, A., & Gupta, M., (2016a). An application of brand personality to green consumers: A thematic analysis. *Qualitative Report, 21*(8), 1531–1545.

Acharya, A., & Gupta, M., (2016b). Self-image enhancement through branded accessories among youths: A phenomenological study in India. *Qualitative Report, 21*(7), 1203–1215.

Allard, E., Mortimer, K., Gallo, S., Link, H., & Wortham, S., (2014). Immigrant Spanish as liability or asset? Generational diversity in language ideologies at school. *Journal of Language, Identity and Education, 13*(5), 335–353.

Baran, M., (2014). Competency models and the generational diversity of a company workforce. *Economics and Sociology, 7*(2), 209–217.

Cawich, S. O., Johnson, P. B., Dan, D., & Naraynsingh, V., (2014). Surgical leadership in the time of significant generational diversity. *Surgeon, 12*(4), 235–236.

Clausing, S. L., Kurtz, D. L., Prendeville, J., & Walt, J. L., (2003). Generational diversity: The Nexters. *AORN Journal, 78*(3), 373–379.

Felgen, J. A., & Kinnaird, L., (2001). Dynamic dialogue: Application to generational diversity. *Seminars for Nurse Managers, 9*(3), 164–168.

Gerhardt, M. W., & Hebbalalu, D., (2018). Mind the gap: Moving from ethnocentric to ethno relative perspectives of generational diversity. *Millennials: Trends, Characteristics and Perspectives* pp. 1–18.

Gupta, M., & Kumar, Y., (2015). Justice and employee engagement: Examining the mediating role of trust in Indian B-schools. *Asia-Pacific Journal of Business Administration, 7*(1), 89–103.

Gupta, M., & Pandey, J., (2018). Impact of student engagement on affective learning: Evidence from a large Indian university. *Current Psychology, 37*(1), 414–421.

Gupta, M., & Ravindranath, S., (2018). Managing physically challenged workers at micro sign. *South Asian Journal of Business and Management Cases, 7*(1), 34–40.

Gupta, M., & Sayeed, O., (2016). Social responsibility and commitment in management institutes: Mediation by engagement. *Business: Theory and Practice, 17*(3), 280–287.

Gupta, M., & Shaheen, M., (2017a). Impact of work engagement on turnover intention: Moderation by psychological capital in India. *Business: Theory and Practice, 18*, 136–143.

Gupta, M., & Shaheen, M., (2017b). The relationship between psychological capital and turnover intention: Work engagement as mediator and work experience as moderator. *Journal Pengurusan, 49*, 117–126.

Gupta, M., & Shaheen, M., (2018). Does work engagement enhance general well-being and control at work? Mediating role of psychological capital. *Evidence-Based HRM, 6*(3), 272–286.

Gupta, M., & Shukla, K., (2018). An empirical clarification on the assessment of engagement at work. *Advances in Developing Human Resources, 20*(1), 44–57.

Gupta, M., (2017). Corporate social responsibility, employee-company identification, and organizational commitment: Mediation by employee engagement. *Current Psychology, 36*(1), 101–109.

Gupta, M., (2018). Engaging employees at work: Insights from India. *Advances in Developing Human Resources, 20*(1), 3–10.

Gupta, M., Acharya, A., & Gupta, R., (2015). Impact of work engagement on performance in Indian higher education system. *Review of European Studies, 7*(3), 192–201.

Gupta, M., Ganguli, S., & Ponnam, A., (2015). Factors affecting employee engagement in India: A study on off shoring of financial services. *Qualitative Report, 20*(4), 498–515.

Gupta, M., Ravindranath, S., & Kumar, Y. L. N., (2018). Voicing concerns for greater engagement: Does a supervisor's job insecurity and organizational culture matter? *Evidence-Based HRM, 6*(1), 54–65.

Gupta, M., Shaheen, M., & Das, M. (2019). Engaging employees for quality of life: mediation by psychological capital. *The Service Industries Journal, 39*(5–6), 403–419.

Gupta, M., Shaheen, M., & Reddy, P. K., (2017). Impact of psychological capital on organizational citizenship behavior: Mediation by work engagement. *Journal of Management Development, 36*(7), 973–983.

Hamlin, L., & Gillespie, B. M., (2011). Beam me up, Scotty, but not just yet: Understanding generational diversity in the perioperative milieu. *ACORN, 24*(4), 36–43.

Hart, S. M., (2006). Generational diversity: Impact on recruitment and retention of registered nurses. *Journal of Nursing Administration, 36*(1), 10–12.

Hendricks, J. M., & Cope, V. C., (2013). Generational diversity: What nurse managers need to know? *Journal of Advanced Nursing, 69*(3), 717–725.

Johnson, S. A., & Romanello, M. L., (2005). Generational diversity: Teaching and learning approaches. *Nurse Educator, 30*(5), 212–216.

Kaminska, R., & Borzillo, S., (2018). Challenges to the learning organization in the context of generational diversity and social networks. *Learning Organization, 25*(2), 92–101.

Kramer, L. W., (2010). Generational diversity. *Dimensions of Critical Care Nursing, 29*(3), 125–128.

McNamara, S. A., (2005). Incorporating generational diversity. *AORN Journal, 81*(6), 1149–1152.

Money, S. R., O'Donnell, M. E., & Gray, R. J., (2014). In the time of significant generational diversity: Surgical leadership must step up!. *Surgeon, 12*(1), 3–6.

Pandey, J., Gupta, M., & Naqvi, F., (2016). Developing decision making measure a mixed method approach to operationalize Sankhya philosophy. *European Journal of Science and Theology, 12*(2), 177–189.

Parry, E., & Urwin, P., (2010). The impact of generational diversity on people management. *Managing an Age-Diverse Workforce*, pp. 95–111.

Parry, E., (2014a). Generational diversity at work: New research perspectives. *Generational Diversity at Work: New Research Perspectives*, pp. 1–238.

Parry, E., (2014b). New perspectives on generational diversity at work: Introduction. *Generational Diversity at Work: New Research Perspectives*, pp. 1–8.

Reid, P., (2013). Generational diversity: Removing barriers to building bridges. *Managing Diversity in the Military: The Value of Inclusion in a Culture of Uniformity*, pp. 174–187.

Sakdiyakorn, M., & Wattanacharoensil, W., (2018). Generational diversity in the workplace: A systematic review in the hospitality context. *Cornell Hospitality Quarterly, 59*(2), 135–159.

Standifer, R. L., & Lester, S. W., (2016). To see ourselves as others see us: How perceptions of generational diversity affect the workplace. *Work Pressures: New Agendas in Communication,* pp. 65–77.

Touchton, D., & Acker-Hocevar, M., (2016). Generational diversity and feminist epistemology for building inclusive, democratic, collaborative community. *Quandaries of the Small-District Superintendency,* pp. 91–111.

Tsai, F. S., Lin, C. H., Lin, J. L., Lu, I. P., & Nugroho, A., (2018). Generational diversity, overconfidence, and decision-making in family business: A knowledge heterogeneity perspective. *Asia Pacific Management Review, 23*(1), 53–59.

Villas-Boas, S., Lima De Oliveira, A., Ramos, N., & Montero, I., (2018). Social support and generational diversity: The potential of the LSNS-6. *Pedagogia Social, 31,* 177–189.

Weingarten, R. M., & Weingarten, C. T., (2013). Generational diversity in ED and education settings: A daughter-mother perspective. *Journal of Emergency Nursing, 39*(4), 369–371.

CHAPTER 7

Managing Married Employees

ANUGAMINI P. SRIVASTAVA

Symbiosis Institute of Business Management Pune, Symbiosis International Deemed University, Maharashtra, India, E-mail: srivastavaanu0@gmail.com

ABSTRACT

Employees are a critical resource for an organization. It is their commitment, loyalty, and involvement in the job that extends towards the organizational productivity and performance. When employees stay for a longer time, it contributes toward the organizational performance and the building up of a strong culture within the organization, with higher trust and productivity and lowers employee turnover and absenteeism. However, with continuous change in the workforce composition, the managerial issues are increasing. It is becoming difficult for the managers to accommodate, organize, control, and direct the ever-changing generations in their workforce. Therefore, this chapter deals with how married employees posit a role towards workforce diversity.

7.1 INTRODUCTION

Employees are the critical resource for an organization (Jaworski, 2018). It is their commitment, loyalty, and involvement in the job which extends towards the organizational productivity and performance (Su, Wright, and Ulrich, 2018). In this line, Srivastava, and Shree (2018) stated that "When employees stay for a longer time, it contributes toward the organizational performance and the building up of a strong culture within the organization, with higher trust and productivity and lowers employee turnover and absenteeism" (p. 1477). However, with continuous change in the workforce composition, the managerial issues are increasing It is becoming difficult for the managers to accommodate, organize, control, and direct the ever-changing generations in their workforce (Mueller and Gopalakrishna, 2016; Wakisaka, 2013).

7.2 NEED FOR DISCUSSING THIS THEME

Marriage is the union of two people with a promise to be with each other through the last breath of life (Loichinger, and Cheng, 2018; Tang, Huang, and Wang, 2017). A healthy married life gives life satisfaction, subjective well-being (Carr, Freedman, Cornman, and Schwarz, 2014) and boosts the career in the long run. Though, an unhealthy relationship leads to stress, burnout, and poor physical and psychological health to both the partners, including the suffering of their children (Ruggles, 2016). A married person has to go through multiple phases of personality changes, physical aging, wax, and wanes in love life, child, family expenses, investments, divorce, retirement planning and many more (Ackerman, 1965). It is more often considered a roller coaster ride when it interacts with jobs and work assignments in an organization. A married employee has to manage, or we may say balance the work and life (Tang, Huang, and Wang, 2017; Simon, 2002). At the same time, managers in an organization have to understand the stage of married employees, both female and male, and plan out the strategies to make them comfortable. Simultaneously, managers need to retain the talented employees and get the best out of them concerning decision making, problem-solving, and achievement of common goals (Scase, and Goffee, 2017). Thus, this chapter will include and present a varied aspect of married employees and how they pose a management dilemma among managers at all levels in an organization.

7.3 REVIEW OF THIS THEME

7.3.1 MANAGING MARRIED WOMEN

Marriage is special for both males and females when it comes to living their life with their loved ones. Although being both partners at work, family life gets affected. Their shifts, work responsibility, over-demanding jobs cause complexities in their personal lives (Loichinger, and Cheng, 2018; Tang, Huang, and Wang, 2017). It not only affects their love life, personal interactions, and physical relationships, it also affects their psychological health. Mental stress due to unhealthy relationship thus affects their work performance also (Hammer et al., 2005) Therefore, organizations need to take care of the work-family balance. Overburdening married employees for a long time not only affects their interest and commitment towards the job but, also reduces their overall productivity, efficiency, and effectiveness (Ruggles, 2016).

Married women in a culture like India, has to face multiple situations all across her life. On the one hand, family, and community are more demanding from her rather than her male counterparts, on the other hand, her own family including kids, husband, and in-laws expect her to fulfill household and customary needs (Lamichhane, Puri, Tamang, and Dulal, 2011; Jain, 2018). This puts married female employees in role conflict. She suffers from a high level of stress and role ambiguity, as to which role she has to put into priority and which one as secondary. Managing married women has to be cautious and diligent, although should not discard her from career growth, promotions, and rewards (Beutell, and Greenhaus, 1982; Akintayo, 2010). As married working women also undergo multiple phases in life like childbearing, hormonal changes, aging, and health dilemmas (Singh, Rai, and Singh, 2012), management of the organization need to consider these facts and evaluate their practices based on it (Lamichhane, Puri, Tamang, and Dulal, 2011).

Along with this payment and incentives are given to married working women should complement her contributions, efforts, and performance level. There should not be any biases or stereotypes regarding assigning the payments just because they are female. The protection of female employees should also be taken care. Provisions should be developed to stop harassment, both psychological and physical, to provide a healthy and positive work culture (Cech, and Blair-Loy, 2010).

As married women are more creative and innovative and can have a broader look towards problem-solving, more opportunity should be given to them to work in a position which suits their talent (Amah, 2010). Retention policies for pregnant women, women with infants, and on maternity leave should also be adopted (Lamichhane, Puri, Tamang, and Dulal, 2011; Cech, and Blair-Loy, 2010). To start the same, managers need to reject Stone-Age thinking. Offer them flexibility, equal, and rewarding pay to their male colleagues and make them feel valuable to the organization (Lucas, 2018).

7.3.2 MANAGING MARRIED MEN

Managing married men is as simple as it is complicated. Men being married have to go through multiple phases as well (Das, and Singh, 2014). From being single, without any responsibility to sole earner or the primary earner in the family; from being independent and ignorant to being responsible and caring; from spending thrift to investment planning; from handling a woman to handling and managing the whole family; they have many shift in their roles across their life. All of these factors affect the work efforts of the employees. The pressure to work well and excel and at the same time stress to handle household expenses has a negative effect on the married men (Gilbert, 2014; Carr, Freedman, Cornman, and Schwarz, 2014). Therefore, managers need to analyses these aspects well before allotting them tasks. Work assignments should not surpass their priorities and responsibility. Simultaneously, they should not be given more priority in comparison to female colleagues. This will enable development of healthy work environment at work. A protective environment should be encouraged. Provisions to avoid false harassment issues and job retention should be adopted. Flexibility in shifts can also be given to them to let work and maintain a tradeoff between work and personal life (Killewald, 2013; Lu et al., 2008).

Although the scenario in the present organizational culture is changing, where married men and women are considered equal concerning responsibility, expenses, and investment, proper human resource (HR) practices are to be developed to suit the requirements of both.

7.3.3 SINGLE EMPLOYEE VS MARRIED EMPLOYEE

With the changes in the workforce composition, the employees working in an organization can be now classified into single and married employees. Where some managers prefer married employee based on stereotype belief of their loyalty, commitment, and performance, many managers are now over allocating work to unmarried employees. Considering them as *lonely, emotional untethered* and *financially untroubled* managers nowadays are preferring employees who are single or are free from family responsibility like wife, parents, and children for higher positions. This is because they can perform the tasks wholeheartedly in the absence of any distractions. High preference is given to those who can devote a majority part of their life in managing, directing, and controlling their demanding assignments. Due to this high level of unfair decisions are being made to celebrate the married couples and their need for leaves and vacations. Single employees are assumed to be available for late-night shifts, free from traveling on weekends, can work on holidays, and take leaves when married employees haven't claimed for. All such beliefs make single employees vulnerable to high workloads and are a highly unfair position.

Opposing the unfair treatment towards single employees, DePaulo (2017) in his article titled *"Single workers aren't there to pick up the slack for their married bosses and colleagues"* published in Quartz, wrote that "There are a lot of misconceptions about single people in the modern-day workplace.... Consider the reaction of former Pennsylvania governor Ed Rendell when Janet Napolitano received the nomination for secretary of homeland security in 2008" (pp. 1, 2). This nomination was given to Janet, as Janet was considered perfect for the job with no personal life and no family. Therefore, she can devote, literally, 19 to 20 hours a day to it.

DePaulo mentioned that managers favoring single employees, sort married employees as less focused, more family-oriented people. They have this opinion that married people require more time to involve in family activities, family time, vacations, leaves, and other benefits, while single employees do not need them. Due to this, they avoid married male and female employees for giving assignments which require higher work orientation. This misconception developed among managers that single and childless employees are free from social involvements is because of the negative stereotypes.

However, research findings provide a contradicting picture. It indicates that once married or were living together actually makes people more insular. They are less focused to community activities and are more focused towards their family events, functions, and like. They have their family commitments and are thus less inclined towards other activities in which single employees are interested.

It is high time to understand that single employees are not self-oriented people. They like to join and meet their friend groups, relatives, and family members more often to their married colleagues. They put more efforts to maintain relationships and are more focused on taking care of their older parents and family, may it be the daughter or the son. They also love enjoying weekends and spend time in their town and communities rather than their married counterparts.

The above discussion provides that considering single employees always available is not fair. They also need their time. Equal opportunities are required for single and married employees for leaves and vacations.

7.3.4 ORGANIZATIONAL FACTORS AFFECTING THE WORK PERFORMANCE OF MARRIED EMPLOYEES

- **Marital Status and Commitment:** With the changing demand of the corporate world, managers are now emphasizing on employee based on their marital status. May it be the finalization of job profile, recruitment, and selection strategies, placement, and relocation decisions or incentives for effective performance, marital status is considered as one of the important criteria.

In recent years, married employees are considered as more committed. Based on their marital status and their inclination towards their spouses or intimate partners are considered as a basis for their decision-making ability, loyalty, and retention rate. In a thesis explaining "The ingredients for a committed workforce," Drenth mentioned the importance of marital status on the commitment of an individual in an organization. By segregating marital status into (1) married; (2) living together; (3) unmarried, with a partner; and (4) unmarried, no partner, he presented that marital status is moderately related to the normative and affective form of commitment towards work and colleagues and organizational citizenship behavior. He further also exhibited that marital status is also corrected to satisfaction with rewards and recognition, communication, openness, work-life balance, and coworkers. However, it does have a negative effect on role clarity, stress, and role conflict. While on the other hand, Camilleri (2002) rejected the hypothesis that marital status affects the organizational commitment (normative, continuance, and affective commitment) of employees. Similarly, few other scholars also presented a similar result towards organizational citizenship behavior, performance, job satisfaction, and the like.

Therefore, this clarifies that marital status cannot be simply considered as a whole sole reason for employee's commitment. The contextual references and cultural differences along with personality does matter while understanding the relationship. Managers need to evaluate this aspect and accordingly develop practices to improve the employee's commitment.

- **Gender-Biased Recruitment:** As per the report published in indianexpress.com, provided that males are dominating the job and salary packages in the organizations. They conducted a survey on Prime Database for the boards of the top 500 companies listed on the National Stock Exchange (NIFTY 500) and verified the difference in remuneration rates of male and female board members. The report mentioned that 86% of board members are male leading to female board members slimming down to merely 14%. This report indicates that organizations still follow "one woman on the board" policy to avoid feminist issues on the one hand and to indicate equal opportunities available for women in the organization. Although the percentage difference indicates the actual picture of gender-based recruitment and appointment policy. A similar report was presented by World Bank which showed that "Six in every ten such jobs prefer male candidates, even as women continue to be preferred for low-quality, low-status, and low-paid informal jobs." The report highlighted that one reason for the female labor participation rate is amongst the lowest in the world in that employers do not want to hire women. 36% of the advertisement given for the blue-collar jobs specified the gender-majority preferred men over women.

- **Gender Wage Gap:** Indianexpress.com report (https://indianexpress.com/article/gender/women-are-paid-less-yes-this-is-old-news/) also specified that not only male employees are preferred for the majority of the jobs, they are also paid higher, multiple times more than the female employees. Where male board members earned Rs. 1.22 crores annually, female board members earned less than Rs. 60 Lakhs per annum in 2016–2017. They provided a median report to avoid skewness issues in averages, which showed that Median earnings for male board members earn more than one and a half times that of women, 1.64 times more to be precise.

The report also showed a comparative analysis of the gender pay gap between 58 public and 442 private sector NIFTY 500 firms. "While earnings for both men and women are greater in the private sector. Overall, the male-female earnings' disparity is greater for PSUs than for private companies. Male board members in PSU organizations earn nearly three times that of women—a male board member earned Rs 16.14 lakh on average vis-à-vis female board members who earned Rs 5.09 lakh on an average. The median earnings for both sexes continued to show similar trends. Men earned 1.33 times that of women in private sector companies; those earnings were four times that of women for PSUs on a median."

Population Survey 2012 Annual Social and Economic Supplement showed the gender wage gap where married women are shown to earn 77 cents per dollar earned by the married male colleagues. They presented that it can be because of two reasons-one that it might not be the case that married men are only considered for higher incomes and second that men are putting off the marriage until they settle themselves with earning more money.

Similarly, it was also shown as married men receive more income than their unmarried counterparts (Thomas, 2018). A study published in 2007 by Cornell University in the American Journal of Sociology examined gaps in wages between different types of workers. It showed that where mother received the lowest pay of any group, married men received the highest income outpacing women and unmarried men.

Therefore, rather than exhibiting themselves as women-oriented organization, it will be better organizations start considering the competencies of female employees, especially the married ones. It should not be restricted to a seat at the table, though essential committees are required to uphold the balance of power.

- **Leaves, Holidays, and Intentions to Absenteeism:** Married employees are also prone to higher absenteeism, leaves, and even exiting the job. It has been sorted that married employees need more leaves especially on special occasions and festivals to fulfill the need of the family. In order to match and balance the needs of the family with work cause distress among them. Distress and overburden of the job cause lower job satisfaction among employees. The higher level of stress leads to burnout causing uninformed absenteeism, sick leaves, and casual leaves. Health issues like musculoskeletal problems are often sorted as the underlying reason to take leaves and absenteeism. Further, the intention to live closer to their family members, with their own family, specifically, kids, and spouse cause unanticipated employee attritions (Lim, and Tai, 2014).

- **Work-Family Conflict:** Married employees are also booked for work-family conflict and its effect on work performance, commitment, and absenteeism. Li, Bragger, and Cropanzano (2017) studied the behavior of married male and female employees towards work-family conflict. They assumed that the flow of conflict from work to family and family to work is based on gender and examined supervisor's perception for the same through statistical analysis they presented association amid employee-rated work-family conflict and supervisor perceptions of employee conflict. The study also indicated that the supervisor's perceptions varied based on both employee gender and the direction of conflict under consideration. Li, Bagger, and Cropanzano, (2017) wrote that "consistent with our expectation, participants who read the WFC statements were more likely to believe that these statements described a man, whereas those who read the FWC statements were more likely to believe that these statements described a woman."

In some studies, it was also mentioned that married employees indulge in extramarital affairs and sexually harass their subordinate to overcome their physical needs and cope up with work-family conflict. These situations not only worsen the work culture but also lead to abusive corporate culture. Additionally, married employees facing such conflict, if they are not able to express it, either suffer psychological issues or cardio-vascular diseases or depression or end up with suicidal intentions, divorces, and may cause physical harm to others (Frone, 2000). Therefore, attempts should be made by the managers to identify the variation in employee's behavior and help them through counseling, proper instructions, healthy discussion or may give some days of relaxation.

- **No-Spouse Policies:** A few organizations supported no spouse policy with a motive that only one of the married partners can work in an organization. Their spouse cannot be considered in the same organization. Although this policy restricts two talents to contribute to the success of an organization, these organizations have their purpose behind it. In these situations, either of the partners is fired from the job just because he or she is married to an employee of the same organization.

Avelenda (1997) stated that "No-spouse and antinepotism policies come in various forms. Some company policies prohibit either relatives or spouses (or both) from working in the same plant, office, branch, department, or company. Other policies prohibit a supervisory relationship between spouses and relatives. Still, further, some companies consider the identity of one's spouse when making hiring and firing decisions. Regardless of form, however, no-spouse, and anti-nepotism policies in the workplace are unjustifiable, antiquated, and unnecessary for running an effective and profitable business." (p. 693).

However, this policy was hugely criticized and was opposed considering it as inaccurate and antiquated notions about spousal relationships, an outdated stereotype causing the inability of spouses to work together. Unable to leave their problems at home, they are competitive, and cannot adequately supervise or manage. Organizations fired employees on having married to the employee of a rival company considering it as a reason for conflict of interest. "They feared that if an employee is married to a competitor, or to a supervisor in the same company, the relationship may create a conflict of interest" (Avelenda, 1997, p. 695).

7.3.5 PERSONAL FACTORS AFFECTING THE WORK PERFORMANCE OF MARRIED EMPLOYEES

- **Marital Status and Psychological Health:** A happy marriage is often considered in studies as the major factor that lead to life satisfaction, job satisfaction, higher commitment, and good physical and psychological health (Mukherjee, 2015; Allen and McCarthy, 2016). However, marital status also has a negative effect on the mental health and physical health. Palner and Mittelmark (2002) reported that controlling the impact of socio-economic status of employees, marriage exerts both main and buffer impact on the perceived mental/physical health relationship. In other words, the stated that married employees are comparatively having lower levels of mental health at different levels of physical fitness (Lewis, 2011).

    Among the factors that connect marital status and psychological health, the most prominent one is job satisfaction. Research sorted that married employees, specifically women, feel less motivated and satisfied with their jobs when end up doing a job different from what they ideally prefer. The gap leads to job discontentment. It also happens, that due to the transfer of husband or family, working women leave their job and for the sake of working they join a particular profession different from what they are either qualified to do or intend to do. Same happens with the male married employees also, though the intensity is somewhat low. Proulx, Helms, and Buehler, (2007) also mentioned the relationship between marriage quality and its effect on physical and psychological health. They indicated that "problematic marriages take an emotional toll, whereas high-quality marriages provide benefits, especially for women and older adults" Simon (2002) stated that as per the Gove (1972); sex-role theory of mental illness, three decades ago, women suffer from *higher rates of psychological distress to their roles in society*. They presented that where on the one hand men have more advantages because of marriage; women are more on the disadvantageous side. Married women, that too working, tend to face emotional problems that affect their work performance.

    The next factor is a balance between emotions, family response, social role accomplishment, and basic need satisfaction. Nelson and Lyubomirsky, (2015) theorized that *"working parents may experience greater wellbeing due to their fulfillment of multiple social roles, greater financial security, opportunities for positive emotions, and satisfaction of basic human needs"* (p.104). They also posited that *"parents are happy when they experience greater meaning and purpose in life, more positive emotions, the fulfillment of their social roles, and satisfaction of their basic human needs. And parents are unhappy when they experience more negative emotions, financial strain, sleep disturbance and fatigue, and troubled marriages"* (p. 102).

Exploring the work-life imbalance issues of working parents, *"disturbing thoughts about a child's illness, an incompetent nanny, or marital disagreements about how to handle a poor report card"* (p. 109) leads to stress and tension at home and negatively impact work experiences. However, on the positive note, Allen, and McCarthy (2015) presented that happy marriages lead to less cardiovascular events, fewer divorces, higher marital satisfaction, and higher income.

In a study conducted on older couples, Rutgers, Cornman, and Schwarz (2014) analyzed the relationship between marital quality and experience wellbeing and life satisfaction. The result provided *"the relative importance of own versus spouse's marital appraisals for well-being, and the extent to which the association between own marital appraisals and well-being is moderated by spouse's appraisals"* (p. 930).

Considering other factors Waite (2000) stated that married couple with higher per capita income, hourly wages, and a lower risk for children of dropping out of school has a positive influence on their work behavior and performance. However, cohabiting, and divorced, separated, and widowed people tend to face psychological traumas and thus suffer from job stress, job strain, and poor mental health. This consequently affects the work performance, commitment, and in-role behavior of employees in the long run (Lundberg and Rose, 2000; Gove and Geerken, 1977).

- **Work from Home and Married Life:** Work from home was introduced as a practice that can enable employees to work from their homes. It had its perceived benefits. Forbes published the results of a survey titled *"Work without Walls"* conducted by Microsoft whitepaper and highlighted the benefits of working from home. They provided that Work from home not only enabled the hiring of the top talent in the organization surpassing geographical limits, it also enabled employers to steal away the top talent from their rivalries, that too without increasing the pay. This practice also enabled the organization to win Best Place to Work awards, as it allowed a more engaged workforce (Bloom, Liang, Roberts, and Ying, 2014; Shehori, 2017).

Since more number of professionals is seeking to have a better balance between their work and life, there has been an increase in companies offering flexible policies to have shifting workplace priorities and realities such as work from home (Gregory, 2019). Indeed, the employees are increasingly finding it the best may for continuing to serve the company thereby maintaining all the tangible benefits of being part of an established company while enjoying all the advantages of being based at home. However, this option seems to have its own pitfalls. Some of the pros and cons are discussed in detail below:

The benefits included responses from different employees and mentioned that through work from home option, they have more opportunity to balance the work and home. They have more time with family, less stressful environment, quieter atmosphere to work less stress of long commute to an office, can avoid traffic, and save gas. They have more options to reduce the overall direct costs like costs of commuting, car wear and tear, fuel, road taxes, parking, and indirect costs such as expensive professional wardrobes and the dry-cleaning of those Simultaneously, as they had fewer distractions in the form of coworker's banter and distractions, unnecessary interruptions, and unimportant meetings, they found themselves more productive. They can set their work environment, setting, mood, and other aspects, which they fail to do in their offices to be more productive (Smith, 2017; Curwen, 2017).

However, this policy also had a few disadvantages. Where on the one hand, professional working from homework in complete isolation, they have difficulties in managing and keeping the work separate from home. Although they manage to avoid office distractions, they have to face other interruptions from children, work, neighbors, friends, and family. As family also calls for their time, it becomes dilemmatic for the employees to decide and priorities. Many times, they also overlooked in the case of promotions, as their physical absence becomes a problem. A problem for married employees is that when they opt for work from home, there work never ends.

On the employer's side, work from home, initially works well, as it makes employees feel that their organization is taking care of them. Further, this motivation helps them to work more and prove their ability. This option reduces attrition rates and enables the selection of employees at lower salaries (Parker, 2008).

7.4 IMPLICATIONS

Based on the above discussion, it is effective through problematic in recruiting married employees in an organization. They have their personal and professional duties, which sometimes make them very efficient, though create stressful scenarios for them. Therefore, while managing married employees, a few points must be kept in mind:

- Recruitment process should verify the location before inviting people to join.
- The selection process should consider the fact, that married employees have duties towards their family as well. Therefore, the appointment and posting should be suitably provided near their hometown. Relocating employees near to their native place will ensure less absenteeism and turnover. Simultaneously, managers need to consider the background of employees and should justify leaving approvals, so that they can visit their native place and celebrate festivities with their families once a year, at least.
- Training process should be effective and productive at the same time. Married employees should not only be trained to accomplish their tasks effectively but also developmental training should be provided on how to manage stress, work-family balance and perform their best on both the platforms (Byron, 2005).
- Appraisals should consider married men and women separately. Married men have less responsibility to take care of their children and old aged family members. Married women should be given grace marks in appraisals. Many a time it happens that married women take leaves to attend their family duties, which costs them in appraisals. Similarly, career breaks should also be taken into consideration, as married women come across multiple phases in life. This shift in phases has a great impact on the psychology and physical body of women. The decision to avoid them during demotion will affect them much, even if they gave their best towards both duties.
- More options to get married women back to the organization should be given. As mentioned, married women have to go through many phases in their life, causing breaks in their careers. Organizations should introduce and encourage the rejoining of married women back into their positions. Further, this break should be considered as experience while counting on appraisals.

- Counseling sessions and exit interviews should be conducted for those who intend to leave the organization due to their household duties. Where some employees become homesick and leave the job, living far away from home and family, also force them to leave the job. Lack of ability to communicate with closed ones affects their mental and physical health. Therefore, such employees should be counseled to reduce their stress, informal activities should be promoted, and exit interviews should be taken to reframe the existing policies.
- Crèches and child care services can also be implemented to enable married women and men to focus on the job and take care of the child at the same time.
- Be thoughtful in giving salaries to make married men and women. It should be fair and rewarding as per the job responsibility and level of risk involved.
- Flexibility should be provided to women to work and complete assignments.
- Pregnancy should not be treated as a reason to expel a married woman.
- The institution of meaningful maternity leave policy will enable married women to return to work. The people's perception and looks towards women coming after career break should be motivating and should not be such that force them to leave the job.

In the subsequent section, a case covering the above concept has been provided (refer Appendix 7.1 for plausible case discussion).

7.5 CASE: SITUATIONS ON MARRIED EMPLOYEE MANAGEMENT

With various human movements taking place across the world, workplace diversity is becoming a critical issue. Many companies found it difficult to manage the changing workplace diversity. This case talks about the four different situations in a company XYZ Ltd.

7.5.1 SITUATION 1

Xyz Ltd is a leading company in the IT industry. The company has a lot of projects which are completed by various groups. Recently one of their groups completed a major project to celebrate the same manager invites their group members after working hours on Friday. Most of the people were happy except Mrs. Smitha. Instead of expressing anything, Mrs. Smitha simply didn't turn up at the party she has to take care of his children. However, the manager presumed that Friday evening must be a good time and all of his team members can show up. Mrs. Smitha thought that a manager must not be able to understand his family commitments or maybe he will not care about it. However, when Mrs. Smitha didn't appear, her team leader concluded that she might not be a good team player. The questions to ponder on include:

- What a manager should do to celebrate the achievement of his team.
- Whether a manager should force Mrs. Smitha to turn up for the party.
- Why Mrs. Smitha did not inform his manager about her inability to attend the party.

7.5.2   SITUATION 2

Similarly, in different scenarios one of the team members Mrs. Andria has put some images of his family as a screen saver on her laptop and had pictures of his families all around his workstation. Another employee of the same team has criticized about this to his team leader. He states that it makes him itchy to see this daily and he wishes it all removed-now. Both the members of the team have words with each other and strains are mounting. In this state, respecting both employees' opinions is challenging. The question to ponder on includes:

- What a manager should do to resolve the conflict among his team members.

7.5.3   SITUATION 3

Similarly, newly married women Mrs. Kaul of a particular cast were wearing various multi-color bangles when she joined the office immediately after her marriage. However, it was against the company dress code, and the manager interrupted her and instructed her not to wear loud bangles. However, when she was hired, she was not informed about these policies. The employee has immediately stopped wearing bangles; however, she resigned one week later. Managers and HR had tried to convince her to stay back, but she didn't agree. The question to ponder on includes:

- What a manager should do in this situation.

7.5.4 SITUATION 4

In another team, one of the people was just married. He was one of the most hardworking employees and often takes extra responsibility to perform the team job. However, since his marriage, he tries to finish his work early and leave for his home. He performed his assigned task but was not willing to take extra responsibilities. However, since he has already set very high expectations, his managers were not happy with him and started scolding him and forced him to take extra work and stay at the company for longer hours. Within a few months, the employee resigned and joined another competing firm. The questions to ponder on include:

- What a manager should do in this situation.
- What was wrong with the manager's behavior.

APPENDIX 7.1: CASE DISCUSSION

SITUATION 1:

Key Issue in this Situation:
- Schedule concerning workplace linked events outside of office hours.
- The schedule that clashes with personal commitments.

Implications:
- In a diverse work-related setting, Acknowledgment, and confirmation of group efforts are vital. Nevertheless, it is also important to give importance to employees personal commitments while scheduling any activity or function outside office hours.

What the Manager Should Do?
- Provide appreciation to your team's hard work and exceptional efforts to finish the project.
- Acknowledge various tasks, duties of and presence of everyone.
- Inform the team about the importance of celebrating as a team.
- Ask the team to suggest the best time for a celebration where all the employees can come. It will show your respect to everyone on the team.

SITUATION 2:

Key Issue:
- Freedom of expression and irritation among co-workers.

Implication:
- The work environment that allows employees to share and show their individual opinions and expressions While preserving a humble and friendly work environment.

What Should a Manager Do?
- The manager should consider the rights of the both employees.
- Balance right to for right of personal expression as well as peers discomfort.

SITUATION 3:

Key Issue:
- Personal beliefs clashes with the grooming policy.

Implications:
- Companies should examine policies and dress codes in light of personal and religious beliefs. It restricts companies of hiring qualified candidates and also they may lose valuable employees.

What Should a Manager Do?
- Make an apology to the employee for not telling her policy while hiring.
- The manager should listen to the concerns of the employee.
- Inspect and remove rules that may not be required.
- The manager should think of providing an exception to the existing policies to take care for a religious need.

SITUATION 4:

Key Issue:
- Personal beliefs clashes with the grooming policy.

Implications:
- Companies should examine policies and dress codes in light of personal and religious beliefs. It restricts companies from hiring qualified candidates and also they may lose valuable employees.

What Should a Manager Do?
Should keep realistic expectations and try not to overburden the employee.
- Make an apology to the employee for shouting at him and keeping unrealistic expectations.
- The manager should understand the changing personal situations of the employee and should not disturb the employee until he is performing his duties with responsibility.

KEYWORDS

- **diversity management**
- **gender**
- **health and marriage**
- **human resource**
- **married men**
- **married women**
- **workforce composition**
- **workforce diversity**
- **working men**
- **working women**

REFERENCES

Ackerman, N. W., (1965). The family approach to marital disorders. In: Greene, B., (ed.), *The Psychotherapies of Marital Disharmony.* New York: The Free Press.

Akintayo, D. I., (2010). Work-family role conflict and organizational commitment among industrial workers in Nigeria. *International Journal of Psychology and Counseling, 2*(1), 1–8.

Allen, M. S., & McCarthy, P. J., (2016). Be happy in your work: The role of positive psychology in working with change and performance. *Journal of Change Management, 16*(1), 55–74.

Amah, O. E., (2010). Family-work conflict and the availability of work-family friendly policy relationship in married employees: The moderating role of work centrality and career consequence. *Research and Practice in Human Resource Management, 18*(2), 35–46.

Avelenda, S. M., (1997). Love and marriage in the American workplace: Why no-spouse policies don't work. *U. Pa. J. Lab. and Emp. L., 1,* 691–700.

Beutell, N. J., & Greenhaus, J. H., (1982). Interrole conflict among married women: The influence of husband and wife characteristics on conflict and coping behavior. *Journal of Vocational Behavior, 21*(1), 99–110.

Bloom, N., Liang, J., Roberts, J., & Ying, Z. J., (2014). Does working from homework? Evidence from a Chinese experiment. *The Quarterly Journal of Economics, 130*(1), 165–218.

Byron, K., (2005). A meta-analytic review of work-family conflict and its antecedents. *Journal of Vocational Behavior, 67*(2), 169–198.

Carr, D., Freedman, V. A., Cornman, J. C., & Schwarz, N., (2014). Happy marriage, happy life? Marital quality and subjective well-being in later life. *Journal of Marriage and Family, 76*(5), 930–948.

Cech, E. A., & Blair-Loy, M., (2010). Perceiving glass ceilings? Meritocratic versus structural explanations of gender inequality among women in science and technology. *Social Problems, 57*(3), 371–397.

Smith, S. (2017). *Seven Reasons Why Home Working is the Future.* https://www.telegraph.co.uk/business/2017/07/24/seven-reasons-home-working-future/ (accessed on 24 February 2020).

Curwen, A. E. (2017). *Hiring Couples Can Inspire Loyalty, Productivity Hiring Couples Can Inspire Loyalty, Productivity.* https://www.shrm.org/resourcesandtools/hr-topics/employee-relations/pages/hiring-couples-can-inspire-loyalty,-productivity.aspx (accessed on 24 February 2020).

Das, A., & Singh, S. K., (2014). Changing men: Challenging stereotypes. Reflections on working with men on gender issues in India. *IDS Bulletin, 45*(1), 69–79.

DePaulo B. (2017). *Single Workers Aren't There to Pick up the Slack for Their Married Bosses and Colleagues* https://qz.com/991030/your-single-coworkers-and-employees-arent-there-to-pick-up-theslack-for-married-people/ (accessed on 24 February 2020).

Frone, M. R., (2000). Work-family conflict and employee psychiatric disorders: The national comorbidity survey. *Journal of Applied Psychology, 85*(6), 888–895.

Gilbert, L. A., (2014). *Men in Dual-Career Families: Current Realities and Future Prospects.* Psychology Press.

Gove, W. R., & Geerken, M. R., (1977). The effect of children and employment on the mental health of married men and women. *Social Forces, 56*(1), 66–76.

Gregory, A. (2019). *The Pros and Cons of Working from Home.* doi: https://www.thebalancesmb.com/thepros-and-cons-of-working-from-home-2951766 (accessed on 24 February 2020).

Hammer, L. B., Cullen, J. C., Neal, M. B., Sinclair, R. R., & Shafiro, M. V., (2005). The longitudinal effects of work-family conflict and positive spillover on depressive symptoms among dual-earner couples. *Journal of Occupational Health Psychology*, *10*(2), 138–154.

Jain, D. (2018). *Indian Companies Often Prefer Men Over Women in Hiring: World Bank Study* https://www.livemint.com/Industry/jRfllDbFXkNJH1itasp8xI/Indian-companies-often-prefemen-over-women-in-hiring-Worl.html (accessed on 24 February 2020).

Jaworski, C., Ravichandran, S., Karpinski, A. C., & Singh, S., (2018). The effects of training satisfaction, employee benefits, and incentives on part-time employees' commitment. *International Journal of Hospitality Management*, *74*, 1–12.

Kaufman, G., & Uhlenberg, P., (2000). The influence of parenthood on the work effort of married men and women. *Social Forces*, *78*(3), 931–947.

Killewald, A., (2013). A reconsideration of the fatherhood premium: Marriage, coresidence, biology, and fathers' wages. *American Sociological Review*, *78*(1), 96–116.

Lamichhane, P., Puri, M., Tamang, J., & Dulal, B., (2011). Women's status and violence against young married women in rural Nepal. *BMC Women's Health*, *11*(1), 19.

Lewis, K. R. (2011). *When Your Spouse is Also Your Coworker*. doi: http://fortune.com/2011/06/09/wheyour-spouse-is-also-your-coworker/ (accessed on 24 February 2020).

Li, A., Bagger, J., & Cropanzano, R., (2017). The impact of stereotypes and supervisor perceptions of employee work-family conflict on job performance ratings. *Human Relations*, *70*(1), 119–145.

Lim, S., Tai, K., (2014). Family incivility and job performance: A moderated mediation model of psychological distress and core self-evaluation. *Journal of Applied Psychology*, *99*(2), 351–359.

Loichinger, E., & Cheng, Y. H. A., (2018). Feminizing the workforce in aging East Asia? The potential of skilled female labor in four advanced economies. *Journal of Population Research*, *35*(2), 187–215.

Lu, L., Kao, S. F., Chang, T. T., Wu, H. P., & Cooper, C. L., (2008). Work/family demands, work flexibility, work/family conflict, and their consequences at work: A national probability sample in Taiwan. *International Journal of Stress Management*, *15*(1), 1–21.

Lucas, S. (2018) *The HR Policies You Need to Hire and Retain More Female Talent*. https://gusto.com/framework/hr/retain-female-talent/ (accessed on 24 February 2020).

Lundberg, S., & Rose, E., (2000). Parenthood and the earnings of married men and women. *Labor Economics*, *7*(6), 689–710.

Mueller, D., & Gopalakrishna, P., (2016). Market-oriented organizations and talent workers: Composition of the workforce and its influence on market orientation. *Journal of Marketing Management*, *4*(2), 1–23.

Mukherjee, M. (2015). *Can Married Couples Work Together?* doi: https://timesofindia.indiatimes.com/lifestyle/relationships/love-sex/Can-married-couples-work-together/articleshow/17609482.cms (accessed on 24 February 2020).

Nelson, S. K., & Lyubomirsky, S., (2015). 6. Juggling family and career: Parents' pathways to a balanced and happy life. *Flourishing in Life, Work, and Careers: Individual Wellbeing and Career Experiences*, pp. 100–120.

Parker, J. (2008). *Legal Dilemma: Relationships at Work*. https://www.personneltoday.com/hr/legal-dilemmarelationships-at-work/ (accessed on 24 February 2020).

Ruggles, S., (2016). Marriage, family systems, and economic opportunity in the USA since 1850. *Gender and Couple Relationships*, *6*, 3–41.

Scase, R., & Goffee, R., (2017). *Reluctant Managers (Routledge Revivals): Their Work and Lifestyles*. Routledge.
Shehori, S. (2017). *Why Employees Should Be Allowed to Work from Home*. https://www.huffingtonpost.com/steven-shehori/why-employees-should-bea_b_14519684.html (accessed on 24 February 2020).
Simon, R. W., (2002). Revisiting the relationships among gender, marital status, and mental health. *American Journal of Sociology, 107*(4), 1065–1096.
Singh, L., Rai, R. K., & Singh, P. K., (2012). Assessing the utilization of maternal and child health care among married adolescent women: Evidence from India. *Journal of Biosocial Science, 44*(1), 1–26.
Srivastava, A. P., & Shree, S., (2018). Examining the effect of employee green involvement on the perception of corporate social responsibility: Moderating role of green training. *Management of Environmental Quality: An International Journal, 301*, 197–210.
Su, Z. X., Wright, P. M., & Ulrich, M. D., (2018). Going beyond the SHRM paradigm: Examining four approaches to governing employees. *Journal of Management, 44*(4), 1598–1619.
Tang, Y., Huang, X., & Wang, Y., (2017). Good marriage at home, creativity at work: Family-work enrichment effect on workplace creativity. *Journal of Organizational Behavior, 38*(5), 749–766.
Thomas, M. (2018). *Indian Companies Often Prefer to Hire Men Over Women—and Pay Them More*. https://qz.com/india/1241434/indian-firms-hiring-and-pay-bias-against-women-according-toa-world-bank-study/ (accessed on 24 February 2020).
Wakisaka, A., (2013). Changes in human resource management of women after the 1985 equal employment opportunity act. *Japan Labor Review, 10*(2), 57–81.
Waite, L. J. (2000). Trends in men's and women's well-being in marriage. *The Ties That Bind: Perspectives on Marriage and Cohabitation, 368*.

FURTHER READING

Acharya, A., & Gupta, M., (2016a). An application of brand personality to green consumers: A thematic analysis. *Qualitative Report, 21*(8), 1531–1545.
Acharya, A., & Gupta, M., (2016b). Self-image enhancement through branded accessories among youths: A phenomenological study in India. *Qualitative Report, 21*(7), 1203–1215.
Adame-Sánchez, C., Caplliure, E. M., & Miquel-Romero, M. J., (2018). Paving the way for coopetition: drivers for work-life balance policy implementation. *Review of Managerial Science, 12*(2), 519–533.
Adisa, T. A., Abdulraheem, I., & Isiaka, S. B., (2019). Patriarchal hegemony: Investigating the impact of patriarchy on women's work-life balance. *Gender in Management, 34*(1), 19–33.
Akanni, A. A., Olayinka, E. O., & Oduaran, C. A., (2018). Work-life balance, job insecurity, and counterproductive work behavior among brewery workers. *North American Journal of Psychology, 20*(2), 289–299.
Alegre, J., & Pasamar, S., (2018). Firm innovativeness and work-life balance. *Technology Analysis and Strategic Management, 30*(4), 421–433.
Alexander, A. G., & Ballou, K. A., (2018). Work-life balance, burnout, and the electronic health record. *American Journal of Medicine, 131*(8), 857–858.
Allam, Z., (2019). An inquisitive enquiry of work-life balance of employees: Evidences from Kingdom of Saudi Arabia. *Management Science Letters, 9*(2), 339–346.

Amutha, R., (2018). Work-life balance among IT industry-An empirical study. *Indian Journal of Public Health Research and Development, 9*(4), 75–78.

Aravinda, K. K. P., & Priyadarshini, R. G., (2018). Study to measure the impact of social media usage on work-life balance. *IOP Conference Series: Materials Science and Engineering, 390*(1).

Azizpoor, P., & Safarzadeh, S., (2016). The relationship between spousal intimacy, perceived equity, and marital quality in married employees. *Asian Social Science, 12*(9), 202–208.

Azizpoor, P., & Safarzadeh, S., (2017). The relationship between perceived individual-couple sacrificial behavior and quality of marital relationship in married employees. *Indian Journal of Public Health Research and Development, 8*(2), 19–23.

Bae, J., Jennings, P. F., Hardeman, C. P., Kim, E., Lee, M., Littleton, T., & Saasa, S., (2019). Compassion satisfaction among social work practitioners: The role of work-life balance. *Journal of Social Service Research.*

Balven, R., Fenters, V., Siegel, D. S., & Waldman, D., (2018). Academic entrepreneurship: The roles of identity, motivation, championing, education, work-life balance, and organizational justice. *Academy of Management Perspectives, 32*(1), 21–42.

Barrett, R., & Holme, A., (2018). Self-rostering can improve work-life balance and staff retention in the NHS. *British Journal of Nursing, 27*(5), 264–265.

Beham, B., Drobnič, S., Präg, P., Baierl, A., & Eckner, J., (2018). Part-time work and gender inequality in Europe: A comparative analysis of satisfaction with work-life balance. *European Societies*, pp. 1–25.

Bertram, J., Mache, S., Harth, V., & Mette, J., (2018). Operational measures for a facilitated work-life balance [Betriebliche Maßnahmen zur Vereinbarkeit verschiedener Lebensbereiche]. *Zentralblatt für Arbeitsmedizin, Arbeitsschutz und Ergonomie, 68*(4), 221–226.

Bittar, P. G., & Nicholas, M. W. (2018). The burden of inbox-messaging systems and its effect on work-life balance in dermatology. *Journal of the American Academy of Dermatology, 79*(2), 361–363.

Blazovich, J. L., Smith, K. T., & Murphy, S. L., (2018). Mother-friendly companies, work-life balance, and emotional well-being: Is there a relationship to financial performance and risk level? *International Journal of Work Organization and Emotion, 9*(4), 303–321.

Braun, B. J., Fritz, T., Lutz, B., Röth, A., Anetsberger, S., Kokemohr, P., & Luketina, R., (2018). Work-life balance: Thoughts of the young surgeon representatives of the German surgical society [Work-Life-Balance: Gedanken des Perspektivforums Junge Chirurgie der Deutschen Gesellschaft für Chirurgie]. *Chirurg, 89*(12), 1009–1012.

Braun, S., & Peus, C., (2018). Crossover of work-life balance perceptions: Does authentic leadership matter? *Journal of Business Ethics, 149*(4), 875–893.

Brown, C., & Yates, J., (2018). Understanding the experience of midlife women taking part in a work-life balance career coaching programme: An interpretative phenomenological analysis. *International Journal of Evidence Based Coaching and Mentoring, 16*(1), 110–125.

Bryan, S. J., & Adair, K. C., (2019). Forty-five good things: A prospective pilot study of the three good things well-being intervention in the USA for healthcare worker emotional exhaustion, depression, work-life balance, and happiness. *BMJ Open, 9*(3).

Caesar, L. D., & Fei, J., (2018). Work-life balance. *Managing Human Resources in the Shipping Industry*, pp. 107–128.

Cain, L., Busser, J., & Kang, H. J., (2018). Executive chefs' calling: Effect on engagement, work-life balance and life satisfaction. *International Journal of Contemporary Hospitality Management, 30*(5), 2287–2307.

Chockalingam, S. M., & Sudarshan, P., (2018). A study on work life balance of it enabled BPO workers in Bangalore city (Karnataka). *International Journal of Engineering and Technology (UAE), 7*(3), 134–137.

Chung, H., & Van Der Lippe, T., (2018). Flexible working, work-life balance, and gender equality: Introduction. *Social Indicators Research.*

Clemen, N. M., Blacker, B. C., Floen, M. J., Schweinle, W. E., & Huber, J. N., (2018). Work-life balance in women physicians in South Dakota: Results of a state-wide assessment survey. *South Dakota Medicine: The Journal of the South Dakota State Medical Association, 71*(12), 550–558.

Cuéllar-Molina, D., García-Cabrera, A. M., & Lucia-Casademunt, A. M., (2018). Is the institutional environment a challenge for the well-being of female managers in Europe? The mediating effect of work-life balance and role clarity practices in the workplace. *International Journal of Environmental Research and Public Health, 15*(9).

Darko-Asumadu, D. A., Sika-Bright, S., & Osei-Tutu, B., (2018). The influence of work-life balance on employees' commitment among bankers in Accra, Ghana. *African Journal of Social Work, 8*(1), 47–55.

Dasgupta, S., Dave, I., McCracken, C. E., Mohl, L., Sachdeva, R., & Border, W., (2018). Burnout and work-life balance among pediatric cardiologists: A single-center experience. *Congenital Heart Disease.*

Denson, N., Szelényi, K., & Bresonis, K., (2018). Correlates of work-life balance for faculty across racial/ethnic groups. *Research in Higher Education, 59*(2), 226–247.

Drenth, J. C., (2009). *The Ingredients for a Committed Workforce.* Master's thesis, University of Twente, The Netherlands.

Dyer, S., Xu, Y., & Sinha, P., (2018). Migration: A means to create work-life balance? *Journal of Management and Organization, 24*(2), 279–294.

Ejlertsson, L., Heijbel, B., Ejlertsson, G., & Andersson, I., (2018). Recovery, work-life balance and work experiences important to self-rated health: A questionnaire study on salutogenic work factors among Swedish primary health care employees. *Work, 59*(1), 155–163.

Emre, O., & De Spiegeleare, S., (2019). The role of work-life balance and autonomy in the relationship between commuting, employee commitment, and well-being. *International Journal of Human Resource Management.*

Gamber, T., Zülch, G., (2018). Approach to improve the work-life balance of employees using agent-based planning and simulation-based evaluation. *Proceedings-Winter Simulation Conference, 2019*, pp. 977–988.

Gawlik, R., (2018). Encompassing the work-life balance into early career decision-making of future employees through the analytic hierarchy process. *Advances in Intelligent Systems and Computing, 594*, 137–147.

George, N., Kiran, P. R., Sulekha, T., Rao, J. S., & Kiran, P., (2018). Work-life balance among Karnataka State Road Transport Corporation (KSRTC) workers in Anekal Town, South India. *Indian Journal of Occupational and Environmental Medicine, 22*(2), 82–85.

Golovina, K., (2018). Gender contract in online commercials in Japan: A critical investigation of the contemporary discourse on the work-life balance. *Sotsiologicheskoe Obozrenie, 17*(1), 160–191.

Gravador, L. N., & Teng-Calleja, M., (2018). Work-life balance crafting behaviors: An empirical study. *Personnel Review, 47*(4), 786–804.

Guilbert, L., Auzoult, L., Gilibert, D., Sovet, L., & Bosselut, G., (2019). Influence of ethical leadership on affective commitment and psychological flourishing: The moderating role of satisfaction with work-life balance [Influence du leadership éthique sur l'engagement affectif et l'épanouissement psychologique: le rôle médiateur de la satisfaction vis-à-vis de l'équilibre entre domaines de vie]. *Psychologie du Travail et des Organizations.*

Gupta, M., & Kumar, Y., (2015). Justice and employee engagement: Examining the mediating role of trust in Indian B-schools. *Asia-Pacific Journal of Business Administration, 7*(1), 89–103.

Gupta, M., & Pandey, J., (2018). Impact of student engagement on affective learning: Evidence from a large Indian university. *Current Psychology, 37*(1), 414–421.

Gupta, M., & Ravindranath, S., (2018). Managing physically challenged workers at micro sign. *South Asian Journal of Business and Management Cases, 7*(1), 34–40.

Gupta, M., & Sayeed, O., (2016). Social responsibility and commitment in management institutes: Mediation by engagement. *Business: Theory and Practice, 17*(3), 280–287.

Gupta, M., & Shaheen, M., (2017a). Impact of work engagement on turnover intention: Moderation by psychological capital in India. *Business: Theory and Practice, 18*, 136–143.

Gupta, M., & Shaheen, M., (2017b). The relationship between psychological capital and turnover intention: Work engagement as mediator and work experience as moderator. *Journal Pengurusan, 49*, 117–126.

Gupta, M., & Shaheen, M., (2018). Does work engagement enhance general well-being and control at work? Mediating role of psychological capital. *Evidence-Based HRM, 6*(3), 272–286.

Gupta, M., & Shukla, K., (2018). An empirical clarification on the assessment of engagement at work. *Advances in Developing Human Resources, 20*(1), 44–57.

Gupta, M., (2017). Corporate social responsibility, employee-company identification, and organizational commitment: Mediation by employee engagement. *Current Psychology, 36*(1), 101–109.

Gupta, M., (2018). Engaging employees at work: Insights from India. *Advances in Developing Human Resources, 20*(1), 3–10.

Gupta, M., Acharya, A., & Gupta, R., (2015). Impact of work engagement on performance in Indian higher education system. *Review of European Studies, 7*(3), 192–201.

Gupta, M., Ganguli, S., & Ponnam, A., (2015). Factors affecting employee engagement in India: A study on offshoring of financial services. *Qualitative Report, 20*(4), 498–515.

Gupta, M., Ravindranath, S., & Kumar, Y. L. N., (2018). Voicing concerns for greater engagement: Does a supervisor's job insecurity and organizational culture matter? *Evidence-Based HRM, 6*(1), 54–65.

Gupta, M., Shaheen, M., & Das, M. (2019). Engaging employees for quality of life: mediation by psychological capital. *The Service Industries Journal, 39*(5–6), 403–419.

Gupta, M., Shaheen, M., & Reddy, P. K., (2017). Impact of psychological capital on organizational citizenship behavior: Mediation by work engagement. *Journal of Management Development, 36*(7), 973–983.

Haar, J. M., Roche, M., & Ten Brummelhuis, L., (2018). A daily diary study of work-life balance in managers: Utilizing a daily process model. *International Journal of Human Resource Management, 29*(18), 2659–2681.

Haar, J. M., Sune, A., Russo, M., & Ollier-Malaterre, A., (2019). A cross-national study on the antecedents of work-life balance from the fit and balance perspective. *Social Indicators Research, 142*(1), 261–282.

Haider, S., Jabeen, S., & Ahmad, J., (2018). Moderated mediation between work-life balance and employee job performance: The role of psychological wellbeing and satisfaction with coworkers. *Journal of Psychology of Work and Organizations, 34*(1), 29–37.

Hampson, S. C., (2018). Mothers do not make good workers: The role of work/life balance policies in reinforcing gendered stereotypes. *Global Discourse, 8*(3), 510–531.

Holden, S., & Sunindijo, R. Y., (2018). Technology, long work hours, and stress worsen work-life balance in the construction industry. *International Journal of Integrated Engineering, 10*(2), 13–18.

Hossen, M. M., Begum, M., & Zhixia, C., (2018). Present status of organizational work-life balance practices in Bangladesh: Employees expectation and organizational arrangements. *Journal of Eastern European and Central Asian Research, 5*(1).

Hua, D. W., Mahmood, N. H. N., Zakaria, W. N. W., Lin, L. C., & Yang, X., (2018). The relationship between work-life balance and women leadership performance: The mediation effect of organizational culture. *International Journal of Engineering and Technology (UAE), 7*(4), 8–13.

Ip, E. J., Lindfelt, T. A., Tran, A. L., Do, A. P., & Barnett, M. J., (2018). Differences in career satisfaction, work-life balance, and stress by gender in a national survey of pharmacy faculty. *Journal of Pharmacy Practice*.

Jackson, L. T. B., & Fransman, E. I., (2018). Flexi work, financial well-being, work-life balance and their effects on subjective experiences of productivity and job satisfaction of females in an institution of higher learning. *South African Journal of Economic and Management Sciences, 21*(1).

James, F. A., & Sudha, S., (2015). An analysis of the effects of shift work on the lives of married employees. Prabandhan: *Indian Journal of Management, 8*(11), 26–33.

Jantzer, A. M., Anderson, J., & Kuehl, R. A., (2018). Breastfeeding support in the workplace: The relationships among breastfeeding support, work-life balance, and job satisfaction. *Journal of Human Lactation, 34*(2), 379–385.

Jerg-Bretzke, L., Krüsmann, P., Traue, H. C., & Limbrecht-Ecklundt, K., (2018). What you will: Results of an empirical analysis of the need to improve work-life balance for physicians ["was ihr wollt," Ergebnisse einer empirischen Bedarfsanalyse zur Verbesserung der Vereinbarkeit von Familie und Beruf bei Ärztinnen und Ärzten]. *Gesundheitswesen, 80*(1), 20–26.

Johari, J., Yean, T. F., & Tjik, Z. Z. I., (2018). Autonomy, workload, work-life balance, and job performance among teachers. *International Journal of Educational Management, 32*(1), 107–120.

Johri, A., & Teo, H. J., (2018). Achieving equilibrium through coworking: Work-life balance in floss through multiple spaces and media use. *Proceedings of the 14th International Symposium on Open Collaboration, Open Sym., 2018*.

Kang, H., & Wang, J., (2018). Creating their own work-life balance: Experiences of highly educated and married female employees in South Korea. *Asian Women, 34*(2), 1–31.

Kaur, S. H., Shankar, A. U., & Nuthan, B. L., (2018). Work-life balance of women working in education sector (With reference to Warangal dist., Telangana). *Journal of Advanced Research in Dynamical and Control Systems, 10*(8 Special Issue), 273–279.

Körber, M., Schmid, K., Drexler, H., & Kiesel, J., (2018). Subjective workload, job satisfaction, and work-life-balance of physicians and nurses in a municipal hospital in a rural area compared to an urban university hospital [Subjektive Arbeitsbelastung, Arbeitszufriedenheit, Work-Life-Balance von Ärzten und Pflegekräften eines Kommunalklinikums im ländlichen Raum im Vergleich zu einem großstädtischen Universitätsklinikum]. *Gesundheitswesen, 80*(5), 444–452.

Kotteswaran, D., & Kala, K., (2018). A study on work life balance and occupational self-efficacy among women entrepreneurs in the Vellore district. *International Journal of Mechanical and Production Engineering Research and Development*, 91–94.

Kowitlawkul, Y., Yap, S. F., Makabe, S., Chan, S., Takagai, J., Tam, W. W. S., & Nurumal, M. S., (2019). Investigating nurses' quality of life and work-life balance statuses in Singapore. *International Nursing Review*, 66(1), 61–69.

Kraak, J. M., Russo, M., & Jiménez, A., (2018). Work-life balance psychological contract perceptions for older workers. *Personnel Review*, 47(6), 1198–1214.

Kumar, K. K., & Premalatha, R., (2018). Conceptual review of work life balance. *Indian Journal of Public Health Research and Development*, 9(12), 2562–2565.

Kumar, K., & Chaturvedi, R., (2018). Role of gender ideology, work-life balance in determining job and life satisfaction among Indian working mothers. *International Journal of Mechanical Engineering and Technology*, 9(8), 898–906.

Kumar, K., & Chaturvedi, R., (2018). Women in construction industry: A work-life balance perspective. *International Journal of Civil Engineering and Technology*, 9(8), 823–829.

Kumar, K., & Velmurugan, R., (2018). A study on the work life balance of generation Y information technology (IT) employees in Cochin. *International Journal of Engineering and Technology (UAE)*, 7(3.6 Special Issue 6), 142–147.

La Barbera, M., & Lombardo, E., (2019). "The long and winding road": A comparative policy analysis of multilevel judicial implementation of work-life balance in Spain. *Journal of Comparative Policy Analysis: Research and Practice*, 21(1), 9–24.

Larsen, T. P., & Navrbjerg, S. E., (2018). Bargaining for equal pay and work-life balance in Danish companies-Does gender matter? *Journal of Industrial Relations*, 60(2), 176–200.

Lederer, W., Paal, P., Von Langen, D., Sanwald, A., Traweger, C., & Kinzl, J. F., (2018). Consolidation of working hours and work-life balance in anesthesiologists: A crosssectional national survey. *PLoS One*, 13(10).

Lee, D. J., & Sirgy, M. J., (2018). What do people do to achieve work-life balance? A Formative conceptualization to help develop a metric for large-scale quality-of-life surveys. *Social Indicators Research*, 138(2), 771–791.

Lee, H., & Ko, G., (2018). Relationship between work-life balance and job engagement. *Indian Journal of Public Health Research and Development*, 9(8), 651–656.

Lindfelt, T., Ip, E. J., Gomez, A., & Barnett, M. J., (2018). The impact of work-life balance on intention to stay in academia: Results from a national survey of pharmacy faculty. *Research in Social and Administrative Pharmacy*, 14(4), 387–390.

Loria, K., & Morris, D., (2018). How to recruit and retain young physicians: Doctors looking for good work/life balance, improved financial stability. *Urology Times*, 46(6), 42–43.

Lucia-Casademunt, A. M., García-Cabrera, A. M., Padilla-Angulo, L., & Cuéllar-Molina, D., (2018). Returning to work after childbirth in Europe: Well-being, work-life balance, and the interplay of supervisor support. *Frontiers in Psychology*, 9.

Manor, U., & Desiana, P. M., (2018). Work-life balance, motivation, and personality of MSE owners on firm performance in Greater Jakarta. *Pertanika Journal of Social Sciences and Humanities*, 26, 127–138.

Marhánková, J. H., (2018). Work-life balance and policies of care. *Gender a Vyzkum/Gender and Research*, 19(1), 143–145.

Matilla-Santander, N., Lidón-Moyano, C., González-Marrón, A., Bunch, K., Martín-Sánchez, J. C., & Martínez-Sánchez, J. M., (2019). Attitudes toward working conditions: are European Union workers satisfied with their working hours and work-life balance? [Actitudes frente a las condiciones laborales: ¿está la población trabajadora de la Unión Europea satisfecha con sus horas de trabajo y su balance trabajo-vida?]. *Gaceta Sanitaria, 33*(2), 162–168.

Mauk, K. L., (2018). Health for the healer: Keeping work-life balance in rehabilitation nursing. *Rehabilitation Nursing, 43*(2), 63–64.

Mazerolle, S. M., & Hunter, C., (2018). Work-life balance in the professional sports setting: The athletic trainer's perspective. *International Journal of Athletic Therapy and Training, 23*(4), 141–149.

Mazerolle, S. M., Eason, C. M., & Goodman, A., (2018). An examination of relationships among resiliency, hardiness, affectivity, and work-life balance in collegiate athletic trainers. *Journal of Athletic Training, 53*(8), 788–795.

Mazerolle, S. M., Pitney, W. A., Goodman, A., Eason, C. M., Spak, S., Scriber, K. C., Voll, C. A., Detwiler, K., Rock, J., Cooper, L., & Simone, E., (2018). National athletic trainers' Association position statement: Facilitating work-life balance in athletic training practice settings. *Journal of Athletic Training, 53*(8), 796–811.

Melo, P. C., Ge, J., Craig, T., Brewer, M. J., & Thronicker, I., (2018). Does work-life balance affect pro-environmental behavior? Evidence for the UK using longitudinal microdata. *Ecological Economics, 145*, 170–181.

Moltz, M. C., (2019). Work-life balance and national context in attraction to public employment. *International Journal of Public Administration, 42*(4), 334–344.

Mousa, M., (2018). Inspiring work-life balance: Responsible leadership among female pharmacists in the Egyptian health sector. *Entrepreneurial Business and Economics Review, 6*(1), 71–90.

Mushfiqur, R., Mordi, C., Oruh, E. S., Nwagbara, U., Mordi, T., & Turner, I. M., (2018). The impacts of work-life-balance (WLB) challenges on social sustainability: The experience of Nigerian female medical doctors. *Employee Relations, 40*(5), 868–888.

Mustapa, N. S., Noor, K. M., & Abdul, M. M., (2018). Why can't we have both? A discussion on work-life balance and women career advancement in Malaysia. *Journal of Asian Finance, Economics and Business, 5*(3), 103–112.

Nayak, P., & Sharma, N., (2018). Managing faculty's work-life balance in Indian business schools [Upravljanje z usklajevanjem dela in družine visokošolskih učiteljev na Indijskih poslovnih šolah]. *Teorija in Praksa, 55*(3), 604–621.

Neumann, J. L., Mau, L. W., Virani, S., Denzen, E. M., Boyle, D. A., Boyle, N. J., et al., (2018). Burnout, moral distress, work-life balance, and career satisfaction among hematopoietic cell transplantation professionals. *Biology of Blood and Marrow Transplantation, 24*(4), 849–860.

Nirmalasari, I., (2019). The analysis of work life balance to nurse performance through work satisfaction and organization commitment (a case study on nurses in Nirmalasuri Hospital, Sukoharjo). *Proceedings of the 32nd International Business Information Management Association Conference, IBIMA 2018–Vision 2020: Sustainable Economic Development and Application of Innovation Management from Regional Expansion to Global Growth*, pp. 6135–6143.

Oludayo, O. A., Falola, H. O., Obianuju, A., & Demilade, F., (2018). Work-life balance initiative as a predictor of employees' behavioral outcomes. *Academy of Strategic Management Journal, 17*(1).

Panday, K. C., & Bhagat, M., (2018). Impact of work-life balance in project-based organizations. *Indian Journal of Labor Economics, 61*(1), 171–180.

Pandey, J., Gupta, M., & Naqvi, F., (2016). Developing decision making measure a mixed method approach to operationalize Sankhya philosophy. *European Journal of Science and Theology, 12*(2), 177–189.

Pedrini, M., Ferri, L. M., & Riva, E., (2018). Institutional pressures and internal motivations of work-life balance organizational arrangements in Italy. *International Journal of Human Resources Development and Management, 18*(43528), 257–281.

Perrigino, M. B., Dunford, B. B., & Wilson, K. S., (2018). Work-family backlash: The "dark side" of work-life balance (WLB) policies. *Academy of Management Annals, 12*(2), 600–630.

Peters, R., (2019). Apprehension shapes employee satisfaction: Intellectual challenge, work/life balance, compensation, and an unclear business outlook create uncertainty among European bio/pharma employees. *Pharmaceutical Technology Europe, 31*(1), 22–26.

Powell, S. K., (2018). Work-life balance: How some case managers do it? *Professional Case Management, 23*(5), 235–239.

Prithi, S., & Vasumathi, A., (2018). The influence of demographic profile on work life balance of women employees in tannery industry: An empirical study. *Pertanika Journal of Social Sciences and Humanities, 26*(1), 259–284.

Proost, K., & Verhaest, D., (2018). Should we tell the recruiter that we value a good work-life Balance? *Journal of Personnel Psychology, 17*(3), 120–130.

Puryer, J., Sidhu, G., & Sritharan, R., (2018). The career intentions, work-life balance and retirement plans of UK dental undergraduates. *British Dental Journal, 224*(7), 536–540.

Raghavendra, H. K., & Raghunandan, M. V., (2018). Women engineers and work life balance a case study of women working in manufacturing industries in Mysore city. *International Journal of Mechanical Engineering and Technology, 9*(1), 752–755.

Rajini, G., (2018). Revisiting the symptoms of work-life balance: A dependency analysis of employees in ICT sector. *Advances in Intelligent Systems and Computing, 632,* 643–653.

Rajya, L. M., Sudhir, R. M., & Satyavathi, M., (2018). Work-life balance of working women professionals-a study of women in different sectors in Warangal region. *International Journal of Civil Engineering and Technology, 9*(4), 446–451.

Ramírez, C. J., (2018). Work-life balance in UK: A comparative study to Spanish system [La conciliación laboral y familiar en Reino Unido: Una visión comparada con el sistema Español]. *Trabajo y Derecho, 38.*

Robertson, K. M., Lautsch, B. A., & Hannah, D. R., (2019). Role negotiation and systems-level work-life balance. *Personnel Review, 48*(2), 570–594.

Salolomo, B., & Agbaeze, E. K., (2019). Effect of work-life balance on performance of money deposit banks in south-south Nigeria. *Management Science Letters, 9*(4), 535–548.

Saternus, Z., & Staab, K., (2018). Towards a smart availability assistant for desired work life balance. *International Conference on Information Systems,* ICIS 2018.

Schnettler, B., Miranda-Zapata, E., Lobos, G., Saracostti, M., Denegri, M., Lapo, M., & Hueche, C., (2018). The mediating role of family and food-related life satisfaction in the relationships between family support, parent work-life balance, and adolescent life satisfaction in dual-earner families. *International Journal of Environmental Research and Public Health, 15*(11).

Schwartz, S. P., Adair, K. C., Bae, J., Rehder, K. J., Shanafelt, T. D., Profit, J., & Sexton, J. B., (2019). Work-life balance behaviors cluster in work settings and relate to burnout and safety culture: A cross-sectional survey analysis. *BMJ Quality and Safety, 28*(2), 142–150.

Shekhar, S. K., Jose, T. P., & Sudhakar, R., (2018). A study on work life balance of female pharmacists working in private hospitals. *Research Journal of Pharmacy and Technology, 11*(1), 107–110.

Sirgy, M. J., & Lee, D. J., (2018). Work-life balance: An integrative review. *Applied Research in Quality of Life, 13*(1), 229–254.

Smalley, M., (2018). Work-life balance versus responsibility to patients. *Veterinary Record, 183*(22), 699.

St John, M., & Bradford, C. R., (2019). Work-life balance among head and neck surgeons – Seeking visionary leadership from everywhere. *JAMA Otolaryngology – Head and Neck Surgery*.

Stavrou, E., & Ierodiakonou, C., (2018). Expanding the work-life balance discourse to LGBT employees: Proposed research framework and organizational responses. *Human Resource Management, 57*(6), 1355–1370.

Talukder, A. K. M., Vickers, M., & Khan, A., (2018). Supervisor support and work-life balance: Impacts on job performance in the Australian financial sector. *Personnel Review, 47*(3), 727–744.

Thiyagarajan, R., & Tamizhjyothi, K., (2018). Job satisfaction of work life balance of women employed in unorganized sector in Kanchipuram district, Tamil Nadu. *Indian Journal of Public Health Research and Development, 9*(1), 226–231.

Valentine, C. M., (2018). Tackling the quadruple aim: Helping cardiovascular professionals find work-life balance. *Journal of the American College of Cardiology, 71*(15), 1707–1709.

Vasumathi, A., & Prithi, S., (2018). Work life balance of women workers in tannery industry, Tamil Nadu – An empirical bivariate study. *International Journal of Services and Operations Management, 29*(3), 420–438.

Vives, A., Gray, N., González, F., & Molina, A., (2018). Gender and aging at work in chile: Employment, working conditions, work-life balance and health of men and women in an aging workforce. *Annals of Work Exposures and Health, 62*(4), 475–489.

Wesarat, P. O., Majid, A. H., Sharif, M. Y., Khaidir, A., & Susanto, P., (2018). Mediating effect of job satisfaction on the relationship between work-life balance and job performance among academics: Data screening. *International Journal of Engineering and Technology (UAE), 7*(4), 214–216.

Wilkinson, K., Tomlinson, J., & Gardiner, J., (2018). The perceived fairness of work-life balance policies: A UK case study of solo-living managers and professionals without children. *Human Resource Management Journal, 28*(2), 325–339.

Witzig, T. E., & Smith, S. M., (2019). Work-life balance solutions for physicians—It's all about you, your work, and others. *Mayo Clinic Proceedings, 94*(4), 573–576.

Yang, J. W., Suh, C., Lee, C. K., & Son, B. C., (2018). The work-life balance and psychosocial well-being of South Korean workers. *Annals of Occupational and Environmental Medicine, 30*(1).

Yu, H. H., (2019). Work-life balance: An exploratory analysis of family-friendly policies for reducing turnover intentions among women in U.S. federal law enforcement. *International Journal of Public Administration, 42*(4), 345–357.

CHAPTER 8

Cross-Border Mergers: The Use of Employment Engagement Tools in Overcoming Challenges of Workforce Cultural Diversity

SHWETA LALWANI

Assistant Professor, School of Management, Sir Padampat Singhania University, Bhatewar, Udaipur (Rajasthan), PIN – 313601, India, Tel.: (+91) 9950396133, E-mail: shweta.lalwani@spsu.ac.in

ABSTRACT

Diversity will increase significantly in the future years. Successful organizations recognize the need for immediate action and be ready and willing to spend resources on managing diversity in the workplace now. Companies today are combining in record numbers. Executives pursue mergers, acquisitions, and joint ventures as a means to create value. Cultural diversity has emerged as one of the dominant barriers to effective integrations. The chapter focuses on the challenges of cultural diversity in firms after merging and addressing them using employee engagement tools. The chapter explores on what are the challenges faced and how human resources (HRs) personnel or leadership through employee engagement interventions mitigate challenges that diversity can spring forth.

8.1 INTRODUCTION

Workplace diversity refers to differences between people in an organization with respect to race, gender, ethnic group, age, personality, cognitive style, tenure, organizational function, education, background, etc. Diversity is about people's perception of self and others. Those perceptions affect their interactions and conversations in an organization. For such a diverse workforce to function effectively in an organization together, human resource (HR) professionals need to deal effectively with issues such as communication, adaptability, and change. Successful organizations recognize the need for immediate action and are ready and willing to spend resources on managing diversity in the workplace now (Josh Greenberg, 2004). Executives pursue mergers, acquisitions, and joint ventures as a means to create value by (1) acquiring technologies, products, and market access, (2) creating economies of scale, and (3) establishing a global brand presence. Cultural diversity has emerged as one of the dominant barriers to effective integrations. In one study, culture was found to be the cause of 30% of failed integrations. Companies with different cultures find it difficult, if not often impossible, to make decisions quickly and correctly or to operate effectively (Miller and Fernandes, 2009). Given that culture will seldom stop a proposed transaction, it becomes the responsibility of the people managing the deal to stop culture from undermining their desired goals. The most widely used approach to managing the cultural issues is to define a set of desirable cultural attributes (a typical set being: customer-focused, innovative, entrepreneurial, decisive, team-oriented, respectful of others) and then to exhort employees to adopt these attributes in their daily behavior (Miller and Fernandes, 2009). It is important for management to respond to changes brought about such as restructuring and transfer of employees in order to address challenges confronted due to different caliber of employees as well as induction of new staff to the organizational culture.

8.2 NEED FOR DISCUSSING THIS THEME

The chapter explores on what are the challenges faced and how HRs personnel or leadership through employee engagement interventions mitigate challenges that diversity can spring forth. Managing diversity enables organizations to derive excellence and talent out of its workforce since the workforce will feel accommodated from diverse countries and cultures. It is, therefore, the responsibility of the management and HRs personnel to introspect and work into their policies of integration of employees from different backgrounds in order to bring about a correct divergence of these cultures. However, challenges brought by diversity can be turned into the strength of enterprises such as transfer of skills and affirmative action. A narration and literature of qualitative study will seek to address issues of diversity in newly established organizations through mergers and acquisitions (Reuben, 2017). If issues of diversity are not carefully redressed, they can end up spiraling to legal frictions which may impact the organization negatively. The chapter covers:

- Mergers & acquisition (M&A)-concept, psychological issues related to M&A.
- Employee engagement-concept, employee engagement during change, employee engagement, and organizational culture.
- Role of employee engagement in merged entities.
- Case studies.

8.3 REVIEW OF THIS THEME

8.3.1 MERGERS & ACQUISITION (M&A)

M&A provides means to acquire expertise, technology, and products. Over the past two decades, M&A have become a global phenomenon and a popular strategic choice for company growth and expansion (Hansen et al., 2004). Many researches were done to examine human and cultural aspects of M&A and it was found that the real problem is not financial issues but the lack of intercultural synergy and sensitivity between the organizations. The organizational cultural issues create communication breakdowns and therefore act as a barrier for successful integration of the two organizations (Epstein, 2004). Cartwright and Cooper (1993) reinforced the previous findings by stating that the financial and other strategic benefits expected from M&A are undermined by the cultural conflicts (Cartwright and Cooper, 1993). Michelle (2006), in her article "Post-merger Culture Clash: Can Cultural Leadership Lessen Casualties?," states that post-merger cultural clashes are often the main reason for the disappointing M&A outcomes and that unfortunately poor research exists to conduct the merged organizations to a suitable cultural integration. Therefore, she underlines that the cultural leadership is the most important and influential factor in order to achieve a sustainable culture. The article includes a qualitative study exploring an analysis of the interviews with 42 post-merger employees in order to put in evidence the role of the leader during the post-merger culture adaptation. The findings of the study have some implication for leaders who are desired to anticipate the post-merger culture clashes (Michelle, 2006). The psychological issues arising out of the merger process is shown in Figure 8.1 and described as under:

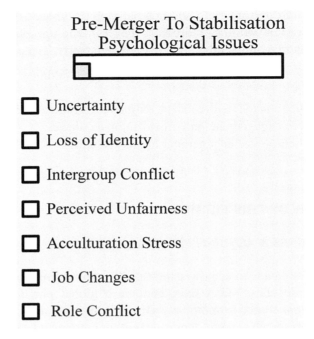

FIGURE 8.1 Pre-merger to stabilization psychological issues.

- **Uncertainty:** Fear in the minds of employees as to what is going to happen next? What will be the merged entity like? What lies in future for them? Are all possible issues of anxiety in the minds of the employees.
- **Loss of Identity:** When the merger takes place, a lot many positions are dissolved and the employees occupying such positions are in a situation of a fix, i.e., which new position will be given to them and whether it would be at a par with the previous one.
- **Intergroup Conflict:** New merged firms face a lot of friction due to lack of communication, understanding between the employees. Various groups are formed based on common interests and such groups take care of their mutual rights.
- **Perceived Unfairness:** The workforce might not agree on suitability of the merger and for the proposed changes. They might fear post-merger adjustment, i.e., changes in the job requirement, duties, salaries, and equitable distribution of these.
- **Acculturation Stress:** Acculturation is the process of social, psychological, and cultural change that stems from mixing between cultures.

The effects of acculturation can be seen at multiple levels in both the original and newly adopted cultures. Acculturative stress refers to the stress response of employees in response to their experiences of acculturation, i.e., blending of new cultures. Stressors can include but are not limited to the pressures of learning a new language, maintaining one's native language, balancing differing cultural values, and brokering between native and host differences in acceptable social behaviors. Acculturative stress can manifest in many ways, including but not limited to anxiety, depression, substance abuse, and other forms of mental and physical maladaptation (Berry, 2006; Davis, 2016). When members of two cultures meet in the stressful situation of a merger each perceives members of the other culture as foreign and intrinsically mistrusts them. Regardless of how much effort is made to welcome the new group on board, working with them leads to a natural tendency to be critical of "their" values, the way "they" work, "their" attitudes, how "their" priorities and methods impact on "our" aspirations, "our" security. Consequently, there are three major cognitive processes that take place when employees consider the "other" culture during the merger process.

- Polarization: People describe the two cultures in a way that highlights the differences or contrasts. Unchecked this becomes "stereotyping," a belief system about the other culture that is difficult to change.
- Evaluation: People place a positive value on their culture and a negative one on the other.
- Ethnocentrism: Unwillingness to see behaviors and events from the point of view of the other.

These distorted perceptions and hostile feelings toward employees from the other organization may become common. As employees experience cultural differences in their daily operations, failures are typically attributed to members of the other company. The result is post-merger conflict or "culture clash" (People and Culture, 2009).

- **Job Changes:** With M&As the positions, duties, work profile, coworkers might change leading to anxiety and tension.

These issues are predictable and manageable. It is important to build the communication and engagement activities around their emergence. It is also important to have employee attitude measures in place that surface when and where these issues are emerging. Transition speed around key decisions is essential to maintain motivation-beware of organizational drift.

Managers need to pay careful attention to signs of stress. Managers need to be aware of "merger syndrome" and have effective methodologies for relieving team stress (People and Culture, 2009). The degree to which these issues are experienced will in part be determined by the cultural congruence of the two organizations, however, inherent cultural differences are exposed and culture-related stress, tension, and resistance are likely to be highest when employees are pushed to abandon their old culture and learn a new one (People and Culture, 2009).

Douglas D. Ross, Managing Director, Square Peg International Ltd (2005) is the author of the article "Culture Management in M&As: A focus on culture and people is critical to make integration strategies work." The author was invited to speak to the Telecom Finance Conference in London, Creating Value through M&A, about managing cultural transition issues in M&As and joint venture situations. During this discussion, the author emphasized that management from the acquiring company usually is unprepared to deal with post-merger politics that can lead to a reduction of the outcomes, the cause are the underestimation of the culture integration challenges or the human factor. So in order to tackle directly with the cultural factors, the author stressed the importance to develop an "integration plan." In this plan, advices are given to the leaders to constitute and implement a new corporate culture. Indeed, once the new organization knows what it wants to be, aligns its systems, processes, and procedures to reinforce the desired culture, the next stage is the most difficult one; it concerns the alignment of the employees and leadership team with the new culture. So, according to Douglas D. Ross once the culture is defined it is important to:

- Obtain individual buy-in from leaders;
- Address the "me" issues;
- Identify integration risk factors;
- Avoid deadly sins of M&A's;
- Learn from best practices.

He concluded his article by pointing out that "the time to make change is limited but the way in which cultural integration is handled will make the difference between success and failure of the deal" (Douglas, 2005). Thus, the organization who addressed cultural issues properly during mergers and acquisitions experienced good results as compare to other organizations who neglected this aspect, for instance, Adidas-Reebok merger, and Nissan-Renault merger. Hence, there exists a strong need to consider the cultural issues as critical during the M&A and address them properly. This may be addresses well using Employee engagement tools. What is Employee engagement? Which tools of engagement may be used during the merging process to overcome various psychological issues?

8.3.2 EMPLOYEE ENGAGEMENT

Employee engagement is an emotional state where employees feel passionate, energetic, and committed to their work. This translates into employees who give their hearts, spirits, minds, and hands to deliver a high level of performance to the organization. Employee Engagement means physical, emotional, cognitive connection of an employee towards his work with commitment and positivity. Engaged employees are motivated, dedicated to their work and exceeds the required efforts to meet business goals.

A few definitions of Employee Engagement are presented in Table 8.1 below:

TABLE 8.1 Employee Engagement Definitions

Year	Name	Definition
1990	William Kahn	The harnessing of organization members' selves to their work roles; in engagement, people employ and express themselves physically, cognitively, and emotionally during role performances.
1993	Schmidt et al.	An employee's involvement with, commitment to, and satisfaction with work. Employee engagement is a part of employee retention.
1997	Bevan et al.	An engaged employee is someone who is aware of business context, and works closely with colleagues to improve performance within the job for the benefit of the organization.
2009	Brad Federman	A roadmap for creating profits, optimizing Performance and Increasing Loyalty

Shuck and Wollard defined EE as "an individual employee's cognitive, emotional, and behavioral state directed toward desired organizational outcomes" described EE as an individual-level variable often measured at the organizational level. The role of meaningful workplace environment and an employee's involvement in contextually meaningful work were examined as antecedents to EE. These antecedents as found were Authentic corporate culture, Clear expectations, Corporate social responsibility, Encouragement, Feedback, Hygiene factors, Job characteristics, Job control, Job fit, Leadership, Level of task challenge, Manager expectations, Manager self-efficacy,

Mission, and vision, Opportunities for learning, Perception of workplace safety, Positive workplace climate, Rewards, Supportive organizational culture, Talent management, Use of strengths (Shuck and Wollard, 2010). Employee engagement is the emotional attachment employees feel towards their place of work, job role, position within the company, colleagues, and culture and the affect this attachment has on wellbeing and productivity. From an employer's point-of-view, employee engagement is concerned with using new measures and initiatives to increase the positive emotional attachment felt and therefore productivity and overall business success. Employee engagement is seen by many to be an example of a competitive advantage. Employee engagement is defined as "the level of an employee's psychological investment in their organization" (Aon Hewitt, 2017).

- **Employee Engagement During Change:** Engagement suffers during periods of uncertainty. Employees can be nervous about change, and feel insecure or demotivated when the goalposts are continuously moving. Their need for information outstrips what management can provide, damaging confidence in the leadership and direction of the firm. And as convergence gathers pace, it will be harder than ever to stay ahead of the innovation curve and to make long term investment decisions. Communicating 'what's next' becomes increasingly difficult. At the same time, companies needs to ask more of employees to get through change. This can harm perceptions of the exchange between contribution and reward. Some breakthroughs will prove controversial. Organizations will need to respect the societal consensus on progress to keep employees engaged behind the strategic vision. High ethical standards and mutual trust will help ensure that staff acts in the best interests of the company, and in line with what's socially acceptable. Collaboration may prove uncomfortable for competing businesses and for technical specialists from different disciplines (who generally prefer to focus on their own area of expertise). Companies will need to establish the right climate, processes, and platforms for collaboration to succeed, and equip people with the right skills and attributes (Haygroup, 2014). The turbulent business environment has caused the need for several organizations to develop the ability to quickly change their strategies and expand into fields which they are not experienced in. Although the academics became interested in the field of organizational culture in the 1950s, it was only in the 1970s when it got truly discovered and in the 1980s, its research became more popular than the research of strategic planning.

- **Employee Engagement and Organizational Culture:** According to Schein organizational culture matters as decisions might have unexpected and damaging results unless the cultural aspects are not taken into account in the decision-making process (Porvari, 2014). These situations could be avoided, if decision-makers would pay more attention to the cultural values of their organizations as changes will face resistance within an existing culture. It is a strong, hidden, and often unrecognized force of different powers which control our daily decisions we make personally and the ones that we make as a group. It affects how we perceive things, how we value different things around us, and how we think. Culture got more attention due to the fact that American companies started paying attention to their Japanese competitors, as it was realized that the differences between them were not only technological or structural (Porvari, 2014). Cross-cultural mergers bring about turmoil of fear, uncertainty, conflicts, and resistance in the merging entities. Various change interventions have been developed and studied which may alleviate the process of change in the new setups. One of these interventions is *Employee Engagement*. Employee engagement is a strong tool to help employees overcome various issues arising amidst the process of M&A. Cross-Cultural adjustments in the new environment may be smoothened out by engagement tools spanning from communication, growth to leadership. An open conversation in between diverse workforce, cross-cultural trainings, advancement opportunities, autonomy, grievance meetings, open-door policy, leader's support, good working environment with enabling infrastructure, benefits/incentives, environment of fairness, justice, and equity, etc. can bring about the more peaceful and cordial merger in an organization.

Thus, it is an imperative to study the management of cultural diversity in merging organizations with the help of effective engagement tools. Figure 8.2 explains various global top employee engagement tools as given by Aon Hewitt in 2017 in their report on Employee engagement, which may be helpful to overcome workforce diversity.

8.3.3 ROLE OF EMPLOYEE ENGAGEMENT IN THE MERGED ENTITIES

The expansion of businesses across the globe led to the involvement of many activities of M&A. Since merging may be done across businesses in different countries diversity in all human aspects as cultures, languages, personalities, and even mannerisms are bound to occur. Failure to align culture in M&A can lead to the collapse of the whole establishment since there are probabilities of delays and misinterpretations (Reuben, 2017). M&A across the globe brings together challenges in the form of mixture of cultures, educational backgrounds, values, and human perceptions as well as technology, with this migration of workers across the globe, especially for multi-national corporations. Managing such diversity is a big challenge for merged organizations today. One of the biggest challenges before global leaders is keeping the diverse workforce engaged. How can a manager build cross-cultural teams? What tools of engagement need to be adopted which also caters to focus on diversity strategy of the company? How a manager can bring about cultural sensitivity? Are some of the questions to be addressed by the global managers today? Why Merger and Acquisitions?

FIGURE 8.2 Global top employee engagement tools.

Primarily value creation or value enhancement is the goal of any merger. These are business combinations and the reasons are based on pecuniary elements. Some of the reasons behind mergers are:

- **Capacity Augmentation:** One of the most common causes of merger is capacity augmentation through combined forces. Usually, companies target such a move to leverage expensive manufacturing operations. However, capacity might not just pertain to manufacturing operations; it may emanate from procuring a unique technology platform instead of building it all over again. Capacity augmentation usually is the driving force in mergers in biopharmaceutical and automobile companies.

- **Achieving Competitive Edge:** Competition is cutthroat these days. Without adequate strategies in its pool, companies will not survive this wave of innovations. Many companies take the merger route to expand their footprints in a new market where the partnering company already has a strong presence. In other situations, attractive brand portfolio lures companies into mergers.
- **Surviving Tough Times:** Global economy is going through a phase of uncertainty and combined strength is always better in tough times. When survival becomes a challenge, combining is the best option. In the crisis period, 2008–2011, many banks took this path to cushion themselves from balance sheet risks.
- **Diversification:** Sensible companies just do not believe in keeping all eggs in one basket. Diversification is the key. By combining their products and services, they may gain a competitive edge over others. Diversification is simply adding products in the portfolio which is not part of current operations. A classic example of this is the acquisition of EDS by HP in 2008 to add services-oriented features in their technology offerings. The combination of HP and EDS will help HP greatly strengthen its IT services and consulting business and allowed it to compete more ably with not only IBM but other emerging-market players in the field. With this acquisition of approximately $13.9 billion value, HP has one of the technology industry's broadest portfolios of products, services, and end-to-end solutions. The combined offerings are focused on helping clients accelerate growth, mitigate risks and lower costs (Alto, 2008).
- **Cost Cutting:** Economies of scale is the soul of most businesses. When two companies are in the same line of business or produce similar goods and services, it makes perfect sense for them to combine locations or reduce operating costs by integrating and streamlining support functions. This becomes a large opportunity to lower costs. The math is simple here. When the total cost of production is lowered with increasing volume, total profits are maximized (Vaidya, 2016).

Post-merger integrations require a high level of employee engagement. Cross-cultural merger is a phase of transition where employees experience shockwaves of anxiety related to job roles, new leadership, diverse new workforce with a cultural mix, different educational background, experience, languages, personality, anxiety on redeployment of duties and responsibilities. The role of a manager to build a united diverse workforce in these times is a humongous challenge. It is the management's challenge to include competitive performance, quality improvement, as well as business processes. It is important to be able to pull together different people from different perspectives in order to achieve a common goal of any enterprise by working together, sharing ideas, solving problems together, in order to be competitive. For instance, Merger of Adidas Reebok could be successful because of the Cultural blend-The culture of Adidas and Reebok effortlessly merged and gave a new identity to the organization. Distinguishing factors were many (Vaidya, 2016). Adidas is originally a German company and Reebok an American entity; Adidas was all about sports, while Reebok redefined lifestyle. However, proper communication, clear strategies, and effective implementation did the job which is the effective tool of engagement (Vaidya, 2016). If leadership can manage the aspect of diversity, then the survival of the enterprise is secure. Failure to address diversity challenges will render the business ineffective as there are bound to be conflicts that will derail the purpose and the mission of the enterprise while competitors will be striving on further expansion and profits (Evan et al., 2003). Trying to balance differences brought about by diversity can bring more conflicts, disappointments, and dissatisfaction among workers because obviously those more skilled and qualified will be earning more salaries than those who are not, this to some can be viewed as favoritism which is not. It then requires the HR manager with sharp skills and special talents in human capital to steer the ship to success (Cox and Beale, 1997). One of the tools which can be effectively used by the leader to smoothen out the integration process can be Employee engagement. Employee engagement is the emotional commitment the employee has to the organization and its goals which mean engaged employees actually care about their work and their company.

In a survey made by AON Hewitt in 2017, employees were asked the following questions to assess employee engagement level:

- If they Say positive things about their organization and act as advocates.
- If they intend to Stay at their organization for a long time.
- If they are motivated to Strive to give their best efforts to help the organization succeed.

Accordingly, a Model of Employee engagement was proposed as per Figure 8.3 (Aon Hewitt, 2017).

As per the model, there are two categories of Engagement drivers:

- Foundation-company practices, the basics and the work; and
- Differentiators-brand, leadership, and performance.

Cross-Border Mergers: The Use of Employment

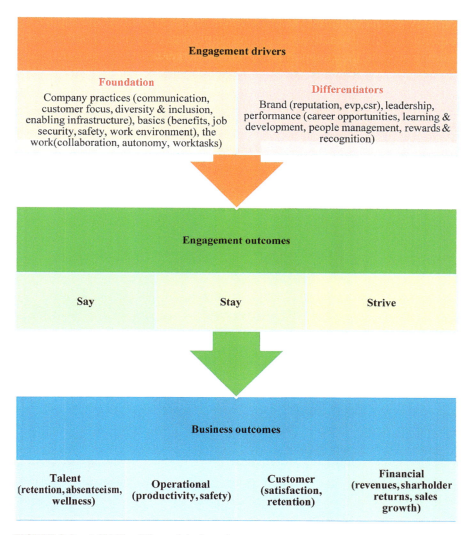

FIGURE 8.3 AON Hewitt's model of employee engagement.

These drivers help in attaining engagement outcomes as good word-of-mouth for the organization, loyalty, and focus and commitment on organizational goals. Hence, business Outcomes as employee retention, reduced absenteeism, wellness, enhanced productivity, more customer satisfaction, customer retention and broader impact on revenues, shareholder returns is achieved. Thus, engaged employees contribute more to the work and try to think out of the box, such employees are willing, able, and actually add to the company's success. Employee engagement brings in positive and supportive corporate culture. By actively engaging employees early as part of the brand-building process, their support for the entire merger becomes that much more enthusiastic. Engagement builds momentum, shapes attitudes and beliefs, generate confidence, sustains loyalty, and, ultimately, creates value (Brew, 2012). There are different methods that can be implemented for an effective employee engagement post-merger at the workplace as shown in Figure 8.4.

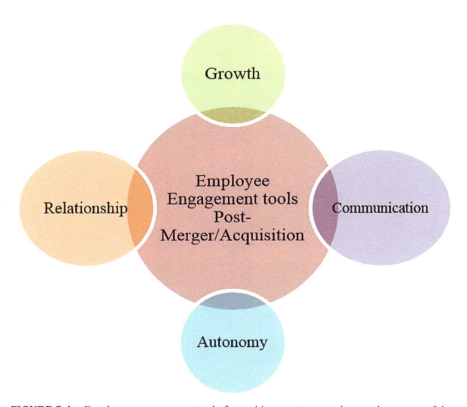

FIGURE 8.4 Employee engagement tools for making post-merger integration successful.

- **Communication:** Clarity on fears, concerns, and anxiety could be given to the employees via communicating with them, making them participate in the events pertaining to change and projecting the positives about the change process. This form of personalized communication will alleviate fears by helping employees adjust more easily to their new environment. Seeking suggestions and ideas from these employees on the new venture will make them feel valuable for the organization. This morale-boosting action will positively benefit the company as happy, engaged employees are bound to contribute more productively to the success of an organization. Thus by eliminating the concepts of fear and apprehension by productive discussions and conversations, an organization can reap the fruits of high employee engagement.

- **Relationship:** Employees are the assets of an organization. During these times of change, the leader can play role of a mentor, a guide, and a philosopher for the apprehensive employees. The strong bonding with the employees can make the process of transition smooth and easily adaptable. A two-way dialogue is a necessity to build a connection with the employees. Employees also need to ensure that their co-workers team-up together, provide reliable support, and make personal sacrifices during these stressful times. HR managers and business leaders must build compatibility through effective training programs, team exercises, and team-building sessions to triumph in bonding and motivating employees (Adhikari, 2016).
- **Growth:** Career opportunities, advancement, support, and enabling infrastructure builds a strong culture of engagement in these changing times. A strong belief in the prosperous future of the organization and eventually their own development helps in bringing a positive environment related to merger. The organization should be able to articulate and deliver on the promises made to its employees which will create a magnetic sense of belonging to the organization thus creating 'Employee Value Proposition.'
- **Autonomy:** Empowering employees to deal with corporate cultural changes requires intelligent planning and clever resource allocation. Business leaders and managers must find ways to set up effective engagement channels for clarifying roles and processes, setting up team meetings, conducting team building sessions and undertaking team management exercises. These channels of engagement are vital to the long-term success of M&A by enabling teams to bond effectively. Employees are the biggest assets for any organization, so finding ways to retain their loyalty and improve their productivity can only lead to positive outcomes. Fostering an engaging environment where employees from both companies can bond with each other is pivotal to the overall success of the corporate merger (Adhikari, 2016).

As per The AON's Hewitt report on 2017 trends in Global Employee engagement, there are four key interventions that meet employees' engagement needs during change:

- **Connection:** Employees want a personal connection to leaders and coworkers.
- **Control:** Employees almost always feel some loss of control with change events.
- The need for a clear career path showed up consistently in our analyses.
- **Capability:** Employees' need for skill building and development is a constant but takes on increased importance in times of change (Aon Hewitt, 2017).

Thus, engaging employees successfully during M&A is always going to be challenging for any business leader or manager. Discomfort, anxiety, and fear are natural emotions to expect from employees, so leaders must find smart ways to assuage these fears for enhanced productivity. Once a company is able to positively mitigate the risks of discontented employees by meaningfully involving them in decisions that affect them, it will witness a positive change in employee behavior. This will give birth to a legion of highly motivated employees committed to ensuring the continued success of the organization (Adhikari, 2016).

Leaders will be required to lead empathetically and courageously through these turbulent and uncertain times. A recipe of employee engagement would be incomplete without these ingredients of communication, growth, autonomy, and congenial relationships in the organization. Table 8.2 shows the various employee engagement interventions used at different merger stages to cope with emotional/psychological impacts on employees due to integration.

There are various instances where the mergers were successful due to the application of engagement interventions during the post-merger stage.

8.4 IMPLICATIONS

Thus, employees may be engaged in the post-merger phase by creating an environment of trust, building cross-company teams, appointing cultural ambassadors from the merged companies, focusing on open communication, a continuous direction from the leader to its employees, clear strategies and its effective implementation and a direct dialogue with the employee unions to maintain good employee relations.

TABLE 8.2 Merger's Psychological Stage and Employee Engagement Interventions Used for Coping-Up

Cycle Stage of Merger	Occurrence of Activities	Cognitive/Behavioral Effects	Employee Engagement Interventions
Pre-stage	• Some probability that M&A will happen • Decision making by the Upper Management on M&A • Active Grapevine communication occurs in the firm • Employees find factors to resist the change	Anxiety Fear of unknown	• Communication • Counseling • Speeding up the transition
Pre-preparations for Merger	• Declaring the Merger decision. Announcement of the merging entities • Clarifications and communicating the reasons, future achievements and the future objectives of the merger • Sharing decisions on succession, new leadership • Establishment of the merged entities	Role Conflict Frustration Role ambiguity Fear of losing social identity	• Disengagement efforts • Suggestion Meetings • Goal setting • Role clarification focus • Cross organizational teams and working
Aftermath	• Exchange of ideas, acceptance, sensitivity, and Interactions between the newly partnered organizations • Role conflicts on duties for employees • Workplace restructuring • Acquaintance with diverse cultures • Knowledge enhancement	Acculturation Accepting changed Leadership Intergroup conflict Stress	• Detailed cultural assessment • Mirroring workshops • Social networking • Post-merger job design
New Equilibrium	• Post-merger adjustments • Employees have new standards of behavior • Employee cohesiveness is sought • Certain issues of conflict continue	Relieved Role clarity New expectations from the leader Reskilling requirement	• Two-way communication • Job transfer training • Counseling • Fair and objective HR policies • Equal participation

Adapted from: http://www.peopleandculture.co.uk/documents/uploads/white_papers_managing_complex_times/Psychology-of-Mergers.pdf.

8.5 CONCLUSION

As the economy becomes increasingly global, our workforce becomes increasingly diverse. Organizational success and competitiveness will depend on the ability to manage diversity in the workplace effectively. A manager needs to evaluate his/her organization's diversity policies, include engagement strategies, and plan for the future. The personal commitment of executive and managerial teams is a must. Leaders and managers within organizations must incorporate diversity policies into every aspect of the organization's function and purpose. Attitudes toward diversity originate at the top and get filtered downwards. Management cooperation and participation are required to create a culture conducive to the success of an organization's plan (Greenberg, 2004).

In the subsequent section, a case covering the above concept has been provided (refer Appendix 8.1 for plausible case discussion).

8.6 CASE: AULWIN BANK AND GOODWILL BANK MERGER

In June 2018, Aulwin Bank decided to expand their business by acquiring Goodwill bank. Even Goodwill bank wished to increase shareholder value, growth opportunities for its employees, and provide latest technology-based services to its customers. The leaders had a difficult time managing the merger. This recent change had brought about resistance from employees in both the banks. Aulwin Bank was one of the most reputed banks in the country. Aulwin Bank's size was four times to that of Goodwill Bank. It has staff strength of 12,000 employees whereas Goodwill bank had a staff strength of 3000 employees. Managing workforce diversity was a challenge for the leaders. Employees differed on age, experience, culture, and gender. There were vast variations in profiles, grades, designations, and salaries of personnel. Because of the multiplicity, there were conflicts amongst the employees as they feared that their positions might change along with the team members. Various interventions were applied by the leaders which brought about a change in the behavioral patterns of the employees leading to accept the changes.

Time Frame	Behavior Patterns	Engagement Interventions
Day 1	Resistance, fear, conflict	Clear communication, regular meetings, diverse teams, environment of trust, focus on benefits of merger, linkage of incentives with performance and feedback.
After two months	Transition, listening	
After a year	Acceptance, new equilibrium	
After one more year	Refreezing, teamwork, development	

Thus, with these tools Aulwin Bank seemed to have successfully managed the HR and transformation aspects of the merger.

APPENDIX 8.1: CASE DISCUSSION

SYNOPSIS OF THE CASE

The case highlights a situation where there is a merger of two banks. There are pre-merger and post-merger HR-related issues which need a planning for its management. The case focuses on various tools and interventions which are used to manage the transition in the merged entities.

TARGET GROUP

Students from undergraduate and postgraduate courses may analyze the case.

LEARNING OBJECTIVES

To learn about various interventions and techniques related to employee engagement and how it can be utilized to manage diversity issues at the workplace.

TEACHING STRATEGY

The students may be asked to have a broad perspective and open discussions may be invited. The phases of change management may be linked with the case situation as Freezing-Movement-Refreezing. Students may be asked to cite and collect a few more examples on Post-merger psychological issues which may be discussed in the groups.

QUESTIONS FOR DISCUSSION

- What are the various employee behavioral issues related to the merger of entities?
- Which were the tools used by these banks to cope-up with the change?
- What other tools of employee engagement may be used by banks to cope up with the psychological issues during mergers?

KEYWORDS

- **autonomy**
- **cognitive impact**
- **communication**
- **cross-cultural diversity**
- **employee engagement interventions**
- **mergers and acquisitions**

REFERENCES

Adhikari, S., (2016). *Business World.* http://businessworld.in/article/Managing-Employees-During-Company-Mergers-Acquisitions-/22-01-2016-90545/ (accessed on 24 February 2020).

Alto, P., (2008). http://www.hp.com/hpinfo/newsroom/press/2008/080826xa.html (accessed on 24 February 2020).

Aon, H., (2017). Whitepaper. http://aonhewitt.co.in/Home/Resources/Whitepapers/2017-Trends-in-Global-Employee-Engagement (accessed on 24 February 2020).

Berry, J., (2006). Stress perspectives on acculturation. In: *The Cambridge Handbook of Acculturation Psychology* (pp. 43–57). Cambridge University Press.

Brew, A., (2012). *Branding Business.* http://www.brandingbusiness.com/blogs/engaged-employees-strong-brands-and-the-secret-of-successful-mergers. (accessed on 24 February 2020).

Cartwright, & Cooper, (1993). The psychological impact of merger and acquisition on the individual: A study of building society managers. *Human Relations, 46*(3), 327–347. http://journals.sagepub.com/doi/abs/10.1177/001872679304600302 (accessed on 24 February 2020).

Cox, & Beale, (1997). *Developing Competency to Manage Diversity: Readings, Cases and Activities* (illustrated ed.). San Francisco: Berrett-Koehler.

Davis, C. B., (2016). The longitudinal associations between discrimination, depressive symptoms, and prosocial behaviors in US Latino/a recent immigrant adolescents. *Journal of Youth and Adolescence, 45*(3), 457–470.

Dheeraj, V. (2005). *Wallstreet Mojo.* https://www.wallstreetmojo.com/successful-mergers-and-acquisitions/ (accessed on 24 February 2020).

Douglas, R. (2005). Culture Management in Mergers and Acquisitions. A focus on culture and people is critical to make integration strategies work. http://www.diva-portal.org/smash/get/diva2:1212/fulltext01 (accessed on 24 February 2020).

Epstein, (2004). The drivers of success in post-merger integration. *Organization Dynamics, 33*(2), 172–189.

Evan, Dean, &Thompson, (2003). *Total Quality Management* (3rd ed.). London: McGraw-Hill.

Hansen, M., & Nohria, F., (2004). How to build collaborative advantage. *MIT Sloan Management Review, 46*(1), 22–30.

Haygroup. *The New Rules of Engagement.* https://www.haygroup.com/downloads/in/Hay%20 Group%20New%20Rules%20of%20Engagement%20Report.pdf. (accessed on 24 February 2020).

Josh, G. (2017). *The Multicultural Advantage.* http://www.multiculturaladvantage.com/recruit/ diversity/diversity-in-the-workplace-benefits-challenges-solutions.asp. (accessed on 24 February 2020).

Josh, G. (2004). www.multiculturaladvantage.com: http://www.multiculturaladvantage.com/ recruit/diversity/diversity-in-the-workplace-benefits-challenges-solutions.asp. (accessed on 24 February 2020).

Michelle, (2004). *Post-Merger Culture Clash: Can Cultural Leadership Lessen Casualties*, *2*(4), pp 395–426.

Miller, & Fernandes, (2009). *Culture Issues in Cross-Border Merger and Acquisitions.* USA: Deloitte Consulting LLP.2009.

People and Culture. *Psychology of Mergers.* http://www.peopleandculture.co.uk/documents/ uploads/white_papers_managing_complex_ti (accessed on 24 February 2020).

Reuben, M., (2017). Managing diversity in enterprises after mergers and acquisitions process. *International Journal of Management Science and Business Administration, 3*(6), 23–27. doi: 10.18775/ijmsba.1849–5664–5419.2014.36.1003.

Shodhganga: *A Reservoir of Indian Theses @* Inflibnet. http://shodhganga.inflibnet.ac.in/ bitstream/10603/174076/7/07_chapter%202.pdf. (accessed on 24 February 2020).

Shuck, & Wollard, (2010). Employee engagement & HRD: A seminal review of the foundations. *Human Resource Development Review, 9*(1), 89–110. doi: 10.1177/153448430935356010.1 177/1534484309353560.

FURTHER READING

Acharya, A., & Gupta, M., (2016a). An application of brand personality to green consumers: A thematic analysis. *Qualitative Report, 21*(8), 1531–1545.

Acharya, A., & Gupta, M., (2016b). Self-image enhancement through branded accessories among youths: A phenomenological study in India. *Qualitative Report, 21*(7), 1203–1215.

Allen, J., Jimmieson, N., Bordia, P., & Irmer, B., (2007). Uncertainty during organizational change: Managing perceptions through communication. *Journal of Change Management, 7*(2), 187–210.

Appelbaum, S. H., Gandell, J., Shapiro, B. T., Belisle, P., & Hoeven, E., (2000). Anatomy of a merger: Behavior of organizational factors and process throughout the pre- during post-stages (Part 2). *Management Decision, 38*(10), 674–684.

Appelbaum, S., & Gandell, J. (2003). A cross method analysis of the impact of culture and communications upon a health care merger. *Journal of Management Development, 22*(5), 370–409.

Appelbaum, S., & Shapiro, B., (2007). Mergers 101 (Part 1): Training managers for communications and leadership. *Industrial and Commercial Training, 39*(3), 128–136.

Appelbaum, S., Gandell, J., Yortis, H., Proper, S., & Jobin, F., (2000). Anatomy of a merger: Behavior of organizational factors and processes throughout the pre-during-post stages (Part 1). *Management Decision, 38*(9), 645–662.

Arai, K., (2004). Daimler Chrysler confronts the challenges of global integration. *Human Resource Management, 12*(2), 5–8.

Ayoko, O. B., & Härtel, C., (2006). Cultural diversity and leadership: A conceptual model of leader intervention in conflict events in culturally heterogeneous workgroups. *Cross Cultural Management: An International Journal, 13*(4), 345–360.

Badubi, R., Botha, P. A., & Swanepoel, S., (2014). Job satisfaction levels of teachers at two high schools in Botswana. *Proceedings of World Business, Finance and Management Conference* pp. 8, 9.

Baptiste, R., (2002). The merger of ACE and CARE two Caribbean banks. *Journal of Applied Behavioral Science, 38*(4), 466–477.

Bijlsma-Frankema, K., (2001). On managing cultural integration and cultural change processes in mergers and acquisitions. *Journal of European Industrial Training, 25*(2–4), 192–207.

Bleijenbergh, I., et al., (2010). Aligning talent management with approaches to equality and diversity: Challenges for UK public sector managers. *Equality, Diversity and Inclusion: An International Journal, 29*(5), 422–435.

Cartwright, S., & Schoenberg, R., (2006). Thirty years of mergers and acquisitions research: Recent advances and future opportunities. *British Journal of Management.*

Cartwright, S., & Cooper, C. L., (1993). The psychological impact of merger and acquisition on the individual: A study of building society managers. *Human Relations, 46,* 327–47.

Chawla, A., & Kelloway, E. K., (2004). Predicting openness and commitment to change. *Leadership and Organization Development Journal, 2*(6), 485–98.

Cheong, Y., (2004). Fostering local knowledge and human development in globalization of education. *International Journal of Educational Management, 18*(1), 7–24.

Covin, T. J., Kolenko, T. A., Sightler, K. W., & Tudor, R. K., (1997). Leadership style and post merger satisfaction. *Journal of Management Development, 16*(1), 22–33.

Eisele, J., (1996). Die helfte geht schief. *Absatzwirtschaft, 5,* 86–96.

Esty, K., Griffin, R., & Schorr-Hirsch, M., (1995). *Workplace Diversity: A Manager's Guide to Solving Problems and Turning Diversity into Competitive Advantage.* Avon, M. A: Adams Media Corporation.

Feldman, M. L., & Murata, D. K., (1991). Why mergers often go 'PFFT.' *ABA Banking Journal, 83,* 34–36.

Fralicx, R. D., & Bolster, C. J., (1997). *Preventing Culture Shock.* Modern Healthcare.

Greenberg, J., & Baron, R. A., (2003). *Behavior in Organizations* (p. 4). Boston: Allyn and Bacon.

Gupta, M., & Kumar, Y., (2015). Justice and employee engagement: Examining the mediating role of trust in Indian B-schools. *Asia-Pacific Journal of Business Administration, 7*(1), 89–103.

Gupta, M., & Pandey, J., (2018). Impact of student engagement on affective learning: Evidence from a large Indian University. *Current Psychology, 37*(1), 414–421.

Gupta, M., & Ravindranath, S., (2018). Managing physically challenged workers at micro sign. *South Asian Journal of Business and Management Cases, 7*(1), 34–40.

Gupta, M., & Sayeed, O., (2016). Social responsibility and commitment in management institutes: Mediation by engagement. *Business: Theory and Practice, 17*(3), 280–287.

Gupta, M., & Shaheen, M., (2017a). Impact of work engagement on turnover intention: Moderation by psychological capital in India. *Business: Theory and Practice, 18,* 136–143.

Gupta, M., & Shaheen, M., (2017b). The relationship between psychological capital and turnover intention: Work engagement as mediator and work experience as moderator. *Journal Pengurusan, 49,* 117–126.

Gupta, M., & Shaheen, M., (2018). Does work engagement enhance general well-being and control at work? Mediating role of psychological capital. *Evidence-Based HRM, 6*(3), 272–286.

Gupta, M., & Shukla, K., (2018). An empirical clarification on the assessment of engagement at work. *Advances in Developing Human Resources, 20*(1), 44–57.

Gupta, M., (2017). Corporate social responsibility, employee-company identification, and organizational commitment: Mediation by employee engagement. *Current Psychology, 36*(1), 101–109.

Gupta, M., (2018). Engaging employees at work: Insights from India. *Advances in Developing Human Resources, 20*(1), 3–10.

Gupta, M., Acharya, A., & Gupta, R., (2015). Impact of work engagement on performance in Indian higher education system. *Review of European Studies, 7*(3), 192–201.

Gupta, M., Ganguli, S., & Ponnam, A., (2015). Factors affecting employee engagement in India: A study on offshoring of financial services. *Qualitative Report, 20*(4), 498–515.

Gupta, M., Ravindranath, S., & Kumar, Y. L. N., (2018). Voicing concerns for greater engagement: Does a supervisor's job insecurity and organizational culture matter? *Evidence-Based HRM, 6*(1), 54–65.

Gupta, M., Shaheen, M., & Das, M. (2019). Engaging employees for quality of life: mediation by psychological capital. *The Service Industries Journal, 39*(5–6), 403–419.

Gupta, M., Shaheen, M., & Reddy, P. K., (2017). Impact of psychological capital on organizational citizenship behavior: Mediation by work engagement. *Journal of Management Development, 36*(7), 973–983.

Hill, C. W. L., & Jones, G. R., (2001). *Strategic Management.* Boston: Houghton Mifflin.

Hovers, J., (1971). *Expansion Through Acquisition.* London, England: Business Books.

Kahn, & William, A., (1990). Psychological conditions of personal engagement and disengagement at work. *Academy of Management Journal, 33(4), 692–724.*

KPMG, (1999). *Mergers and Acquisitions: Global Research Report.* London: KPMG.

Kwoka, J. E. Jr., (2002). Mergers and productivity. *Journal of Economic Literature, XL, 540,* 541.

Langford, R., Brown, III. C., & Mathieu, B., (2004). Making M&A pay: Lessons from the world's most successful acquirer. *Strategy and Leadership, 32*(1), pp5–14.

Lichtenberg, F. R., & Siegel, D., (1987). Productivity and changes in ownership of manufacturing plants. *Brookings Papers on Economic Activity, 3,* 643–673.

MacDonald, R., (2005). A template for shareholder value creation on M&As. *Strategic Direction, 21*(5), 3–10.

Morrell, K. M., Loan-Clarke, J., & Wilkinson, A. J., (2004). Organizational change and employee turnover. *Personnel Review, 33*(2), 161–173.

Nguyen, H., & Kleiner, B., (2003). The effective management of mergers. *Leadership and Organization Development Journal, 24*(8), 447–454.

Nikandrou, I., Papalexandris, N., & Bourantas, D., (2000). Gaining employee trust after acquisition. *Employee Relations, 22*(4), 334–355.

Pandey, J., Gupta, M., & Naqvi, F., (2016). Developing decision making measure a mixed-method approach to operationalize Sankhya philosophy. *European Journal of Science and Theology, 12*(2), 177–189.

Papadakis, V., (2007). Growth through mergers and acquisitions: How it won't be a loser's game. *Business Strategy Series, 8*(1), 43–50.

Pollitt, D., (2006). Raise a glass to HR at Wolverhampton and Dudley Breweries. *Human Resource Management International Digest, 14*(1), 9–12.

Presley, J., Solaro, N., Tsui, C., & Wong, H., (2002). Deutsche Bank & bankers trust. A merger going global. *North American Penetration a Business Organization Presentation.*

Shelton, D., Hall, R., & Darling, J., (2003). When cultures collide: The challenge of global integration. *European Business Review*, *15*(5), 312–323.
Want, J., (2003). When worlds collide: culture clash. *Journal of Business Strategy*, *24*(4), 14–21.
Weber, Y., (1996). Corporate cultural fit and performance in mergers and acquisitions. *Human Relations*, *49*(9), 1181–95.
Yin, R. K., (1984). *Case Study Research: Design and Methods*. Newbury Park, CA: Sage.

CHAPTER 9

Engaging People with Physical Disability at Work

ANITHA ACHARYA

Marketing and Strategy Department, ICFAI Business School (IBS), Hyderabad, The ICFAI Foundation for Higher Education (IFHE) (Deemed to be University u/s 3 of the UGC Act 1956), Hyderabad, India, Tel.: (+91) 8712290557, E-mail: anitha.acharya@ibsindia.org

ABSTRACT

A physical disability is the long-term impairment or loss of part of body's physical function. According to the International Labor Organization (ILO) estimates as of July 2017, there are roughly 650 million people worldwide who are classified as disabled. Most of the countries around the world are working to give these individuals a better opportunity at finding and retaining jobs. The chapter highlights the benefits of having employees with physical disability at work. The chapter also discusses about how to engage employees with physical disability at workplace. It also explains about what are the different laws which are available for disabled people.

9.1 INTRODUCTION

Disabled literary characters usually remain on the margins of fiction as uncomplicated figures or exotic aliens whose bodily configurations operate glasses, triggering response from other characters. The disparity between disabled as an attributed, decontextualizing identity and the perceptions and experiences of real people living with disabilities suggests that this figure of otherness emerges from, interpreting, conferring, and positioning meaning upon bodies (Bricout and Bentley, 2000). The rhetorical effect of representing disability derives from social relations between people who assume the normate position and those who are assigned the disabled position.

Whether one lives with a disability or meets someone who has one, the real experience of disability is more multifaceted and more energetic than depiction usually suggests. For example, the expertise which people with disability have to learn is managing social encounters. The initial interaction between normal and disabled people varies markedly from the usual relations between readers and disabled people. In the first encounter with other individuals, a tremendous amount of information must be organized and interpreted simultaneously, each member probe the explicit for the implicit, and then decide what is significant for particular purposes, and prepares a reply that is guided by many cues, both delicate and evident. If someone has disability, however, it almost always dominates and skews the normate's process of sorting out perceptions and forming a reaction. The interaction is usually stressed because the nondisabled person may feel dread, disgust, or might be surprised, all of which according to social protocol is difficult to express them. In addition, the disconfirming dissonance amid experienced and expressed reaction, a person with disability often does not know how to act toward a disabled person: how or whether to offer help; whether to accept the disability; what expressions, gestures to use or avoid. Possibly most destructive to the potential for continuing relations is the normate's recurrent postulation that a disability cancels out other qualities, reducing the complex person to a single attribute. This uncertainty and dissension make the encounter especially stressful for the nondisabled person who is unacquainted to disabled people. The disabled person may be anxious about whether the encounter will be uncomfortable for either of them to maintain and may feel the ever-present danger of denial. People with disability must learn to manage relationships from the beginning (Williams and O'Reilly, 1998). In other words, disabled people must use ardor, charm, or humor to relieve people without disability of their discomfort (Goggin and Newell, 2003; Schartz et al., 2006; Jaeger and Bowman, 2005).

9.2 NEED FOR DISCUSSING THIS THEME

Physical disability can entail difficulties with sitting and standing, bladder control, sight, speech, hearing, walking, and mobility, muscle control, sleeping, fits, and seizures. A physical disability may be genetic. It can also come about through something that happened before or during birth or later in life through an illness or injury. A physical disability may be apparent, such as loss of a limb, or less obvious, for instance, epilepsy. However, there are some advantages and disadvantages about the disabled population, which often keep them out of job even if they are qualified workers. They are as follows:

- **Tax Benefits:** If companies want to reduce their tax liability a little can recruit disabled person so that they can claim tax benefits which are only given to those companies who have recruited major or any number of employees who are disabled. The cost of having a barrier-free environment or preparing ramps is reduced by the government to support companies to hire a disabled employee and also helps the disabled people to have a better access to the area around them. To increase employment opportunity government organizations in many countries provide funds for recruiting disabled workers in a small scale business. Therefore, employers can take advantage of such opportunities to reduce the financial burden of the company and increase ways of gathering funding for small business. Government funds differ from direct salary tax reduction or indirectly through providing subsidies for companies who provide appropriate infrastructure and creating a healthy work environment for disabled people.
- **Talented Workforce:** Irrespective of the company size, talented workers can be recruited. Individuals with disability are very hardworking people because they have reached such a capable level that they can do a work independently. Disabled people put in lot of effort to reach certain level and to better their level at something than the normal people. So if organizations recruit a good professional person then it will pay them off in the long run. By recruiting disabled workers, an organization will be having another large set of the population who, although may not be normal in all aspects but can show to be equally talented when it comes to the job requirement. In reality, prior studies (Thomson, 2017; Greening and Turban, 2000) have shown that given equal opportunities, disabled people can perform on par with people who do not have any disability. So, appropriate investment should be made in recruiting disabled people by an organization.
- **Brand Image:** Recruiting disabled workers can incalculably increase the brand image of the company. It will enhance the goodwill of the company and it will create a positive mindset among the citizens in the society. Although, that does not mean that the company should start to select such workers only for enhancing the image. The selection of the employees, nevertheless, should be based on the eligibility criteria of the job. But, the organization should ensure that the job opportunity is equally available to all kinds of people. This will prove to be

highly useful for the organization's assets as it can increase positive feeling about the organization among existing employees and other stakeholders.

- **Benefits in a Legal Matter:** Having disabled people in an organization gives an extra edge to an organization in legal matters and gives an advantage in any cases associated to discrimination. In addition, the added social support towards the organization helps to gather a good support towards the organization's cause. If organizations treat disabled-employees as they treat any other worker, then it will help in taking help of various acts and legal documents to prevent any legal problems from arising and in addition, have the benefit of some extra payback the government has to offer.
- **Trust:** Many employers believe that just because some people behave in some special way means that they cannot be assigned with jobs. But, many prior research works (Suter et al., 2007) have proved that disabled persons can be trusted more on confidential matters and that they do serve up to the expectation of the manager. They are very loyal and hard working in every matter. They try to put in their best efforts at anything.
- **Diversity:** In today's industry, the key to success is diversity. Diverse workforce ensures that organizations have a talented pool of workers that will be able to handle workloads of various intensity and types at any point of time. Furthermore, disabled workers provide a different type of diversity in the workplace which is the inimitable ability in both mental and physical ways. Although, they may lack in their physical abilities, most of the mentally sound disabled employees make up for it with positive mental abilities which others can only think about.
- **Loyalty:** Disabled workers are found out to be more loyal than their normal counterparts. In spite of, many twisted opinions that many managers might be having about these people, researchers have shown that disabled worker is loyal to their bosses and have more probability of meeting the deadline than their other counterparts. They are very focused on their work.
- **Lower Turnover:** According to Jain and Lobo disabled workers are usually subject to lower turnover rate and they prefer to do their job efficiently without complaining much to their supervisor (2012). Disabled people, being loyal in nature, prefer to stay in their current job and they are less likely to change their position, post, or job on frequent basis.

- **Less Payment of Paid-Leave:** Disabled people are not only loyal and focused, but they even take fewer holidays or unprepared leaves from their office. This can help organizations run flawlessly where the presence of each person on every working day counts. Additionally, it also helps the organization to save additional money on paid-leave which, they would otherwise have to bear for the normal counterparts. In a survey conducted by Accenture (2008), it is found that the physically disabled people take fewer pay sick days compared to normal average employee. This will always be advantageous for an organization.
- **Productivity and Speed:** Disabilities at times mean that an employer is not able to work at a velocity comparable to other employees. For instance, it may take longer for someone with a physical disability to move office equipment, whereas someone with a mental disability may take longer to read and interpret official documents. This sometimes makes a lot of difference in jobs where the individual's ability to keep pace affects other functions of the organization, such as in a factory with an assembly line. Nevertheless, it is not necessarily a disadvantage in other sectors, such as an art restoration of arts business where the quality, and not the speed, of each project is more important.
- **Discrimination:** Although the Rights of Persons with Disabilities Bill-2016 and other laws in India have opened doors for the disabled in the workplace and many employers strive to comply, disabled workers still can face discrimination. For instance, coworkers may refuse to accept having a disabled worker as a member of their team if the specific disability affects the progress on the project. In some cases, HRs department employees intervene to settle conflicts, and the employer may need to take time and resources to train all employees about the consequences of workplace discrimination.
- **Technology and Infrastructures:** In few countries (for example America, Canada, and India), employers are required to make rational accommodations to infrastructures at workplace so that a disabled individual may work in an efficient manner if recruited. For example, they may have to construct a wheelchair ramp. In addition, in few instances, other co-workers might not be familiar with the specific technologies a disabled individual might require. For example, a specialist in software who is an expert in company rules and regulations about software programs may not be as familiar with a speech-recognition program and therefore might take some time to find solutions for the problems.

9.3 REVIEW OF THIS THEME

9.3.1 ENGAGEMENT AND DISABILITY

Organizations are recognizing the need for employing diverse workforce in their organization. As per the World Health Organization (WHO), 18% of people with disabilities are employed. This means that a majority of people with disabilities who are qualified are not employed and are looking for job. Since organizations benefit a lot if they recruit people with disabilities, they have to ensure that they recruit more people with disabilities in their organizations. The first step which organizations have to take is to create a disabled-friendly workplace. The workplace should engage, support, and encourage people with disability to perform to their best.

Some of the steps which organizations can adopt to ensure that employees with disability are engaged to their work are as follows:

- **Awareness and Training:** Organizations should build awareness among its employees. Since an aware workforce is also an empowered workforce. In order to engage employees with disability they have to be fully integrated into the workplace, this is possible only if all the employees are familiar with the stated dedication of their organization to being disabled-friendly. Imparting training to the employees will help them to gain more insight into how to deal with coworkers who are disabled. Some employees may possibly be knowingly or reflexively biased about their disabled co-workers. Training employees will help them to overcome these notions. Organization should also provide its employees with all the information regarding how to help disabled co-workers in case of emergency.
- **Social Support:** Organizations can also take outside support to impart training to their employees. Several government and non-profit organizations are working towards new comprehensive workplaces that can hold awareness initiatives and seminars at the organizations. This will not only improve employee participation and engagement but will also help to bring new ideas to solve any issue.
- **Technology:** Providing assistive technology will help disabled employees to be more engaged. Assistive technology will help employees with disability to carry out their work without any hindrance. For example, organizations can provide its employees with disability computer keyboards which are color-coded, sign language applications, assistive devices for listening, and software to read the screen.

- **Appraisal:** Organizations should be honest and impartial while giving feedback to employees with disability. Honest feedback will help employees with disability to be responsible for their work as other normal employees. If managers are lenient towards disabled employees then this will affect the work culture of the organization and will also affect the performance of both abled and disabled employees.
- **Accessibility:** One of the priorities of any organizations is to ensure that there is free accessible workplace to its employees with disability. This will help them to move around the organization and communicate with their coworkers more freely. Organizations should also ensure that employees with disability have access to all the communication materials and also the company website. This will help employees to be more engaged to their work.
- **Well-Being:** Organizations should focus on employee's health and well-being. Managers can incorporate physical activities and other recreational activities at their workplace. Managers can also consider involving the family members of disabled employees during the training sessions, this will help employees with disability to understand and get used to their new roles much faster.

9.3.2 INTERNATIONAL LAW

International covenant on economic, social, and cultural rights (ICESCR) is applicable on the above theme. The provisions of the covenants are as follows:

- Article 2: Right to non-discrimination.
- Article 6: Right to work.
- Article 7: Right to just and favorable conditions of work.
- Article 9: Right to social security.
- Article 11: Right to an adequate standard of living.

The central point in the acquiring of economic security is the capability to engage in gainful employment on the basis of equality with others. Articles 6 and 7 of the ICESCR guarantee the right to work. Apart from the above points, the provision needs to be read in conjunction with the guarantee of non-discrimination in the enjoyment of rights contained in Article 2. Apart from ICESCR, there are other important Protocols relating to the right of individuals with disability to work. These include the ILO Protocol on Discrimination in Occupation and Employment 1958, which was ratified by the Indian government on 3 June 1960, and the

ILO Protocol on Vocational Rehabilitation and Employment (Disabled Persons) 1983.

International covenant on civil and political rights (ICCPR) is also applicable. The following articles of the covenants are relevant from the point of view of the employment rights of the disabled:

- Article 6: Right to life.
- Article 7: Freedom from torture, cruel, inhuman, or degrading treatment, and punishment.
- Article 9: Right to liberty and security of persons.
- Article 2, 3, and 26: Right to equality and the right to take part in public affairs.

9.3.3 PHYSICAL DISABILITY IN INDIA

The rights which should not be taken by the government or an individual are the fundamental. They consist within its gamut the right to liberty and life, right to freedom of expression and speech, equality, freedom of movement, right to education, etc., these rights not only cover political but also social, cultural, and economical dimensions. The main aim is to promote social progress and better standard of living. The human rights movement has intrepidly and firmly shifted the interest of policymakers from the mere rider of charitable services to enthusiastically protecting their basic right to dignity and self-respect. In this new scenario, the individuals with disability are viewed as individuals with a wide range of abilities and each one of them willing and capable to utilize his/her talents (Albrecht and Snyder, 2005; Siperstein et al., 2006). On the other hand, the society is seen as the real cause of the misery of individuals with disabilities since it continues to put numerous barriers as expressed in education, employment, architecture, transport, health, and dozens of other activities.

Advances in surgical and medical sciences, greater understanding of the causes of disability, improved methods of coping with disability, breakthroughs in technology, and increasing consciousness of civil rights and the emergence of people with disabilities displaying skills and knowledge to improve their own lives, are some of the factors which have contributed to the new thinking that the individuals with disability also deserve a decorous status in society as the normal individuals.

Human rights have become a matter of international concern and their advocates are exercising helpful influence across cultural and geographical boundaries. Lot of countries is opting for laws which are in favor to individuals with a disability. Any direct/indirect discrimination against individuals with disability is punishable by the legislation. A logical and broad framework for the encouragement of just and fair policies and their efficient implementation are established by the legislation. The legislation has drafted formal procedures which hasten the process of full and total integration of the disabled in the society. Individuals with disability are an enormous minority group, and have been subjected to direct and indirect discrimination for centuries in most countries of the world, including India (Barnartt and Scotch, 2001; Friedner and Osborne, 2015).

In India, the number of individuals with disability is so large, they have complex problems, limited scarce resources and social attitudes so detrimental, it is only legislation which can ultimately bring about a substantial change in an unvarying manner (Dorodi, 2013). Though legislation cannot alone radically change the fabric of a society in a short span of time, it can however, augment ease of access of the disabled to education and employment and means of transport and communication. The impact of well-planned legislation, in the long run, would be insightful and liberating.

In order to ensure better protection of human rights of individuals with disability, The individuals with Disabilities (Equal Opportunities, Protection of Rights and Full Participation) Act, 1995 was formed in the year 1995 in order to give effect to the proclamation on the Full Participation and Equality of the People with Disabilities in the Asian and Pacific Region. Chapter 6 of the Act deals with employment and contains various incidental provisions related thereto. The relevant laws are mentioned in the following subsections.

9.3.4 INDIAN LAWS

- **Constitutional Provisions and Human Rights:** Though the fundamental rights stress on the existing rights, the Directive Principles provide the energetic movement towards the goal of providing Human Rights for all. Right to employment under the Indian Constitution can be considered in the light of relevant provisions of parts III and IV of the Constitution and in particular the following:
    - **Article 14:** Equality before the law and equal protection of laws within India.
    - **Article 7:** Declaration of Human Rights.
    - **Article 16:** Equal opportunity in the matters of public employment (Article 16).
    - **Article 43:** Living wage, benefits, etc. for workers (Article 43).

- ➢ **Article 41:** Directs the State to ensure the people within the limit of its economic capacity and development, right to work, to education and to public assistance in certain cases.
- **The Mental Health Act, 1987:** This act was enacted to regulate admissions to hospitals for mentally-ill individuals who do not have sufficient knowledge to seek medical treatment on a voluntary basis and also to protect the rights of such individuals if they are being detained.
- **The Rehabilitation Council of India Act, 1992:** The act was passed to normalize the manpower development programs in the field of education of persons with special needs. The main aim is to regulate the training policies and programs in the field of rehabilitation of people with disabilities, to recognize institutions/universities running degree/diploma/certificate courses in the field of rehabilitation of the disabled, to recognize and equalize foreign degree/diploma/certificate courses, and to standardize training courses for rehabilitation professionals.
- **The National Policy on Education 1986:** The aim of this act was to provide education, to all the citizens, including people with disabilities. The objective of this policy is to integrate the physically and mentally handicapped with the general community as equal partners and to prepare them for growth and to assist them to face life with confidence.
- **The Workmen's Compensation Act, 1923:** As per Section 3 of the act, if personal injury is caused to a workman by accident arising out of or in the course of his employment, his employer shall be liable to pay compensation.
- **Employees' State Insurance Act, 1948, Section 46(c):** According to this act, periodical payment shall be made to an insured person suffering from disablement as a result of an employment injury sustained as an employee and certified to be eligible for such payments by an authority specified in this behalf by the regulations.
- **The Payment of Gratuity Act, 1972:** As per Section 4 of this Act, gratuity shall be payable to an employee on the termination of his employment on his death or disablement due to accident or disease, if he has rendered continuous service of five years.

9.3.5 RIGHTS OF DISABLED PERSONS IN INDIA

Some of the basic rights that the Person and with disabilities are entitled to:

- **Certificate:** It is the most basic document that a disabled person should possess in order to avail certain benefits and concessions. The State Medical Boards established under the State governments can issue a disability certificate to any person with more than 40% disability. The concerned person can visit the nearest hospital and will be issued the certificate after the checkup and determination of percentage of disability. The certificate is valid for five years and can be renewed if the disability is temporary and valid for the entire lifetime in case of permanent disability.
- **Travel Concession:** Persons with disability are entitled to certain concession in the amount of train tickets bought at the railway counter or online. In some states, such persons can also avail bus ticket concession under State government bus service by showing the disability certificate to the conductor.
- **Pension:** People who are above 18 years of age, living below the poverty line and suffering with more than 80% disability and are entitled to the disability pension under the Indira Gandhi National Disability Pension Scheme. Various NGOs are dedicated to this cause and they help disabled people to get their disability pension.
- **Guardianship Certificate:** Persons with certain disabilities like mental retardation, cerebral palsy, etc., are in a special situation and would not be able to take important legal decisions once they become adults. Hence a Legal Guardianship Certificate is issued to such disabled person which makes another person his legal guardian who thereby is entitled to take all legal decisions on behalf of the disabled person even after he becomes an adult i.e., completes 18 years of age.
- **Income Tax Concession:** Under sections 80DD and 80U of Income Tax Act, 1961, persons with disabilities are also entitled to certain income tax concessions.
- **Employment:** In government jobs, 4% of the seats are reserved for persons with disabilities.

9.3.6 IMPEDIMENTS IN THE EMPLOYMENT PROCESS OF PEOPLE WITH DISABILITY

Some of the impediments that limit employment opportunities for people with physical disabilities:

- **Adjustment:** Employers and employees need ongoing direction in providing and requesting adjustments, respectively.

- **Self-Identification:** employees are hesitant to self-identify; however, as federal contractors need this information for compliance with the Department of Labor (DOL) regulations.
- **Discomfort:** Perceptions subsist that people with disabilities are less qualified, which creates discomfort among employers to employ them in their organizations.
- **Access to Information:** Employers are besieged by the amount of information available, but concurrently need more information.
- **Motivation:** Employers need greater visible Acknowledgment for their efforts to employ more individuals with physical disabilities, to persuade continued buy-in and contribution in disability and insertion efforts.

9.3.7 STRATEGIES TO EASE THE IMPEDIMENTS

- **Change:** The management should change perceptions and stereotypes, the manager's focus should be on building a rational, economic business where there is no place for bias, since biases influence perceptions. For example, Lemon tree hotel have addressed this impediment through programs and awareness-raising activities that work to change these perceptions.
- **Education:** The management should educate and inform all the employees and create awareness among employees of the set of practices for increasing disability employment.
- **Incentives:** Motivate with incentives, companies must be acknowledged and rewarded for their progress, as well as individual employees at all levels since they will also respond positively due to incentives.

9.4 IMPLICATIONS

Some of the efforts which companies need to take to engage the employees with disabilities are:

- **Involvement:** Top management involvement.
- **Policies:** Sound and supportive human resource (HR) policies.
- **Environment:** Efforts to create an inclusive environment to support disabled employees among other employees.
- **Collaboration:** Collaboration with non-governmental organizations to update their skills regularly.

- **Infrastructure:** Modifications in the infrastructure to support the disabled employees.
- **Commitment:** Build up leadership commitment honest commitment towards leadership requires leaders to institute the employment of people with disabilities as a priority. They should place people with disabilities in leadership positions, communicate their actions and commitment inside and outside of their company, and, preferably, find their fervor.
- **Accountability:** Create a position and identify a person who will be in charge of attracting, engaging, and advancing people with disabilities and make sure that person gets the required support and resources which they need to be successful. That person should be made accountable for achieving objectives of the company. Before giving the final offer, consider the character of the candidate along with their competence in hiring decisions.
- **Identity:** Managers should make it safe to identify people with disabilities. Managers should develop an organizational climate that makes it safe for individuals to disclose any disabilities. The managers should also inform the individuals at the time of orientation what organizations have to offer for individuals with disabilities, for example, flexible work options, technology, access to accommodations, and various other facilities.
- **Skill Levels:** Managers should raise understanding and skill levels of their employees. Training should be imparted to all employees on manners and consideration. Managers should be aware of their role in level playing field, particularly their responsibility in eliminating unintended biases and inequities, they should know how to interview impartially and fully understand their accountabilities and legal responsibilities.
- Although existing results for employing people with disabilities have not been astral, there are three trends which will affect the future employment opportunities of people with disabilities:
- As the occurrence of disability increases with age, the proportion of the talent pool with disabilities will grow as the population ages. Apart from age, technological innovations that have widened the scope for disabled people to have access to the workplace. Advancement in health care has extended and improved the life of people with disabilities. Due to the above-stated reasons, the proportion of the talent pool composed of people with disabilities will be on the rise.

- The existing incentives and proposed quotas, especially for those doing business in the domestic market and globalization. The motivation to hire people with disabilities will increase.
- The wider application of universal design will further improve access due to changing attitudes toward remote work and the increasing accessibility of technologies that enable remote work, increase access to communication and information technologies.

9.5 CONCLUSION

According to India's philosophy of *Vasudhaiv Kutumbakamb* which means the world is one family and all the family members deserve equal care and treatment. Members who are defective, physically weak, should not be abandoned and should not be treated differently. The ideas and aspirations of disabled members should align with the Millennium Development Goals and this is possible when there is decentralization of rehabilitation planning and implementation. As a consequence, all of us should come together and extend our support and assistance to create a world without chauvinism and favoritism, and disgrace where sign language is also treated as a common medium as spoken and written words with another medium of expression that is through empathy. We have to create a world where disabled can participate in cultural activities, have easy access to public places and utilities, and are capable to enjoy their lives in a meaningful way.

Physical disability is not the end of the world. Many individuals with disabilities live their lives just like any other normal person. They enjoy shopping, sports, movies, etc., and also involve in lot of other activities. They work as very hard as the normal person and support their family. Nevertheless, they do need help and acceptance from the social community to help them to integrate into the mainstream society. Giving them opportunities at work and treating them like any normal individual are some of the ways that can help them to live life to the fullest.

9.5.1 BEST PRACTICES

- **SAP Labs India:** SAP was founded in the year 1972. They are the market leader in enterprise application software, SAP helps its clients in fighting the effects of complexity, generates new opportunities for growth and innovation. SAP started its operations in India in the year

1998. It hires individuals with a disability. In the year 2016, they hired an additional 10% of disabled people. They have special evacuation chairs for people with disability and they also have incorporated Braille signs across all buildings in their organization. In the year 2011, they started working with the Autism Society of Indian to choose candidates with autism spectrum disorder (ASD). The SAP took advantage of the individual's photographic reminiscence to execute complex tests, which would involve about 50 steps. When they were tested against regular engineers, it was found that the regular engineers were right only on 80% of the occasions, while those with autism were right every time. By 2020, SAP is aiming Aims to have 1% of its global workforce consisting of autistic employees who can be fit into relevant information technology roles.

- **Accenture India:** Accenture PLC was founded in the year 1989 by Arthur E. Andersen and Clarence DeLany is a management consulting firm and has a presence globally. It provides services like consulting, technology, strategy, operations, and digital to its clients. Since 2010, Accenture has been organizing the *Skills to Succeed* initiative program in India. The program empowers persons with disabilities, who receive training in the skills needed in some high-demand job sectors, for instance, business process outsourcing, facilities management, and hospitality, retail, construction, and micro-enterprise. The company has a network called 'PwD Champions Network' that includes PwD employees. The company also provides business cards printed in Braille to all its inclusion and diversity sponsors to sensitize and impact people. The company also encourages reasonable modifications to a job or the work environment to ensure that a qualified individual with a disability has privileges and rights equal to those of employees without disabilities.
- **Tata Group:** It is an Indian multinational conglomerate holding company headquartered in Mumbai, Maharashtra, India. It was founded in 1868 by Jamsetji Tata. Titan Industries, Tata group's watch-making division started employing the disabled in the 1980s. The differently-abled constitutes about 4% of Tata's employees. They are given tasks with low levels of body risk, the least physical movement, and minimum verbal communication. The Tata Group employs the differently-abled in, Tanishq, Tata Motors Titan, and Industries.
- **Lemon Tree Hotels:** It was founded in the year 2002 and is headquartered in New Delhi. It is India's largest hotel chain in the mid-priced hotel sector, and the 3rd largest overall, in terms of controlling

interest in leased and owned rooms. As of 2017, 22% of their staff consisted of people with disabilities. It aims to increase the number to 10% by the end of 2025. Lemon Tree has also developed a standard process to induct people with disabilities into its entire hotels pan India. It believes in being an equal opportunity employer and has been acknowledged on various platforms for its work towards employing individuals with a disability. The disabled employees consist of down symptoms, speech, and hearing impairment, and orthopedically handicapped. In fact, one employee with a disability is working as a part of the President and Executive Director's team in the corporate office. Disabled employees are spread across the management roles, supervisory roles, and semi-skilled in Food and Beverage services.

- **National Association for the Blind (Employment and Training)-(Nabet):** Nabet was formed in the year 2009 by two brothers, Abhishek, and Arjun Mishra as a social enterprise that loops in corporate who in turn offer training and subsequent employment to the individuals who are disabled. The employment opportunities provided by the organization includes data entry, contact center, software testing, and accessibility testing.

9.5.2 MINISTRY OF RAILWAYS AND AHMED

- **Problem:** Ahmed was paid less salary by the Indian Railways because of his disability.
- **Verdict:** The Allahabad High Court has held that the government cannot accommodate an employee in a post with a lower pay scale just because he or she acquired some disability during service and became unfit for the current post.
- **Compensation:** Ahmed was paid all the dues with a 7% interest from the due date.

9.5.3 KUNAL SINGH AND UNION OF INDIA

- **Problem:** The Apex Court in India considered the issue as to whether the services of an employee who incurs any disability during such service could be dispensed from work on the ground of such disability.
- **Verdict:** The court was considering the case of the appellant, a constable for 17 years in Special Service Bureau of Union of India suffering

amputation of a leg on account of an injury caused while on the duty. The medical Board had declared him to be permanently debilitated and therefore appellant was terminated from service. His petition before the High Court challenging the action of the government and seeking relief by way of alternative duty, which he could discharge, was dismissed by the High Court. Allowing the appeal the Apex Court held that merely because the appellant got invalidity pension is no ground to deny mandatory protection made available under the Act. Once he was found not suitable for the post he was holding, he could be shifted to some other post with same pay scale and service benefits and if that was not possible he could be kept on a supernumerary post until a suitable post is available or he attains the age of superannuation.

- **Compensation:** Union of India was asked to compensate Kunal Singh. In the subsequent section, a case covering the above concept has been provided (refer Appendix 9.1 for plausible case discussion).

9.6 CASE: CHALLENGES OF ALEX

Alex suffers from right hip early osteoarthritis. He believes that this condition amounts to a disability within the connotation of the Equality Act 2010. Alex is currently employed as a Personal Trainer at "OZO" gymnasium (his "Employer") and has worked with them for the past 6 years. Alex was diagnosed with hip early osteoarthritis 3 years ago. Alex feels that he has been treated less favorably by his Employer due to his disability divergent to the Equality Act 2010. Alex has on numerous instances made his Employer aware that he is suffering from a disability for which he requires sound accommodations to his working practices. Alex has requested the following changes:

- Reduction in his working hours. He wants to work for 30 hours per week.
- Regular breaks from his shifts so that he can take some rest to reduce the hip pain.
- Adjustment to the shift pattern for personal trainers.
- Although Alex's Employer knew of his disability for over 3 years now, it has indefatigably botched to make any adjustments to accommodate his disability. His manager's view is that Alex's disability does not create a positive image for its Personal Trainers and company. The subjecting of Alex to disability discrimination has meant that Alex

has been prevented from working the reduced hours he requested and this has had a detrimental effect on his current health which has exacerbated the effects of his disability. Three months ago, Alex raised a formal grievance as he felt that he had no alternative but to do so in circumstances where all of his previous concerns raised verbally had been ignored. Alex's Employer did not uphold his grievance and denied all liability for discrimination. Alex's Employer did, however, agree to reduce his hours to 25 hours per week (with no adjustment or flexibility to allow him to work in excess of that should the need arise), requesting that he works on all the days of the week when the crowd is less and preventing him from working on busiest days and times of the week. He has also been allowed to take a 10-minute break when he feels in pain provided that his boss authorizes such breaks in order to ensure that his boss is aware of his whereabouts.

Alex's Employer wishes to change Alex's terms and conditions of employment to reflect his new working hours (25 hours per week) and days of work to include working every weekend. Alex was told that he will face legal proceedings if he does not accept the proposed varied terms.

Alex contemplates that his Employer has botched to give any good reason for not agreeing to make the accommodations he requested and that the recommended accommodation that it is willing to make are unreasonable in the situation he is in. Alex is aware that his employer is hiring and new staffs are being recruited or being asked to cover during peak hours and they have been asked to be personal trainers for their clients instead of allowing Alex to be personal trainer.

Alex went to see an advocate for legal advice to see if he had any impending employment claims against his Employer. He was advised by the advocate that the Equality Act 2010 requires employers to make reasonable adjustments for employees who have a disability. In addition, that employee with a disability should not be treated less favorably because of a disability. In Alex's case, his Employer did not provide any reasons as to why it could not allow Alex to work 30 hours per week. Alex's Employer had not sought a medical opinion. In all the circumstances, therefore, Alex's Employer had failed to make reasonable adjustments. Adding up to this, Alex's employer subjected him to less favorable treatment by insisting he works at the quietest times every weekend (when his colleagues who did not suffer from a disability did not have to work every weekend) and by insisting that he seeks his boss's approval before taking breaks.

In addition to a claim for disability discrimination, Alex could also claim victimization under the Equality Act 2010 because he was subjected to further less favorable treatment because he made a complaint about disability discrimination, as his Employer threatened that he will face legal proceedings if he does not accept the proposed changes to his terms and conditions of employment.

Alex was advised that if he were to pursue a claim in the Employment Tribunal for disability discrimination, he would be entitled to compensation for his injury to feelings, his future loss of income (if he were to resign from the current job), and also the personal injury he had suffered due to his condition becoming worse as a result of his Employer's failure to accommodate his disability. It was also explained to Alex that the Employment Tribunal would make a recommendation about reasonable adjustments for his continued employment.

While interacting with his advocate Alex was worried about the costs involved in the legal proceedings. However, when his advocate discussed the matter with him it became clear that he had Legal Expenses Insurance which would fund legal assistance. Alex's advocate helped him to apply to his insurers for funding and then issued an Employment Tribunal claim on his behalf.

9.6.1 EMPLOYER LAW

The Equality Act 2010 is the law that prohibits unfair treatment and helps achieve equal opportunities in the workplace and in wider society. Answer the following questions based on the case:

- Explain the rights of Alex.
- How can Alex's employer keep him engaged?

APPENDIX 9.1: CASE DISCUSSION

ABSTRACT

The case describes the discrimination faced at the workplace by an employee who is physically challenged. Even though there is a lot of laws, which say that physically challenged employees should not be discriminated. The organization is not following the rules.

Teaching Objectives and Target Audience

The case is designed to enable students to identify the laws which are available for employees in the organization in case they face any discrimination. The case also helps students to explore strategies that the employee might adopt to overcome them.

This case is meant for BBA and MBA students as part of their human resource (HR) ethics curriculum.

Teaching Scheme

As a short case, it can be used as a basis for plenary discussion instead of group discussion.

Case Analysis

By law, all employees have a number of rights that have been cautiously documented to ensure that all individuals are treated fairly by their employers. Students can discuss in detail some of the rights like:

- Whistleblower rights;
- Right to a safe workplace;
- Right to fare wages;
- Right to be free from harassment and discrimination of all types;
- Medical leave.

KEYWORDS

- **Accenture India**
- **advantages**
- **engagement**
- **physical disability**
- **rights**
- **workforce**

REFERENCES

Albrecht, G. L., & Snyder, S. L., (2005). *Encyclopedia of Disability* (Vol. 5). Sage Publications.

Anuradha, M., Meera, P., & Pratiti, R., (2006). *Rights of the Disabled*. National Human Rights Commission. https://nhrc.nic.in/sites/default/files/DisabledRights.pdf (accessed on 24 February 2020).

Barnartt, S. N., & Scotch, R. K., (2001). *Disability Protests: Contentious Politics 1970–1999*. Gallaudet University Press.

Bricout, J. C., & Bentley, K. J., (2000). Disability status and perceptions of employability by employers. *Social Work Research, 24*(2), 87–95.

Commission on Social Determinants of Health, (2008). *Closing the Gap in a Generation: Health Equity Through Action on the Social Determinants of Health*. Geneva, World Health Organization.

Dorodi, S., (2013). *The Disconnect with Disability*. The Hindu. http://www.thehindu.com/opinion/op-ed/the-disconnect-with-disability/article4667332 (accessed on 24 February 2020).

Friedner, M., & Osborne, J., (2015). New disability mobilities and accessibilities in urban India. *City and Society, 27*(1), 9–29.

Goggin, G., & Newell, C., (2003). *Digital Disability: The Social Construction of Disability in New Media*. Rowman and Littlefield.

Greening, D. W., & Turban, D. B., (2000). Corporate social performance as a competitive advantage in attracting a quality workforce. *Business and Society, 39*(3), 254–280.

Jaeger, P. T., & Bowman, C. A., (2005). *Understanding Disability: Inclusion, Access, Diversity, and Civil Rights*. Greenwood Publishing Group.

Jain, S., & Lobo, R., (2012). Diversity and inclusion: A business imperative in global professional services. In: *Globalization of Professional Services* (pp. 181–187). Springer Berlin Heidelberg, (2012).

Peter, C., (2015). *Employer Engagement Strategy Office of Disability Employment Policy*. https://www.dol.gov/odep/pdf/20150201EESFinalReport.pdf (accessed on 24 February 2020).

Schartz, H. A., Hendricks, D. J., & Blanck, P., (2006). Workplace accommodations: Evidence-based outcomes. *Work, 27*(4), 345–354.

Schneider, A. L., & Ingram, H. M., (2005). *Deserving and Entitled: Social Constructions and Public Policy*. SUNY Press.

Siperstein, G. N., Romano, N., Mohler, A., & Parker, R., (2006). A national survey of consumer attitudes towards companies that hire people with disabilities. *Journal of Vocational Rehabilitation, 24*(1), 3–9.

Suter, R., Scott-Parker, S., & Zadek, S., (2007). Realizing potential: Disability confidence builds better business. *GLADNET Collection*, p. 463.

Thomson, R. G., (2017). *Extraordinary Bodies: Figuring Physical Disability in American Culture and Literature*. Columbia University Press.

Williams, K. Y., & O'Reilly, III, C. A., (1998). *Demography and Research in Organizational Behavior, 20*, 77–140.

FURTHER READING

Acharya, A., & Gupta, M., (2016a). An application of brand personality to green consumers: A thematic analysis. *Qualitative Report, 21*(8), 1531–1545.

Acharya, A., & Gupta, M., (2016b). Self-image enhancement through branded accessories among youths: A phenomenological study in India. *Qualitative Report, 21*(7), 1203–1215.

Ahamed, N. U., Sundaraj, K., Ahmad, B., Rahman, M., Ali, M. A., Islam, M. A., & Palaniappan, R., (2013). Rehabilitation systems for physically disabled patients: A brief review of sensor-based computerized signal-monitoring systems. *Biomedical Research (India), 24*(3), 370–376.

Ahmad, M., (2013). Health care access and barriers for the physically disabled in rural Punjab, Pakistan. *International Journal of Sociology and Social Policy, 33*(3), 246–260.

Alingh, R. A., Hoekstra, F., Van Der Schans, C. P., Hettinga, F. J., Dekker, R., & Van Der Woude, L. H. V., (2015). Protocol of a longitudinal cohort study on physical activity behavior in physically disabled patients participating in a rehabilitation counseling programme: ReSpAct. *BMJ Open, 5*(1), e007591.

Alzahrani, M. S., Kammoun, J. S., Ali, M. S., & Ben-Abdallah, H., (2018). Watchful-eye: A 3D skeleton-based system for fall detection of physically-disabled cane users. *Lecture Notes of the Institute for Computer Sciences, Social-Informatics and Telecommunications Engineering, LNICST, 247*, 107–116.

Aragão, J. S., De França, I. S. X., Coura, A. S., Medeiros, C. C. M., & Enders, B. C., (2016). Vulnerability associated with sexually transmitted infections in physically disabled people. *Ciencia e Saude Coletiva, 21*(10), 3143–3152.

Aykara, A., (2015). Factors affecting social adaptation of physically disabled students during inclusive education, rights-based approach, and school social work. *Turkish Online Journal of Educational Technology, 2*(2), 204–207.

Balaji, L., Nishanthini, G., & Dhanalakshmi, A., (2015). Smart phone accelerometer sensor based wireless robot for physically disabled people. *2014 IEEE International Conference on Computational Intelligence and Computing Research, IEEE ICCIC 2014*, pp. 1–4.

Banik, P. P., Azam, M. K., Mondal, C., & Rahman, M. A., (2015). Single channel electrooculography based human-computer interface for physically disabled persons. *2nd International Conference on Electrical Engineering and Information and Communication Technology, ICEEICT 2015*, pp. 1–6.

Blutinger, J. C., (2018). The fate of mentally and physically disabled children in Nazi Germany. *Plight and Fate of Children During and Following Genocide, 10*, 105–121.

Bombom, L. S., & Abdullahi, I., (2016). Travel patterns and challenges of physically disabled persons in Nigeria. *Geo Journal, 81*(4), 519–533.

Boon, E. T., (2015). Everybody is a musician, everybody is an orchestra: Musical and bodily dialogues with physically disabled children in Turkey. *International Journal of Community Music, 8*(2), 149–161.

Borren, I., Tambs, K., Idstad, M., Ask, H., & Sundet, J. M., (2012). Psychological distress and subjective well-being in partners of somatically ill or physically disabled: The Nord-Trøndelag health study. *Scandinavian Journal of Psychology, 53*(6), 475–482.

Botticello, A. L., Chen, Y., & Tulsky, D. S., (2012). Geographic variation in participation for physically disabled adults: The contribution of area economic factors to employment after spinal cord injury. *Social Science and Medicine, 75*(8), 1505–1513.

Bouabdellah, L., Kharbache, H., Mokdad, M., & Mebarki, B., (2018). Fitting school buildings to the requirements of physically disabled students in Algeria: An ergonomic study. *Advances in Intelligent Systems and Computing, 585*, 470–478.

Bozkir, Ç., Özer, A., & Pehlivan, E., (2016). Prevalence of obesity and affecting factors in physically disabled adults living in the city center of Malatya. *BMJ Open, 6*(9), e010289.

Breugelmans, J., & Lin, Y., (2011). Biosensor based video game control for physically disabled gamers. *ASME 2011 Dynamic Systems and Control Conference and Bath/ASME Symposium on Fluid Power and Motion Control, DSCC 2011, 2*, 65–71.

Bualar, T., (2012). Physically disabled women and social acceptance in non- disabled community: Evidence from rural Thailand. *European Journal of Social Sciences, 29*(3), 366–376.

Bualar, T., (2014). Barriers to employment: Voices of physically disabled rural women in Thailand. *Community, Work and Family, 17*(2), 181–199.

Buller, M. K., Andersen, P. A., Bettinghaus, E. P., Liu, X., Slater, M. D., Henry, K., Fluharty, L., Fullmer, S., & Buller, D. B., (2018). Randomized trial evaluating targeted photographic health communication messages in three stigmatized populations: Physically-disabled, senior, and overweight/obese individuals. *Journal of Health Communication, 23*(43749), 886–898.

Chattoraj, S., Vishwakarma, K., & Paul, T., (2017). Assistive system for physically disabled people using gesture recognition. *2017 IEEE 2nd International Conference on Signal and Image Processing, ICSIP 2017*, pp. 60–65.

Collins, B., & O'Mahony, P., (2015). Physically disabled adults' perceptions of personal autonomy: Impact on occupational engagement. *OTJR Occupation, Participation and Health, 35*(3), 160–168.

Cornaton, J., Schweizer, A., Ferez, S., & Bancel, N., (2018). The divisive origins of sports for physically disabled people in Switzerland (1956–1968). *Sport in Society, 21*(4), 591–609.

Cornell, C., (2012). Health care disparities: Not just for the physically disabled. *Narrative Inquiry in Bioethics, 2*(3), 163–165.

Culduz, A., Şencan, N., İğnec, D., Kaspar, Ç., Özgören, R., & Çelİk, E., (2014). Opinion of pharmacists towards accessibility of physically disabled people to pharmacy [Fiziksel engelli hastaların eczane ve eczacılık hizmetlerine erişimi]. *Marmara Pharmaceutical Journal, 18*(2), 73–78.

Da Silva, G. A. I., Dos Santos, V. P., Mainenti, M. R. M., De Figueiredo, F. M., Ribeiro, B. G., & De Abreu, S. E., (2014). Basal and resting metabolic rates of physically disabled adult subjects: A systematic review of controlled cross-sectional studies. *Annals of Nutrition and Metabolism, 65*(4), 243–252.

De Mesquita-Guimarães, K. S. F., Ferreira, D. C. A., Da Silva, R. A. B., Díaz-Serrano, K. V., De Queiroz, A. M., Mantovani, C. P. T., & De Rossi, A., (2016). Development of an intraoral device for social inclusion of a physically disabled patient. *Special Care in Dentistry, 36*(1), 53–56.

Ellappan, V., Upadhayay, D., & Yadav, R., (2016). Vandana hand gesture recognition for physically disabled people. *International Journal of Pharmacy and Technology, 8*(4), 25640–25647.

Enajeh, S. M. A., Cavus, N., & Ibrahim, D., (2018). Development of a voice recognition based system to help physically disabled people use the Facebook. *Quality and Quantity, 52*, 1343–1352.

Endress, P., (2011). Analysis of a development assistance project organized by a local association of physically disabled people in Ouahigouya: The supportive Handirace [Analyse d'un projet d'aide au développement organisé par une association locale d'handicapés moteurs de Ouahigouya: la Handicourse solidaire]. *Staps, 93*(3), 7–23.

Faroom, S., Ali, M. N., Yousaf, S., & Deen, S. U., (2018). Literature review on home automation system for physically disabled peoples. *2018 International Conference on Computing, Mathematics and Engineering Technologies: Invent, Innovate and Integrate for Socioeconomic Development, iCoMET 2018 - Proceedings*, pp. 1–5.

Ferez, S., Ruffié, S., Issanchou, D., & Cornaton, J., (2018). The manager, the doctor, and the technician: political recognition and institutionalization of sport for the physically disabled in France (1968–1973). *Sport in Society, 21*(4), 622–634.

Finnvold, J. E., (2018). Will my child ever go to a university? The link between school segregation practices and Norwegian parents' expectations for their physically disabled child. *Journal of Research in Special Educational Needs, 18*(2), 103–113.

Gaba, G. S., Singh, P., & Arora, S. K., (2016). Electronic wheelchair for physically disabled persons. *Indian Journal of Science and Technology, 9*(21).

Graf, P., (2015). The Educational Center HTL HAK Ungargasse (SZU): Competence center for integration of physically disabled and sensory impaired youths [Schulzentrum HTL HAK Ungargasse (SZU): Kompetenzzentrum für Integration von körper- und sinnesbehinderten Jugendlichen]. *Elektrotechnik und Informationstechnik, 132*(6), 329–331.

Gupta, A., Chitranshi, G., Sharma, V., & Singh, H., (2017). Controlling devices using facial expression identification and recognition for assisting physically disabled. *Proceedings of the 7th International Conference Confluence 2017 on Cloud Computing, Data Science and Engineering*, pp. 574–576.

Gupta, M., & Kumar, Y., (2015). Justice and employee engagement: Examining the mediating role of trust in Indian B-schools. *Asia-Pacific Journal of Business Administration, 7*(1), 89–103.

Gupta, M., & Pandey, J., (2018). Impact of student engagement on affective learning: Evidence from a large Indian university. *Current Psychology, 37*(1), 414–421.

Gupta, M., & Ravindranath, S., (2018). Managing physically challenged workers at micro sign. *South Asian Journal of Business and Management Cases, 7*(1), 34–40.

Gupta, M., & Sayeed, O., (2016). Social responsibility and commitment in management institutes: Mediation by engagement. *Business: Theory and Practice, 17*(3), 280–287.

Gupta, M., & Shaheen, M., (2017a). Impact of work engagement on turnover intention: Moderation by psychological capital in India. *Business: Theory and Practice, 18*, 136–143.

Gupta, M., & Shaheen, M., (2017b). The relationship between psychological capital and turnover intention: Work engagement as mediator and work experience as moderator. *Journal Pengurusan, 49*, 117–126.

Gupta, M., & Shaheen, M., (2018). Does work engagement enhance general well-being and control at work? Mediating role of psychological capital. *Evidence-Based HRM, 6*(3), 272–286.

Gupta, M., & Shukla, K., (2018). An empirical clarification on the assessment of engagement at work. *Advances in Developing Human Resources, 20*(1), 44–57.

Gupta, M., (2017). Corporate social responsibility, employee-company identification, and organizational commitment: Mediation by employee engagement. *Current Psychology, 36*(1), 101–109.

Gupta, M., (2018). Engaging employees at work: Insights from India. *Advances in Developing Human Resources, 20*(1), 3–10.

Gupta, M., Acharya, A., & Gupta, R., (2015). Impact of work engagement on performance in Indian higher education system. *Review of European Studies, 7*(3), 192–201.

Gupta, M., Ganguli, S., & Ponnam, A., (2015). Factors affecting employee engagement in India: A study on off shoring of financial services. *Qualitative Report, 20*(4), 498–515.

Gupta, M., Ravindranath, S., & Kumar, Y. L. N., (2018). Voicing concerns for greater engagement: Does a supervisor's job insecurity and organizational culture matter? *Evidence-Based HRM, 6*(1), 54–65.

Gupta, M., Shaheen, M., & Das, M. (2019). Engaging employees for quality of life: mediation by psychological capital. *The Service Industries Journal, 39*(5–6), 403–419.

Gupta, M., Shaheen, M., & Reddy, P. K., (2017). Impact of psychological capital on organizational citizenship behavior: Mediation by work engagement. *Journal of Management Development, 36*(7), 973–983.

Gupta, S., (2017). Rights of physically disabled persons: An inclusive approach. *Marginalities in India: Themes and Perspectives* (pp. 247–260).

Hantke, S., Sagha, H., Cummins, N., & Schuller, B., (2017). Emotional speech of mentally and physically disabled individuals: Introducing the EmotAsS database and first findings. *Proceedings of the Annual Conference of the International Speech Communication Association, Interspeech*, pp. 3137–3141.

Hong, Y., Bruniaux, P., Zeng, X., & Curteza, A., (2016). Simulation based customized garment design for the physically disabled people of scoliosis type. *14th International Industrial Simulation Conference, ISC 2016*, pp. 50–56.

Hong, Y., Bruniaux, P., Zeng, X., Liu, K., Curteza, A., & Chen, Y., (2018). Visual-simulation-based personalized garment block design method for physically disabled people with scoliosis (PDPS). *Autex Research Journal, 18*(1), 35–45.

Hong, Y., Bruniaux, P., Zhang, J., Liu, K., Dong, M., & Chen, Y., (2018). Application of 3D-TO-2D garment design for atypical morphology: A design case for physically disabled people with scoliosis. *Industria Textila, 69*(1), 59–64.

Hong, Y., Zeng, X., Bruniaux, P., Curteza, A., & Chen, Y., (2018). Movement analysis and ergonomic garment opening design of garment block patterns for physically disabled people with scoliosis using fuzzy logic. *Advances in Intelligent Systems and Computing, 590*, 303–314.

Hong, Y., Zeng, X., Bruniaux, P., Curteza, A., Stelian, M., & Chen, Y., (2018). Garment opening position evaluation using kinesiological analysis of dressing activities: Case study of physically disabled people with scoliosis (PDPS). *Textile Research Journal, 88*(20), 2303–2318.

Ikeda, T., Nagai, T., Kato-Nishimura, K., Mohri, I., & Taniike, M., (2012). Sleep problems in physically disabled children and burden on caregivers. *Brain and Development, 34*(3), 223–229.

Iqbal, M. A., Asrafuzzaman, S. K., Arifin, M. M., & Hossain, S. K. A., (2016). Smart home appliance control system for physically disabled people using kinect and X10. *2016 5th International Conference on Informatics, Electronics and Vision, ICIEV 2016*, pp. 891–895.

Izsó, L., Székely, I., & Dános, L., (2015). Possibilities of the ErgoScope high fidelity work simulator in skill assessment, skill development, and vocational aptitude tests of physically disabled persons. *Studies in Health Technology and Informatics, 217*, 825–831.

Jaeger, L., Kroenung, J., & Kupetz, A., (2013). Me vs. cyber-me-analyzing the effects of perceived stigma of physically disabled people on the disguise of the real self in virtual environments. *International Conference on Information Systems (ICIS 2013): Reshaping Society Through Information Systems Design, 2*, 1004–1022.

Jaffer, S., & Ma, L., (2014). Preschoolers show less trust in physically disabled or obese informants. *Frontiers in Psychology, 5.*
Jai, G. J., Subha, A. T. V., & Krishna, M. N. R., (2015). Design of a detection system for the tracking of physically disabled humans. *International Journal of Applied Engineering Research, 10*(2), 3937–3945.
Jamain-Samson, S., (2014). The world games for the physically disabled of Saint-Etienne (1970–1975): A testimony to an unavailable split within the young French Sports Federation for the disabled [Les Jeux mondiaux pour handicapés physiques de Saint-Étienne (1970–1975), témoins d'une inéluctable scission au sein du jeune mouvement handisport français]. *European Studies in Sports History, 7,* 169–196.
Jumma, B. J., & Çerkez, Y., (2017). Parents' mistreatment towards physically disabled children. *International Journal of Economic Perspectives, 11*(1), 527–532.
Kamei, S., Ohashi, M., & Hori, M., (2015). Consideration of embodied expertise transfer process in manufacturing-site-through the development example of wheelchair for the physically disabled users. *Procedia Computer Science, 64,* 125–131.
Kassah, B. L. L., Kassah, A. K., & Agbota, T. K., (2014). Abuse of physically disabled women in Ghana: Its emotional consequences and coping strategies. *Disability and Rehabilitation, 36*(8), 665–671.
Kim, M., (2016). The effect of filial therapy program on physically disabled mothers' self-esteem and parenting stress. *Advanced Science Letters, 22*(11), 3586–3588.
Klaphajone, J., Thaikruea, L., Boontrakulpoontawee, M., Vivatwongwana, P., Kanongnuch, S., & Tantong, A., (2013). Assessment of music therapy for rehabilitation among physically disabled people in Chiang Mai province: A pilot study. *Music and Medicine, 5*(1), 23–30.
Knight, K. H., Porcellato, L., & Tume, L., (2014). Out-of-school lives of physically disabled children and young people in the United Kingdom: A qualitative literature review. *Journal of Child Health Care, 18*(3), 275–285.
Kraus, T., Singer, G., Wegmann, H., Tschauner, S., Svehlik, M., Steinwender, G., & Sorantin, E., (2014). Injuries in physically disabled children. *Seminars in Musculoskeletal Radiology, 18*(5), 513–522.
Krops, L. A., Dekker, R., Geertzen, J. H. B., & Dijkstra, P. U., (2018). Development of an intervention to stimulate physical activity in hard-to-reach physically disabled people and design of a pilot implementation: An intervention mapping approach. *BMJ Open, 8*(3).
Krops, L. A., Folkertsma, N., Hols, D. H. J., Geertzen, J. H. B., Dijkstra, P. U., & Dekker, R., (2018). Target population's requirements on a community-based intervention for stimulating physical activity in hard-to-reach physically disabled people: An interview study. *Disability and Rehabilitation,* pp. 1–8.
Krops, L. A., Hols, D. H. J., Folkertsma, N., Dijkstra, P. U., Geertzen, J. H. B., & Dekker, R., (2018). Requirements on a community-based intervention for stimulating physical activity in physically disabled people: A focus group study amongst experts. *Disability and Rehabilitation, 40*(20), 2400–2407.
Lagalle, M., Ruet, A., Villart, M., Azouvi, P., & Michelon, H., (2015). Use of psychotropic drugs in physically disabled patients: One-shot prevalence and medical practice assessment in a physical and rehabilitation medicine ward. *Annals of Physical and Rehabilitation Medicine, 58*(6), 357–358.
Lappeteläinen, A., Sevón, E., & Vehkakoski, T., (2017). Forbidden option or planned decision? Physically disabled women's narratives on the choice of motherhood. *Scandinavian Journal of Disability Research, 19*(2), 140–150.

Lappeteläinen, A., Sevón, E., & Vehkakoski, T., (2018). 'Celebrating diverse motherhood': Physically disabled women's counter-narratives to their stigmatized identity as mothers. *Families, Relationships and Societies*, 7(3), 499–514.

Lawang, W., Horey, D., Blackford, J., Sunsern, R., & Riewpaiboon, W., (2013). Support interventions for caregivers of physically disabled adults: A systematic review. *Nursing and Health Sciences*, 15(4), 534–545.

Lee, Y., & Kim, M., (2015). The effect of filial therapy on physically disabled mothers' empathy ability and their non-disabled children's behavior problems. *Indian Journal of Science and Technology*, 8(25).

Lee, Y., Park, J., & Jang, M., (2018). Perceived effects of home renovation on independence of physically disabled Koreans living at home. *Disability and Rehabilitation*, 40(20), 2380–2387.

Linse, K., Aust, E., Joos, M., & Hermann, A., (2018). Communication matters-pitfalls and promise of high-tech communication devices in palliative care of severely physically disabled patients with amyotrophic lateral sclerosis. *Frontiers in Neurology*, 9.

Luo, J., Guo, H., & Wang, J., (2013). Demand and guarantee of internet use for physically disabled children in Wenchuan Earthquake. *Asian Social Science*, 9(3), 112–119.

Malik, M., Bilal, F., Ali, K. M. S., Jabeen, F., Fatima, D. S., & Munir, N., (2013). Quality of life and its relationship with demographic variables among physically disabled patients with artificial limb. *Rawal Medical Journal*, 38(2), 134–138.

Martins, C. P., Mendes, A. K., & Cardoso, F. L., (2011). Capacity of adaptation and sexual esteem in physically disabled athletes [Capacidade de Adaptação e estima sexual em atletas deficientes físicos]. *Revista da Educacao Fisica*, 22(4), 547–554.

Mathus-Vliegen, E. M. H., Karin Van Der Vliet, R. N., Inge, J., Wignand-Van Der Storm, R. N., John, S., & Stadwijk, R. N., (2018). Efficacy and safety of sodium picosulfate/magnesium citrate for bowel preparation in a physically disabled outpatient population: A randomized, endoscopist-blinded comparison with ascorbic acid-enriched polyethylene glycol solution plus bisacodyl (The PICO-MOVI Study). *Diseases of the Colon and Rectum*, 61(2), 239–249.

Mirin, S. N. S., Annuar, K. A. M., & Yook, C. P., (2018). Smart wheelchair using android smartphone for physically disabled people. *International Journal of Engineering and Technology (UAE)*, 7(2), 453–457.

Mirza, R., Tehseen, A., & Joshi, K. A. V., (2012). An indoor navigation approach to aid the physically disabled people. *2012 International Conference on Computing, Electronics and Electrical Technologies, ICCEET 2012*, pp. 979–983.

Mitchell, W., Beresford, B., Brooks, J., Moran, N., & Glendinning, C., (2017). Taking on choice and control in personal care and support: The experiences of physically disabled young adults. *Journal of Social Work*, 17(4), 413–433.

Moradi, A., (2013). Share of self-efficacy, achievement motivation, and self-esteem in predicting entrepreneurial behavior in physically disabled females. *Advances in Environmental Biology*, 7(8), 1795–1803.

Mouget, A. C., (2016). Recreational sexuality of physically disabled men [Sexualité récréative des hommes handicapés moteurs]. *Dialogue*, 212(2), 65–78.

Nagarajan, R., Hariharan, M., & Satiyan, M., (2012). Luminance sticker based facial expression recognition using discrete wavelet transform for physically disabled persons. *Journal of Medical Systems*, 36(4), 2225–2234.

Nasor, M., Rahman, K. K. M., Zubair, M. M., Ansari, H., & Mohamed, F., (2018). Eye-controlled mouse cursor for physically disabled individual. *Advances in Science and Engineering Technology International Conferences, ASET 2018*, pp. 1–4.

Niedbalski, J., (2015). Transformation in the perception of their own bodies among physically disabled persons practicing sport [Przemiany Percepcji Wlasnego Ciala Przez Osoby Z Niepelnosprawnoścía Fizyczna Uprawiajace Sport]. *Studia Socjologiczne, 3*(218), 221–240.

Noda, T., & Kamata, H., (2012). Disabilities and lifestyle of junior and senior high school students attending special-needs schools for the physically disabled. *Kitakanto Medical Journal, 62*(3), 261–270.

Nowshin, N., Rashid, M. M., Akhtar, T., & Akhtar, N., (2019). Infrared sensor controlled wheelchair for physically disabled people. *Advances in Intelligent Systems and Computing, 881*(2), 847–855.

Ohmine, S., Kimura, Y., Saeki, S., & Hachisuka, K., (2012). Community-based survey of amputation derived from the physically disabled person's certification in Kitakyushu City, Japan. *Prosthetics and Orthotics International, 36*(2), 196–202.

Oku, H., Matsubara, K., & Booka, M., (2015). Usability of PDF based digital textbooks to the physically disabled university student. *Studies in Health Technology and Informatics, 217*, 3–10.

Öznacar, B., & Erdağ, D., (2018). Physical education and sports education candidate students' awareness and knowledge status about physical education lessons designed for physically disabled individuals. *Quality and Quantity, 52*, 1365–1370.

Pandey, J., Gupta, M., & Naqvi, F., (2016). Developing decision making measure a mixed method approach to operationalize Sankhya philosophy. *European Journal of Science and Theology, 12*(2), 177–189.

Papasotiriou, M., & Windle, J., (2016). The social experience of physically disabled Australian university students. *Disability and Society, 27*(7), 935–947.

Payne, D. A., Hickey, H., Nelson, A., Rees, K., Bollinger, H., & Hartley, S., (2016). Physically disabled women and sexual identity: A photovoice study. *Disability and Society, 31*(8), 1030–1049.

Phua, K. L., Chong, J. C., Elangovan, R., Liew, Y. X., Ng, H. M., & Seow, Y. W., (2014). Public and private hospitals in Kuala Lumpur and Selangor, Malaysia: How do they fare in terms of accessibility for the physically disabled? *Malaysian Journal of Medicine and Health Sciences, 10*(1), 43–50.

Ravikumar, C. P., & Dathi, M., (2016). A fuzzy-logic based Morse code entry system with a touch-pad interface for physically disabled persons. *2016 IEEE Annual India Conference, INDICON 2016.*

Rostron, J., (2018). Planning and housing the physically disabled. *Journal of Planning and Environment Law,* 1200–1206.

Rostron, J., (2019). Development of the law concerning property and accessibility for the physically disabled. *Journal of Planning and Environment Law,* 126–133.

Satyanarayana, P., Sai Prajwal, K., Chandra, N. V. T., Sri Manojna, E., & Sitara, S., (2016). Advanced motion tracking based mobility assistance for physically disabled. *ARPN Journal of Engineering and Applied Sciences, 11*(15), 9545–9552.

Scott J. A., (2012a). "Cripped" heroes: An analysis of physically disabled professionals' personal narratives of performance of identity. *Southern Communication Journal, 77*(4), 307–328.

Scott, J. A., (2011). Attending to the disembodied character in research on professional narratives: How the performance analysis of physically disabled professionals' personal stories provides insight into the role of the body in narratives of professional identity. *Narrative Inquiry, 21*(2), 238–257.

Scott, J. A., (2012b). Stories of hyper embodiment: An analysis of personal narratives of and through physically disabled bodies. *Text and Performance Quarterly, 32*(2), 100–120.

Scott, J. A., (2015). Almost passing: A performance analysis of personal narratives of physically disabled femininity. *Women's Studies in Communication, 38*(2), 227–249.

Senghor, D. B., Diop, O., & Sombié, I., (2017). Analysis of the impact of healthcare support initiatives for physically disabled people on their access to care in the city of Saint-Louis, Senegal. *BMC Health Services Research, 17*.

Shamsul, A. S., Mohd, R., H., Muholan, K., Noor, Z. H., Ang, W. C., Sei, F. S., et al., (2013). Quality of life and its influencing factors among physically disabled teenagers in Kuala Lumpur, Malaysia. *Malaysian Journal of Public Health Medicine, 13*(2).

Shaw, R. B., & Ginis, K. A. M., (2017). Physical activity among physically disabled populations. *Physical Activity in Diverse Populations: Evidence and Practice*, pp. 226–243.

Shimada, M., (2014). The dental care project for the severely mentally and physically disabled persons under general anesthesia in Okinawa prefecture and its future course. *Journal of Japanese Dental Society of Anesthesiology, 42*(2), 187–189.

Shirahama, N., Sakuragi, Y., Watanabe, S., Nakaya, N., Mori, Y., & Miyamoto, K., (2014). Development of input assistance application for mobile devices for physically disabled. *2014 IEEE/ACIS 15th International Conference on Software Engineering, Artificial Intelligence, Networking and Parallel/Distributed Computing, SNPD 2014 - Proceedings*.

Silva, A. L. A., Reise, S. A. N., De Jesus Viana Sá, E., & Da Silva, T. R., (2019). Game for the digital inclusion of the physically disabled with reduced mobility. *Advances in Intelligent Systems and Computing, 776*, 98–104.

Singh, A., & Kaur, A., (2016). Case study of touch technology Used for teaching physically disabled students. *Proceedings of the 2015 IEEE 3rd International Conference on MOOCs, Innovation and Technology in Education, MITE 2015*, pp. 392–395.

Siva, K. B. G., & Sudhagar, K., (2016). Development of stair climbing intelligent wheelchair for physically disabled people. *International Journal of Control Theory and Applications, 9*(5), 2717–2725.

Skučas, K., (2013). The influence of water therapy on the psychological and physical health of physically disabled people. *Specialusis Ugdymas, 29*(2), 189–193.

Soman, S., & Murthy, B. K., (2015). Using brain-computer interface for synthesized speech communication for the physically disabled. *Procedia Computer Science, 46*, 292–298.

Soto, P. A. B., & Pérez, C. L., (2014). Design and validation of a body image questionnaire for physically disabled persons [Diseño y Validación de un Cuestionario de Imagen Corporal para Personas en Situación de Discapacidad Física]. *Revista Colombiana de Psicologia, 24*(1), 219–233.

Suhas, K. S., Dhal, S., Shankar, P. V., Hugar, S. H., & Tejas, C., (2018). A controllable home environment for the physically disabled uses the principles of BCI. *2018 9th International Conference on Computing, Communication and Networking Technologies, ICCCNT 2018*.

Tai, Y. H., Tian, Y. J., Huang, T. W., & Sun, K. T., (2013). Brainwave technology gives internet access to the physically disabled. *Proceedings - 2013 4th Global Congress on Intelligent Systems, GCIS 2013*, pp. 331–335.

Thomann, G., Magnier, C., Villeneuve, F., & Palluel-Germain, R., (2016). Designing for physically disabled users: benefits from human motion capture: A case study. *Disability and Rehabilitation: Assistive Technology, 11*(8), 695–700.

Tonak, H. A., Kitis, A., & Zencir, M., (2016). Analysis of community participation levels of individuals who are physically disabled and working in industrial environments. *Social Work in Public Health, 31*(7), 638–645.

Van Amsterdam, N., Knoppers, A., & Jongmans, M., (2015). 'It's actually very normal that I'm different'. How physically disabled youth discursively construct and position their body/self. *Sport, Education and Society, 20*(2), 152–170.

Van Der Putten, G. J., Brand, H. S., De Visschere, L. M. J., Schols J. M. G. A., & De Baat, C., (2013). Saliva secretion rate and acidity in a group of physically disabled older care home residents. *Odontology, 101*(1), 108–115.

Var, T., Yeşiltaş, M., Yayli, A., & Öztürk, Y., (2011). A study on the travel patterns of physically disabled people. *Asia Pacific Journal of Tourism Research, 16*(6), 599–618.

Vasanthan, M., Murugappan, M., Nagarajan, R., Ilias, B., & Letchumikanth, J., (2012). Facial expression based computer cursor control system for assisting physically disabled person. *Proceeding - COMNETSAT 2012: 2012 IEEE International Conference on Communication, Networks and Satellite*, pp. 172–176.

Villanueva-Flores, M., Valle-Cabrera, R., & Ramón-Jerónimo, M. A., (2015). Perceived compensation discrimination against physically disabled people in Andalusia. *International Journal of Human Resource Management, 26*(17), 2248–2265.

Wu, L., & Akgunduz, A., (2017). Bio-signal-based geometric modeling application for physically disabled users. *Journal of Intelligent Manufacturing, 28*(7), 1667–1678.

Yen, S. M., Kung, P. T., & Tsai, W. C., (2014). Factors associated with free adult preventive health care utilization among physically disabled people in Taiwan: Nationwide population-based study. *BMC Health Services Research, 14*(1).

Younesi, S., Parsian, H., Hosseini, S. R., Noreddini, H., Mosapour, A., Bijani, A., & Halalkhor, S., (2015). Dyshomeostasis of serum oxidant/antioxidant status and copper, zinc, and selenium levels in elderly physically disabled persons: An AHAP-based study. *Biological Trace Element Research, 166*(2), 136–141.

Zafani, M. D., Omote, S., & Baleotti, L. R., (2015). Observation protocol of physically disabled children's performance: Construction, application and data analysis [Protocolo de Observação do Desempenho de Crianças com Deficiência Física: Construção, Aplicação e Análise de Dados]. *Revista Brasileira de Educacao Especial, 21*(1), 23–38.

CHAPTER 10

Engaging Different Income Classes at Work

ANITHA ACHARYA

Marketing and Strategy Department, ICFAI Business School (IBS), Hyderabad, The ICFAI Foundation for Higher Education (IFHE) (Deemed to be University U/S 3 of the UGC Act 1956), Hyderabad, India, Tel.: (+91) 8712290557, E-mail: anitha.acharya@ibsindia.org

ABSTRACT

Income is defined as the salary/wage that is paid to employees. The job requirements for each level differ in terms of skills, education, and prior work experience. The employees at each level would be promoted to the next depending on their performance and the number of years they have served the organization. The income for each level differs. The chapter illustrates how different income slabs among the employees of an organization have an impact on engagement. It also describes how the income is decided by the organization. The chapter also lists various methods to engage employees who have different income slabs.

10.1 INTRODUCTION

One apparently unrelated workplace trend which could have a substantial conjunctive impact on management is the challenge of engaging employees in organizations (Halbsleben, 2011; May et al., 2004; Pech and Slade, 2006; Gupta, 2017, 2018). According to Kahn, engagement refers to harnessing the members of organization selves to their work roles (1990). In spite of its apparent conceptual overlap with existing constructs such as job involvement and organizational commitment, evidence suggests that engagement is a distinct construct (Hallberg and Schaufeli, 2006). According to a survey of 656 Chief Executive Officers hailing from different countries around the world which was carried out by Wah (1999), the results revealed that engaging employees is the fourth most important management challenge, behind creating customer loyalty, supervising mergers and acquisitions, and reducing the operating cost. Additionally illustrating the scale of this challenge, the Gallup Organization recently found that nearly 20% of U.S. employees were disengaged and in addition, 54% were in effect neutral about their work. The impact of these one trend—the growing challenge of engagement—might prove challenging for many employers.

There is a multi-tiered hierarchical structure of job levels, within the multifaceted system of a traditional business organization. The job levels can range from 'C' level officers to entry-level clerk. The success of any business depends on how employees at different levels contribute towards the smooth operation of the company. The chain of command of job levels is like a vertical pyramid where most of the information and power are with the top-level executives. A lot of organizations is moving toward a few-layered structure with a strong emphasis on employee engagement and collaboration. Yet in these flatter organizations, there are at least five job levels which provide structure to the organization. The five levels are termed as: (a) Entry Level; (b) Experienced Level or Intermediate; (c) First-Level Management; (d) Middle-Level Management; and (e) Executive or Top-Level Management and Chiefs.

The income within each level also differs due to the salary spread. The amount of range spread which refers to the distance from the minimum range to the maximum range, also increase as employees move up higher in the grade structure. At the entry levels, the employees are typically unskilled; in that case, the amount of range spread can be as low as 15% plus or minus the range midpoint. The amount of disparity in work behavior is small at entry-level jobs. The range spread increases at higher job levels and it would range between 20 to 30% plus or minus the range midpoint for middle and senior-level management roles.

In the case of mid-size and large organizations, the compensation systems provide a pay schedule grades which reflect the job complexity. However, in the case of small organizations, the income is not fixed and the employees have the option of negotiating. Designing an overall pay structure for the employees is one of those core building blocks for a human resource (HR) manager, but even the most veteran compensation specialist might want to have an eye on the viable background to authenticate their approach. The common theme for any organization for formulating the compensation structure is the alignment with the organization's compensation philosophy and human capital strategy. Implementing a compensation structure is a significant undertaking that can have lasting impacts.

The primary objective of an organization's compensation function is to help recruit, retain, and engage employees who are talented and required by the organization. Nevertheless, they should also ensure that there are adequate management controls, not just cost controls for planning and organizing, but also cost controls. If the employees are not compensated properly then the HR manager finds himself in the conundrum of frustrated employees who feel that their work has not been: (a) represented properly; (b) properly appreciated; and (c) evaluated properly. This places the HR manager in an uncomfortable position and organizations might find it difficult to engage these employees.

10.2 NEED FOR DISCUSSING THIS THEME

The term employee engagement is catching the attention of both academicians and industry. Prior studies have indicated that employee engagement predicts the outcomes of the employees, helps organizations to improve their performance, and also helps them to be successful in their business (Bakker et al., 2011; Harter et al., 2002; Richman, 2008). Even though there are lots of benefits of employee engagement it has also been observed that employee engagement is on the decline and today there is a deepening disengagement among employees (May et al., 2004; Rynes et al., 2004; Richman, 2008; Genicot and Ray, 2017). It has even been reported that the majority of employees today, are not fully engaged or they are disengaged leading to what has been referred to as an engagement gap that is costing organizations more than $300 billion a year in lost productivity (Johnson, 2004). The purpose of this chapter is to state the various methods which the organizations can adopt to ensure that the employees are engaged irrespective of inequality in income.

10.3 REVIEW OF THIS THEME

According to Robinson et al. the term employee engagement is very popular (2004). Nevertheless, most of what has been written on the topic of employee engagement is originated only in practitioner journals where it has its basis in practice instead of empirical research and theory (Robinson et al., 2004). In addition, employee engagement has been defined in many different ways and the definitions and measures are often similar to other constructs like organizational citizenship behavior and organizational commitment (Saks, 2006). Authors like Richman (2006) and Shaw and Bastock (2005) have defined it as emotional and intellectual commitment to the organization. According to Frank et al. employee engagement refers to the amount of voluntary effort showed by employees towards their jobs (2004). In the academic literature, a number of definitions have been provided. Kahn (1990, p. 694) defines personal engagement as "the harnessing of organization members' selves to their work roles; in engagement, people employ and express themselves physically, cognitively, and emotionally during role performances." Personal disengagement refers to "the uncoupling of selves from work roles; in disengagement, people withdraw and defend themselves physically, cognitively, or emotionally during role performances" (p. 694).

According to Kahn, engagement refers to be psychologically present when occupying and performing an organizational role (1990). Rothbard defined engagement as psychological presence consisting of two components namely absorption and attention (2001). Attention was defined as cognitive availability and the amount of time one spends thinking about a role. Absorption meant being engrossed in a role and refers to the intensity of one's focus on a role. Researchers like Maslach et al. in burnout define engagement as the positive or opposite antithesis of burnout (2001).

According to Maslach et al. the components of engagement are energy, involvement, and efficacy (2001), the direct opposite of the three burnout dimensions of exhaustion, cynicism, and inefficacy (Bakker et al., 2014; Collins, 2017; Schaufeli et al., 2009; Hakanen et al., 2008). Research on burnout and engagement has found that the core dimensions of burnout (exhaustion and cynicism) and engagement (vigor and dedication) are opposites of each other (Bakker et al., 2008; Gonzalez-Roma et al., 2006; Rich et al., 2010). Schaufeli et al. (2002, p. 74) define engagement "as a positive, fulfilling, work-related state of mind that is characterized by vigor, dedication, and absorption." They also stated that engagement is not a momentary and specific state, but relatively, it is "a more persistent and pervasive affective-cognitive state that is not focused on any particular object, event, individual, or behavior" (p. 74).

10.3.1 BENEFITS OF EMPLOYEE ENGAGEMENT

An employee who is engaged is considered to be a productive employee. Employee engagement is an integral part of organizations operations. They are the best resource that a company can have and without that it, there could be negative impact on the operations of the organization. Some of the benefits are discussed below:

- **Profitability:** Organizations that focus on employee engagement will have committed workforce. This would result in higher profits. Higher profits will benefit the shareholders since they will be in a position to get better returns. According to Harter et al. if the employees are engaged they tend to become more productive and efficient and this will help organizations to lower their operating cost and would also result in better profit margins (2002).
- **Employee Retention:** Retaining good employees is very important for the success of the organization. Engaged employees usually don't leave the organization as they are committed to their work. This will help the organizations to attract more qualified individuals to join their organizations. The job satisfaction of the employees will be higher if they are engaged. This will help the organizations to reduce their costs which they will incur if they recruit new employees and train them.
- **Wellbeing:** Happy employees are those who are engaged, and happy employees are productive employees. A clear focus on workplace happiness will help the organizations to unlock the employee's true potential. Additionally, an engaged and happy workforce can turn out to be loyal advocates for the organization.
- **Efficiency:** Employees with the highest levels of commitment perform 20% much better in comparison to employees who are not engaged. Engaged employees will be the most dedicated and productive, which will give organizations a positive boost. Employees who are engaged with their role and who align their goals with the organization's culture are more productive as they are looking beyond personal benefits. In other words, they will work towards the overall success of the organization in mind and this will enable increased performance.
- **Innovation:** Employee engagement helps employees to be innovative. Employees who are not engaged will not have the desire to work innovatively. On the other hand, employees who are engaged will have higher level of satisfaction and interest in their role. This often leads to employees being innovative and creative. Innovation leads to better products and services and can help organizations to generate more profit.

10.3.2 INCOME AND EMPLOYEE ENGAGEMENT

Income is defined as the salary/wage which is paid to the employee's. Although there is a limited empirical work that substantially links employee engagement to income. It is therefore important to find out how different income slabs among the employees of an organization have an impact on engagement. Income and engagement are arguably closely related on the ground that avariciousness and income has a strong connection with employee intention to remain employed with an organization (Torlak and Koc, 2007; Kuznets, 1955; Horn et al., 2017). According to White, income refers to the promise of payment made against the work which is delivered by the employee (1981). In other words, employees articulate their intent to remain with the organization by taking up responsibilities which is given to them and in return expect organizations to fulfill the guarantee of passing on the guarantee into a asset which are liquid in nature and is termed as wage/salary which is paid to the employee as a consequence of the expressed involvement to work. Thus, a bond is created between the organization and the employee continued by the continuous promise of income-earning, which forces employees to be loyal to their organization (Meyer and Allen, 1991; Piketty and Saez, 2003). Income acts as an instrument for employees continued expression of commitment to their organizations. In other words, the more employees are willing to express their dedication to the organization, by implication this means the higher the income, the higher the commitment. Even though, this view may not be complete due to the different impacts of one's environment on their behavior, it nevertheless shows that income does, to some extent, persuade an employee to be engaged to their organizations (Atkinson, 1983; Piketty, 2015).

The growth in income inequality over the last few decades is one of the most extensively researched topics (Bridgman, 2017; Croft et al., 2015; Haines et al., 2016; Katz, 1999). According to prior researchers, one important part of the change in income inequality has been associated to the growth in the college high school income premium since the late 1970s (Katz and Murphy, 1992; Greenwood and Jovanovic, 1990). However, the reasons for the growth in income inequality linked to standard human capital variables like experience and education are limited and these variables contribute to only about a third of the variance of income. The most imperative contributor of income inequality is within-group wage inequality also termed as residual wage—i.e., income dispersion among employees who have the same education and experience (Juhn et al., 1993; Galor and Zeira, 1993). Understanding the sources of growth in residual wage inequality is problematic, as there are many reasons why employees with the same level of education and work experience may report varied income. Perhaps these employees have different levels of valuable but unobserved skills which are linked to, intrinsic ability, individual effort, and the quality of the educational institute from where they have received there degree. Some of the possible reasons for income inequality may be: (a) the value or return to unnoticed skills may be on the rise because of an increase in the demand for skill; (b) the distribution in unobserved skills may be growing over time. For instance, if unnoticed skills are more dispersed among more educated workers and older people, dispersion in unnoticed skills could increase because of composition effects related to the aging and increasing educational achievement of the workforce.

Studies relating to the income structure are as old as the profession of economics. Economist Adam Smith in his book titled "The Wealth of Nations" provided a complete and neat analysis of the determinants of differences in income among individuals and employments. He emphasized that income differences were determined by competitive factors like: (a) compensating differentials for differences in costs of training; (b) probability of success; (c) steadiness of work; (d) and other workplace amenities. He also stated that shifts in demand across occupations and space could generate transitory wage differentials, but that highly elastic supply responses would tend to equalize the merits and demerits of different employments over the long-run in the absence of rigid entry barriers. The gap found in Smith's analysis along with the roles of supply and demand factors and those of institutional forces in affecting income remains through today a key theme of research on the income structure.

Prior studies on the income structure examined changes and differences in wages by occupation (Douglas, 1930; Ober, 1948; Greenwood and Jovanovic, 1990), and industry (Slichter, 1950; Cullen, 1956). According to Douglas who studied the advancement of the wages of white-collar (white-collar refers to managers and clerical workers) and blue-collar (blue-collar refers to manual labor) workers in the United States from 1890 to 1926 revealed substantial decline in the income premium to white-collar work during World War I and argued that the rapid access to education had led the growth in the supply of qualified workers to outshine the growth in demand (1930). Tinbergen stated that the evolution and advancement of technology tend to increase the demand for more-educated employees (1975). The reasons for this increased research emphasis on understanding income structure changes are clear.

The wage structures of some of the Organization for Economic Co-operation and Development (OECD) nations have changed during the past few years. During 1970s, the educational and occupational wage differentials (especially the relative earnings of college graduates) lessened noticeably in almost all advanced nations. Nevertheless, there is development due to the divergent patterns in the evolution of the wage structure. Overall wage inequality and educational wage differentials have expanded greatly in the United Kingdom and the United States since the end of the 1970s (Barro, 2000; Belabed et al., 2017). A lot of effort has been mounted to comprehend these labor market changes, because widening income structure means widening family income and inequality in consumption and also the social problems associated with income inequality.

The overall income distribution is composed of: (a) differences in income between groups (typically defined by demographic or skill categories); (b) and within group income dispersion (residual income inequality). The minimum income which is set by the organizations depends on factors like employment changes among different demographic and skill groups, changes in the market forces of supply and demand, and government institutions (for example local government and labor unions which mandate minimum income). Changes in within-group inequality may be due to changes in the market forces or because of the government institutions which might decide the minimum income which the organizations have to pay to its employees within jobs and across firms and/or industries. According to Freeman and Katz, the demand-supply institutions reasons for income structure changes have three reasons (1994). The first reason is that diverse skill groups and demographic are understood to be imperfect substitutes in production. Therefore, any changes in the demand and supply for labor skills can show a discrepancy in employment and income outcomes. The demand for skill-based individuals could be due to changes in the technology, changes in the input price, shift in the market, and also due to globalization. On the other hand, the supply of individuals could change due to education, immigration, and incentives for educational investments.

The second reason could be due to government institutions. If the role of the government is stronger then they will be less responsive to changes in market forces. The third reason is the changes in the organization structure for example deregulation of the product market, degree of centralization, and changes in the degree of unionization.

Greater equality in income leads to trust. Equality in income and higher levels of trust are two pathways to employee engagement. In case there is inequality in income then the engagement level of the employees will drop.

If the inequality in income is high, the employees might feel powerless. They will perceive that their views are not considered by the organization and this might lead to disengagement. Trust on the organization rests on a foundation of economic equality. When resources (income) are distributed inequitably, people at the bottom and top will not see each other as facing a shared fate. Consequently, the employees will have fewer reasons to trust people of different backgrounds. Also, trust rests on an emotional foundation of confidence and control over one's environment. Where inequality is high, employees will be less likely to consider that the growth opportunity in the organization looks bright, and they will have even fewer reasons to believe that they are the masters of their own fate. Inequality leads to lower levels of trust and thus may also have an indirect impact on engagement. The present job market is highly transparent, and attracting skilled workers is a highly competitive activity for any HR manager since employees have increased bargaining power. Organizations are now investing in analytics tools to find out the reasons as to why employees leave the job and how to engage the employees who have different pay structure.

The compensation of employees is decided based on the following factors:

- The moral and historical standard of average workers of average skills, which is determined by development in technology, world capitalist system, the position of national economy and other factors.
- The cost which is incurred to produce the skills.
- Prevailing market conditions in the job market.
- Discrimination forms and subjugation that may push the salary of certain categories of employees.
- Social control imperatives within production, which may add an increment of inducement-income to the salary of workers, particularly employees who require more on-the-job training.

Even though compensation is necessary to recruit and retain employees, it is not compensation alone that drives employee engagement. Rather, managers must build on a solid compensation foundation in order to persuade the extent that employees will go the extra mile and put discretionary effort into their work by putting more energy, passion, and creativity to the job which is assigned to them. The organizations must deliver the basics, which refers to internally equitable and externally competitive pay. This is particularly important given that as the economy improves and job opportunities increase, a lack of competitive position can create disengagement among employees, which can lead to both troubles recruiting critical-skill talent and unwanted employee turnover. The manager must ensure employees understand where they stand. How do employees know if they are paid competitively, either within or outside the organization? A well-structured performance management system and communication program are vital to creating alignment and understanding between employees and pay program outcomes. A well-structured performance management process can provide managers with the tools not only to deliver competitive base pay, but also to communicate the same to their employees and how the job fits within the organization overall. Understanding where they stand and how pay decisions are made can increase employees' line of sight, which is a key component of employee engagement.

10.3.3 STRATEGIES FOR INCREASING EMPLOYEE ENGAGEMENT

- **Identify Achievements:** If employees are recognized and appreciated then their engagement level will be higher compared to others (Saks, 2008). If employees feel that they are not being appreciated then it will hamper their performance and can also affect the profit of the organization. The manager should take time and effort to recognize the good works the employees are doing and he should reward them accordingly.
- **Communication:** Since all the team members have valuable suggestions it is important that the team leader listens to them effectively. In case the team leader refuses to listen to them then this will lower the engagement level of the employees and this might affect the organization and their culture (Halbesleben and Wheeler, 2008; Saks, 2006; Mishra et al., 2014). Communication should be open in nature and employees should know whom they have to contact if they face any problem. This will create a platform for the employees to share new innovations and ideas. It will also help in bridging the gap between the senior management and others.
- **Growth Opportunities:** Career development is vital for employee engagement. If the employees feel that there is no growth for them and that their hard work and emotional investment aren't being reciprocated, then their engagement level will drop. Managers should meet their team members on a regular basis and discuss their respective targets and the time frame by when they should finish it and build a road map for their future. They should also clearly highlight how they fit into the wider plans of the organization. This will show that the efforts and hard work of the employees aren't going unnoticed.

- **Culture:** To keep employees engaged it is important to build a culture that reflects the organization's brand and creates a productive and lively working environment. It will also help managers to retain their employees. If the organization's culture is good then the employees will get motivated to work and then can also act as brand ambassadors. An engaged and dedicated workforce is a huge contributor to any organization's growth. The right culture can act like a catalyst to help organization achieve their growth.

Following is the list of essential capabilities which were identified by Taylor (2004) for the manager to engage employees at their workplace:

- To build trust among the manager and the employees;
- Building admiration in employees;
- Effective communication;
- Building a work climate that is pleasant and pleasing;
- Flexibility in identifying, sympathetic, and adapting to individual viewpoints;
- To develop the talent of the employees and to train the employees to attain their target;
- High performance-building to strengthen high levels of employee's performance, particularly with regard to top-performing employee;
- To impart the engagement knowledge to all the employees;
- Monitoring engagement team member issues so that preventive action can be taken before someone becomes disengaged;
- To ensure that the employees stay in the organization for a longer period of time;
- Some of the other strategies which were proposed Frank and Taylor (2004) include:
  - Holding managers answerable for the engagement of employees;
  - Selecting and hiring managers on the basis of engagement leader talents;
  - Using metrics to text the impact of managers on engagement.

10.3.4 STRATEGIES FOR IMPROVING ORGANIZATIONAL CULTURE

- **Empowerment:** Employees have to be empowered. Empowered employees will solve problems on their own; they will take possession of their responsibilities, and do whatever it takes to help the organization succeed. This will drive the organizations culture forward. Managers

should demonstrate that they trust their employees. Trusting the employees will ensure them that they are being valued, which can lead to empowerment.
- **Set Expectations:** Managers have to set, manage, and communicate expectations. Employees will find it difficult to understand the cultural vision of the organization initially. If the manager sets clear and expected expectations and communicates them to their employees on a regular basis then it will prevent confusion and limit deviation from the organizations desired vision.
- **Consistency:** To maintain a consistent culture, organization must show uniformity with their actions and communications. The manager should ensure efforts to have consistent expectations and standards for all their workers, and communicate the same to their employees. By focusing on employee, engagement, and investing in them will help organizations in repaying the efforts. Repaying the efforts leads to increase in, productivity performance, and the overall profit.

10.3.5 THE BUDDING ROLE OF THE TEAM MEMBER IN ENGAGEMENT

As engagement is very vital for the success of the organization, engagement agents will take on greater value in the workplace. Some initiatives include:
- **Training:** Training team members as engagement agents. Training them to deal with their own way of thinking of agitation and to listen to and advise fellow team members who are thinking of leaving the organization.
- **Suggestions:** Taking suggestions from employees regarding how to encourage peak performance, so as to engage employees.
- **Commitment (Mutual Commitment Building):** Each member of the team can indicate their engagement talents he or she most values. These values can be documented in a questionnaire, and the leader can commit to express these talents in their everyday behaviors. Team members can, in turn, commit to convey any feelings of disengagement on their part to their managers and they can also encourage their colleagues to do the same thing (Bakker and Bal, 2010; Harris, 2004).
  - Orientation processes which are innovative to the employees (Watson Wyatt, 2003).

- Innovative intra-team engagement processes (Tims et al., 2011). For example, Pfizer Inc. which is an American pharmaceutical corporation headquartered in New York City. Introduced a concept termed as "self-reinforcing peer groups of four to seven senior reps" throughout the company which was communicated via e-mail and phone when many of Pfizer's senior sales representatives had become less motivated.
- **Programs:** Introducing mentoring programs focusing on engagement (Byrne, 2003). Pfizer formed a mentoring program using mentors for younger colleagues, termed as the Master's Group, which was named after the major golf tournament.
- **Creativity:** Ensuring that employees are given opportunities to use their creativity (Babcock-Roberson and Strickland, 2010).
- **Technology:** Use of technology within the framework of virtual teams, allowing more employees to work from home, appreciating the need for greater moldability in the workplace.
- **Modifications:** Making some structural modifications, like, job rotation, flattening of the organization in order to engage workers (De Lange et al., 2008).
- **Environment:** Creating more engaging and welcoming environments for employees (Attridge, 2009; Halbesleben and Wheeler, 2008). For instance, using music and other design features. For example, Starbucks which was founded in 1971 Seattle, Washington is an American coffee company and coffeehouse chain. The company is able to create more holistic experiences for their customers and employees.

10.3.6 ORGANIZATIONAL EFFORTS TO ENCOURAGE EMPLOYEE ENGAGEMENT

Top managers should not only be trained to focus on the strengths of the staff but they should also master the art of preventing a lack of employee engagement. Some of the tools identified by Attridge (2009) are discussed below:

- **Leadership Style:** One of the important factors for encouraging employee engagement is the leadership style. Research on health psychology has revealed that a transformational leadership style is effective for employee engagement (Barling, 2007; Hoon Song et al., 2012). Barling states that a transformational leader provides a clear

vision, motivates, and inspires, offers intellectual challenges, and shows real interest in the needs of the workers (2007).
- **Corporate Culture:** The culture of the organization should also be changed in order to reduce the organizational conditions that lead to stress. Grawitch et al. have identified five factors for the healthy workplace they are: (a) supporting work-life balance; (b) encouraging health and safety on the job; (c) employee involvement/engagement; (d) fostering employee growth development; and (e) praise and recognition (2006). The organization guidelines should be on ethics and principles instead of castigation and stringent rules.
- **Job Design:** Better job design improves employee engagement. The employee strengths should be identified and they should be given the job which matches their abilities. They should also be given the autonomy to take decisions (Shuck, 2011; Wollard and Shuck, 2011).
- **Proper Resources and Support from Managers:** Prior researchers have indicated that there is low productivity and disengagement among employees if they don't get proper resources and support from their managers (Rhoades and Eisenberger, 2002). Thus, it is the duty of the managers to ensure that the employees get the required amount of resources at the right time in order to keep them engaged to their work.
- **Working Conditions:** The working conditions should not be very stressful. Prior researchers have indicated that stressful job leads to employee burnout and exhaustion (Xanthopoulou et al., 2007; Shuck et al., 2011).
- **Projects:** Employees should be allowed to pursue projects which passionate them (McCombie and Spreafico, 2017). For example, Google an American multinational technology company which specializes in Internet-related products and services introduced the concept termed as since-retired 20% time which offers employees flexibility and also gives them free time to pursue passion projects, like volunteer work or community service. The company found that the performance of the company was very good post that.
- **Diversity:** Organizations should give importance to inclusion and diversity. They should realize that innovation is derived from diverse stakeholders. By giving importance to inclusion and diversity, the organizations would be giving a platform for employee's concerns to be heard. For example, Accenture PLC which was founded in the year 1989 is a global management consulting and professional services firm headquartered in Dublin, Republic of Ireland, pays the cost to

its employees who are transgender if they would like to go for gender reassignment surgery.
- **Information Gathering:** Organizations should facilitate the gathering of information in an informal way by encouraging employees to form a community within the organization. This will improve the morale and productivity of the employees. Forming community will also help employees in getting to know who their co-workers are. For example, Gimbel the founder of LaSalle Network which is a HR, recruiting, and staffing firm based in Chicago encourages his employees to participate in wellness activities. The firm also has various clubs (viz. book club) where employees get a platform to get to know one another.
- **Reviews:** Organizations should ensure that they give their employees review of their performance on a regular basis. For example, Illuminate education which was founded in 2009 has implemented a concept termed as collaborative approach. Wherein, seven to twelve colleagues give feedback on a monthly basis through an employee survey form.
- **Promotions:** Eligible employees should be promoted immediately. If organizations wait for a long time until their employees hit arbitrary milestones (one year) then it will demonstrate that the organization is more concerned with following the rule rather than employee engagement.
- **Ergonomic Workspaces:** Organizations should provide Ergonomic workspaces like mediation center, standing desk, chairs which can be adjusted, etc. for example Uber Technologies Inc. which was established in the year 2009 is a peer-to-peer ridesharing, food delivery, and transportation network firm headquartered in San Francisco, California, pays its employees up to 10,000 rupees to spend on anything they want for their workstation.

10.4 IMPLICATIONS

- **Coca Cola Company:** The Coca-Cola Company, which was founded in the year 1886 is an American multinational beverage corporation and is headquartered in Atlanta, Georgia, but it is incorporated in Wilmington, Delaware. Its main business is manufacturing, retailing, and marketing of nonalcoholic beverage syrups and concentrates. The Coca-Cola system is created by more than 700,000 associates. Each associate brings his or her unique talents and ideas to work every day to help the Coca-Cola system achieve the goals outlined in their 2020 vision. The associates also embody Coca-Cola in their communities and act as ambassadors of their brands to the world. The company's core business philosophy is to ensure that their associates are content, vigorous, and treated fairly and with admiration. The company strives very hard to create open work environments as varied as the markets it serves, where associates are encouraged and inspired to create superior results. They also aim to create environments where people are fully engaged and where the company is viewed both within and outside the company as an employer of choice. Some of the measure which the company adopts to engage its associates are:

- **Communication:** The Company encourages a work environment where communication is open so that the associates can effectively implore and leverage innovative ideas, the company engages in frequent dialogue with their associates around the world. Such frequent dialogue provides them with valuable information, promotes business strategies, shares successes, and opportunities, increases awareness, and solicits opinions of associates (Gupta and Pandey, 2018). For example, the company's global associates and bottling partners have contributed ideas during the 2008 Beijing Summer Olympic Games. The inputs given by the employees were the key ingredient to their Company's mission, vision, and values. Another example of their frequent dialogue with their associates was during the global Employee Insights Survey. In 2010, the results of the global Employee Insights Survey showed improvement across almost all survey categories, including an 84% associate engagement score which was a 2% increase compared to 2008 results.
- **Rewards:** The Company's reward and remuneration packages are considered the best in the world, benchmarked against other global, high-performing employers. They also offer a variety of developmental opportunities for their associates, for example, Coca-Cola University, a learning program for high performers.
- **Southwest Airlines:** Southwest Airlines Co. is a major U.S. airline. The airline was established in 1967 by Herb Kelleher and is headquartered in Dallas, Texas; it is also the world's largest low-cost carrier. Southwest Airlines is a company admired for their employee engagement practices. The levels of employee engagement have been high over the years, their employees are fully committed and are passionate

about the company's values and vision, and enthusiastic to help the company achieve success. They've set the bar high-they to allow their employees across all departments to design their own uniform and have given them full autonomy over aspects of their work life. They have set a good example of providing superior customer service due to their employees who are committed to their work.

The company allowed employees from any department to apply to collaborate on new uniform designs, with results really reflecting the personality and company culture in a way that wouldn't have been achieved had employees not been given a say. Employees were receptive to this, describing it as a memorable experience. The company encourages employees to stay motivated to do things differently. One of the important factors of company's engagement practices is it recognizes those employees who go the extra mile. Each week, the chief executive officer of the company gives a "shout out," publicly praising those employees who have gone above and beyond at work. They also have a monthly recognition in Southwest's magazine, featuring those employees who have shined that particular month.

This kind of acknowledgment keeps employees responsive that they are being valued and that their commitment and hard work towards the company doesn't go unobserved. Praising the employees is as important as providing constructive feedback. Employees love to feel motivated and appreciated. It aids them to continue going that extra mile.

According to Herb Kelleher (Founder Southwest Airlines), "competitors can't simply adopt the levels of engagement and commitment found in the company-it takes a special kind of employee and company culture. They can buy all the physical things. The things you can't buy are dedication, devotion, loyalty—the feeling that you are participating in a crusade."

- **Screwfix:** It was founded in the year 1979 and it is the United Kingdom's largest multi-channel retailer of trade tools, accessories, and hardware products. The company keeps their employee engagement levels up by keeping an honest and open culture. Every two weeks, employees are encouraged to provide feedback without any rules to their respective managers. They are encouraged to give feedback on matters like how they think things are managed, how the company interacts with its customers, how things are going, and new ideas for improvement. Among other initiatives, one outcome of this is the implementation of a new customer card, which speeds up the in-store process, identifying customers and allowing them to make quicker purchases. Like many other initiatives now in place, this would never have been successful had the employees not been asked for their valuable suggestions?

Having this kind of regular, 360-degree feedback in place means that important things don't get unnoticed but it keeps the discussion going and ensures a company culture where people really feel that they are contributing to the company. According to Andrew Livingstone (Chief Executive, Screwfix) "Many of the improvements can come from an engaged staff team who understand the business objective and are given a voice."

- **FullContact:** FullContact Inc. is a privately held technology company founded in the year 2010. It is headquartered in Denver, Colorado, United States. FullContact provides a suite of cloud-based contact management solutions for individuals, businesses, and developers. Every year, the company offers their employees $7,500 to take a paid vacation. The company pays its employees literally to go on holiday anywhere they like. During vacation, the employees are requested not to answer work-related messages or calls. The idea behind doing is when the employees return back from their vacation they will be in a better state to work and will be fully committed to fulfill the objectives of the company. The employees also return from paid vacation with a diverse, new outlook.

The other objective of paid vacations is also to purge the issue of employees thinking they're the only one who can solve a problem in the organization. Once employees return from their paid holiday peaceful and find things functioning efficiently in the organization, they will feel less pressure to handle everything themselves and will also develop a sense of trust for their coworkers.

10.5 CONCLUSION

The chapter highlights how income helps in engaging employees at the workplace. Employees whose income is as per the industry standards will stay back with the company. Retaining talented employees helps the organization to have a competitive edge over others.

In the subsequent section, a case covering the above concept has been provided (refer Appendix 10.1 for plausible case discussion).

10.6 CASE: DIVERSE INCOME CLASSES AT ABC

Michael is presently working for a multinational company ABC which was established in the year 2012 and headquartered in Moscow as Project Manager. He has been with the company since 3 years. Prior to joining ABC firm, he was with RNS Company for two years. He came to know about ABC Company from his friend from college. He joined ABC Company because of the position and the pay which was offered to him at the time of the final interview. His friend had told him that the company's culture is very good and that they treat all their employees fairly. Michael was enjoying his work which was assigned to him and had no complaints.

During the annual meet in the year 2017, he met his classmate Tom. He was surprised to know that Tom had joined ABC Company last week as a project manager. During the casual discussion, Tom discussed about the pay which was offered to him. Michael realized that the pay package which was offered to him was 20% more than what he was getting. Michael felt bad and was very disappointed since Tom's and his educational qualification and work experience were the same but still Tom was getting 20% more income compared to him.

After returning home from the annual meet, he discussed the same with his wife. He wanted to discuss about the inequality in pay with his reporting boss. His wife advised him to keep quiet and told him not to discuss the same with his reporting manager since it could affect his promotion which was due in two months. Michael decided not to discuss with his reporting manager and continued doing the work which was assigned to him.

After fifteen days, he got a call from his reporting manager. His manager wanted to discuss about the progress in the project which was assigned to him. The deadline for the submission of the project was fast approaching. His manager felt that the progress was not up to the mark and he was worried that if they don't complete the project within the deadline then it would affect the image of the company. Michael's reporting manager discussed the same with Michael and casually asked him whether there is any concern which he is facing. Michael thought that this is the right time to discuss his concern. Michael told his reporting manager that he has not been treated fairly by his Employer.

Michael's reporting manager told him not to worry and focus on his work and that he would discuss with the management about his concern and assured him that the outcome would be positive. Michael felt relieved after discussing his concern with his reporting manager and completed the task within the deadline.

During the annual review which was conducted after two months, Michael was called by his reporting manager. Michael though that he will be receiving his promotion letter from his manager but was shocked to know that he was not promoted and also there was no increment in his pay. Michael wanted an explanation from his manager; the explanation which he received was the management felt that he is not eligible for promotion this time. The following questions from the case may be answered:

- How would you react if you were in Michael's shoes?
- Explain in detail what steps the manager should have taken when he came to know about Michael's concern.

APPENDIX 10.1: CASE DISCUSSION

ABSTRACT

The case describes the disparity in income faced by an employee whose grade and designation is similar to one of his co-workers who has joined the organization recently.

TEACHING OBJECTIVES AND TARGET AUDIENCE

The case is designed to enable students to identify the laws which are available for employees in the organization in case they face any disparity in their income.

This case is meant for BBA and MBA students as part of their Company Law curriculum.

TEACHING SCHEME

As a short case, it can be used as a basis for plenary discussion instead of group discussion.

CASE ANALYSIS

All employees of a company are eligible for appropriate and fair remuneration. According to Article 39(d) of the constitution, employees have to be paid equally for equal work. The students can discuss about the laws under the Equal Remuneration Act. Some of the highlights of the law are as follows:

- The employer should not reduce the income of the employees.
- Employer should pay equal remuneration to men and women workers for work of a similar nature or same work.
- No discrimination to be made while recruiting women and men employees for the same work.
- The organization has an advisory committee.

KEYWORDS

- **commitment**
- **employee engagement**
- **income**
- **inequality**
- **remuneration**
- **trust**

REFERENCES

Aburto, K. H., Rioseco, M. M., & Moyano-Díaz, E., (2017). Concept of happiness in adults from low-income class. *Paideia, 27*, 386–394.

Acharya, A., & Gupta, M., (2016a). An application of brand personality to green consumers: A thematic analysis. *Qualitative Report, 21*(8), 1531–1545.

Acharya, A., & Gupta, M., (2016b). Self-image enhancement through branded accessories among youths: A phenomenological study in India. *Qualitative Report, 21*(7), 1203–1215.

Acharyya, R., García-Alonso, M. D. C., (2008). Parallel imports, innovations, and national welfare: The role of the sizes of income classes and national markets for health care. *Singapore Economic Review, 53*(1), 57–79.

Atkinson, A. B., (1983). *The Economics of Inequality* (Vol. 2). Oxford: Clarendon Press.

Attridge, M., (2009). Measuring and managing employee work engagement: A review of the research and business literature. *Journal of Workplace Behavioral Health, 24*(4), 383–398.

Babcock-Roberson, M. E., & Strickland, O. J., (2010). The relationship between charismatic leadership, work engagement, and organizational citizenship behaviors. *The Journal of Psychology, 144*(3), 313–326.

Bakker, A. B., & Bal, M. P., (2010). Weekly work engagement and performance: A study among starting teachers. *Journal of Occupational and Organizational Psychology, 83*(1), 189–206.

Bakker, A. B., Albrecht, S. L., & Leiter, M. P., (2011). Key questions regarding work engagement. *European Journal of Work and Organizational Psychology*, *20*(1), 4–28.

Bakker, A. B., Demerouti, E., & Sanz-Vergel, A. I., (2014). Burnout and work engagement: The JD-R approach. *Annu. Rev. Organ. Psychol. Organ. Behav.*, *1*(1), 389–411.

Bakker, A. B., Schaufeli, W. B., Leiter, M. P., & Taris, T. W., (2008). Work engagement: An emerging concept in occupational health psychology. *Work and Stress*, *22*(3), 187–200.

Barro, R. J., (2000). Inequality and growth in a panel of countries. *Journal of Economic Growth*, *5*(1), 5–32.

Belabed, C. A., Theobald, T., & Van Treeck, T., (2017). Income distribution and current account imbalances. *Cambridge Journal of Economics*, *42*(1), 47–94.

Bresnahan, T. F., Brynjolfsson, E., & Hitt, L. M., (2002). Information technology, workplace organization, and the demand for skilled labor: Firm-level evidence. *The Quarterly Journal of Economics*, *117*(1), 339–376.

Bridgman, B., (2017). *Falling Stars: Exports and Income Inequality in Golden Age Hollywood.*

Caron, J., Cole, J., Goettle, R., Onda, C., McFarland, J., & Woollacott, J., (2018). Distributional implications of a national CO_2 tax in the U.S. across income classes and regions: A multi-model overview. *Climate Change Economics*, *9*(1).

Choi, R., & Hwang, B. D., (2014). Medical expense factors and trend of income classes in South Korea. *International Journal of Applied Engineering Research*, *9*(24), 29677–29684.

Cohen, J. R., (1998). Combining historic preservation and income class integration: A case study of the butchers hill neighborhood of Baltimore. *Housing Policy Debate*, *9*(3), 663–697.

Collins, J., (2017). *The General Theory of Employment, Interest, and Money*. Macat Library.

Cramer, J. S., (1986). Estimation of probability models from income class data. *Statistica Neerlandica*, *40*(4), 237–250.

Croft, A., Schmader, T., & Block, K., (2015). An underexamined inequality: Cultural and psychological barriers to men's engagement with communal roles. *Personality and Social Psychology Review*, *19*(4), 343–370.

Da Silva, M. V., Ometto, A. M. H., Furtuoso, M. C. O., Pipitone, M. A. P., & Sturion, G. L., (2000). Access to day-care centers and the nutritional status of Brazilian children: Regional differences by age group and income class [Acesso á creche e estado nutricional das crianças brasileiras: Diferenças regionais, por faixa etária e classes de renda]. *Revista de Nutricao*, *13*(3), 193–199.

De Lange, A. H., De Witte, H., & Notelaers, G., (2008). Should I stay or should I go? Examining longitudinal relations among job resources and work engagement for stayers versus movers. *Work and Stress*, *22*(3), 201–223.

De Martino, J. G., (1989). Residential energy demand in Brazil by income classes issues for the energy sector. *Energy Policy*, *17*(3), 254–263.

Enders, W., & Sandler, T., (2006). Distribution of transnational terrorism among countries by income class and geography after 9/11. *International Studies Quarterly*, *50*(2), 367–393.

Feenberg, D. R., & Rosen, H. S., (1986). The deductibility of state and local taxes: Impact effects by state and income class. *Growth and Change*, *17*(2), 11–31.

Frank, F. D., Finnegan, R. P., & Taylor, C. R., (2004). The race for talent: Retaining and engaging workers in the 21st century. *People and Strategy*, *27*(3), 12–25.

Freeman, R. B., & Katz, L. F., (1994). *Rising Wage Inequality: The United States* (pp. 29–63). Working under different rules, New York: Russell Sage.

FURTHER READING

Galor, O., & Zeira, J., (1993). Income distribution and macroeconomics. *The Review of Economic Studies, 60*(1), 35–52.

Gambini, A., & Marianera, M., (2013). Trends in private consumption in China: The emergence of the Chinese high-income class and its global relevance. *The Chinese Economy: Recent Trends and Policy Issue*s, pp. 235–253.

Genicot, G., & Ray, D., (2017). Aspirations and inequality. *Econometrica, 85*(2), 489–519.

Ghazanfar, S. M., (1978). Sales tax equity again: By age groups and income classes. *Public Finance Review, 6*(3), 343–357.

González-Romá, V., Schaufeli, W. B., Bakker, A. B., & Lloret, S., (2006). Burnout and work engagement: Independent factors or opposite poles? *Journal of Vocational Behavior, 68*(1), 165–174.

Greenwood, J., & Jovanovic, B., (1990). Financial development, growth, and the distribution of income. *Journal of Political Economy, 98*(5, Part 1), 1076–1107.

Gupta, M., & Kumar, Y., (2015). Justice and employee engagement: Examining the mediating role of trust in Indian B-schools. *Asia-Pacific Journal of Business Administration, 7*(1), 89–103.

Gupta, M., & Pandey, J., (2018). Impact of student engagement on affective learning: Evidence from a large Indian University. *Current Psychology, 37*(1), 414–421.

Gupta, M., & Ravindranath, S., (2018). Managing physically challenged workers at microsign. *South Asian Journal of Business and Management Cases, 7*(1), 34–40.

Gupta, M., & Sayeed, O., (2016). Social responsibility and commitment in management institutes: Mediation by engagement. *Business: Theory and Practice, 17*(3), 280–287.

Gupta, M., & Shaheen, M., (2017a). Impact of work engagement on turnover intention: Moderation by psychological capital in India. *Business: Theory and Practice, 18*, 136–143.

Gupta, M., & Shaheen, M., (2017b). The relationship between psychological capital and turnover intention: Work engagement as mediator and work experience as moderator. *Journal Pengurusan, 49*, 117–126.

Gupta, M., & Shaheen, M., (2018). Does work engagement enhance general well-being and control at work? Mediating role of psychological capital. *Evidence-Based HRM, 6*(3), 272–286.

Gupta, M., & Shukla, K., (2018). An empirical clarification on the assessment of engagement at work. *Advances in Developing Human Resources, 20*(1), 44–57.

Gupta, M., (2017). Corporate social responsibility, employee-company identification, and organizational commitment: Mediation by employee engagement. *Current Psychology, 36*(1), 101–109.

Gupta, M., (2018). Engaging employees at work: Insights from India. *Advances in Developing Human Resources, 20*(1), 3–10.

Gupta, M., Acharya, A., & Gupta, R., (2015). Impact of work engagement on performance in Indian higher education system. *Review of European Studies, 7*(3), 192–201.

Gupta, M., Ganguli, S., & Ponnam, A., (2015). Factors affecting employee engagement in India: A study on offshoring of financial services. *Qualitative Report, 20*(4), 498–515.

Gupta, M., Ravindranath, S., & Kumar, Y. L. N., (2018). Voicing concerns for greater engagement: Does a supervisor's job insecurity and organizational culture matter? *Evidence-Based HRM, 6*(1), 54–65.

Gupta, M., Shaheen, M., & Das, M. (2019). Engaging employees for quality of life: mediation by psychological capital. *The Service Industries Journal*, 39(5–6), 403–419.

Gupta, M., Shaheen, M., & Reddy, P. K., (2017). Impact of psychological capital on organizational citizenship behavior: Mediation by work engagement. *Journal of Management Development*, 36(7), 973–983.

Haines, E. L., Deaux, K., & Lofaro, N., (2016). The times they are a-changing or are they not? A comparison of gender stereotypes, 1983–2014. *Psychology of Women Quarterly*, 40(3), 353–363.

Hakanen, J. J., Schaufeli, W. B., & Ahola, K., (2008). The job demands-resources model: A three-year cross-lagged study of burnout, depression, commitment, and work engagement. *Work and Stress*, 22(3), 224–241.

Halbesleben, J. R., & Wheeler, A. R., (2008). The relative roles of engagement and embeddedness in predicting job performance and intention to leave. *Work and Stress*, 22(3), 242–256.

Halbesleben, J. R., (2011). The consequences of engagement: The good, the bad, and the ugly. *European Journal of Work and Organizational Psychology*, 20(1), 68–73.

Harter, J. K., Schmidt, F. L., & Hayes, T. L., (2002). Business-unit-level relationship between employee satisfaction, employee engagement, and business outcomes: A meta-analysis. *Journal of Applied Psychology*, 87(2), 268–279.

Hashimoto, K., & Heath, J. A., (1995). Income elasticity's of educational expenditure by income class: The case of Japanese households. *Economics of Education Review*, 14(1), 63–71.

Hoon, S. J., Kolb, J. A., Hee, L. U., & Kyoung, K. H., (2012). Role of transformational leadership in effective organizational knowledge creation practices: Mediating effects of employees' work engagement. *Human Resource Development Quarterly*, 23(1), 65–101.

Horn, B. P., Maclean, J. C., & Strain, M. R., (2017). Do minimum wage increases influence worker health? *Economic Inquiry*, 55(4), 1986–2007.

Ishi, H., (1979). Individual income tax erosion: By income class in Japan. *Public Finance Review*, 7(3), 303–322.

Jayadev, A., (2009). Income, class, and preferences towards anti-inflation and anti-unemployment policies. *Beyond Inflation Targeting: Assessing the Impacts and Policy Alternatives*, pp. 71–92.

Johnson, M., (2004). *The New Rules of Engagement: Life-Work Balance and Employee Commitment*. CIPD Publishing.

Jones, E., (1997). Consumer demand for carbohydrates: A look across products and income classes. *Agribusiness*, 13(6), 599–612.

Jones, G., (1982). Another approach to estimating different income elasticities for different income classes. With detailed results for southern Brazil and Greece. *Oxford Agrarian Studies*, 11, 124–138.

Jubb, M., (1987). Income, class and the taxman: A note on the distribution of wealth in nineteenth-century Britain. *Historical Research*, 60(141), 118–124.

Juhn, C., Murphy, K. M., & Pierce, B., (1993). Wage inequality and the rise in returns to skill. *Journal of Political Economy*, 101(3), 410–442.

Kahn, W. A., (1990). Psychological conditions of personal engagement and disengagement at work. *Academy of Management Journal*, 33(4), 692–724.

Katz, L. F., & Murphy, K. M., (1992). Changes in relative wages, 1963–1987: Supply and demand factors. *The Quarterly Journal of Economics*, 107(1), 35–78.

Katz, L. F., (1999). Changes in the wage structure and earnings inequality. In: *Handbook of Labor Economics* (Vol. 3, pp. 1463–1555), Elsevier.

Kim, C. W., Lee, S. Y., & Hong, S. C., (2005). Equity in utilization of cancer inpatient services by income classes. *Health Policy, 72*(2), 187–200.

Klein, L. R., Straw, K. H., & Vandome, P., (1956). *Savings and Finances of the Upper Income Classes* (Vol. 18, No. 4, pp. 293–319). Bulletin of the Oxford University Institute of Economics and Statistics.

Kolwicz, P., (2003). Property (β) and orthogonal convexities income class of Köthe sequence spaces. *Publicationes Mathematicae, 63*(4), 587–609.

Kuznets, S., (1955). Economic growth and income inequality. *The American Economic Review*, pp. 1–28.

Lee, Y., & Shin, D., (2016). Measuring social tension from income class segregation. *Journal of Business and Economic Statistics, 34*(3), 457–471.

Lexchin, J., (1996). Income class and pharmaceutical expenditure in Canada: 1964–1990. *Canadian Journal of Public Health, 87*(1), 46–50.

Malik, M. H., & Us Saqib, N., (1989). Tax incidence by income classes in Pakistan. *Pakistan Development Review, 28*(1), 13–25.

Mann, A. J., (1982). The Mexican tax burden by family income class. *Public Finance Review, 10*(3), 305–331.

May, D. R., Gilson, R. L., & Harter, L. M., (2004). The psychological conditions of meaningfulness, safety and availability and the engagement of the human spirit at work. *Journal of Occupational and Organizational Psychology, 77*(1), 11–37.

McCombie, J., & Spreafico, M., (2017). On income inequality: The 2008 great recession and long-term growth. In: *The Crisis Conundrum* (pp. 41–63). Palgrave Macmillan, Cham.

McDowell, D. R., Allen-Smith, J. E., & McLean-Meyinsse, P. E., (1997). Food expenditures and socioeconomic characteristics: Focus on income class. *American Journal of Agricultural Economics, 79*(5), 1444–1451.

Mellen, L. W., (1956). The growth of the middle-income class in the United States. *South African Journal of Economics, 24*(4), 290–296.

Meyer, J. P., & Allen, N. J., (1991). A three-component conceptualization of organizational commitment. *Human Resource Management Review, 1*(1), 61–89.

Mishra, K., Boynton, L., & Mishra, A., (2014). Driving employee engagement: The expanded role of internal communications. *International Journal of Business Communication, 51*(2), 183–202.

Nocum, A. A., Baltao, J. M., Agustin, D. R., & Portus, A. J., (2015). Ergonomic evaluation and design of a mobile application for maternal and infant health for Smartphone users among lower-income class Filipinos. *Procedia Manufacturing, 3*, 5411–5418.

Oguri, Y., Watanabe, N., & Saeda, S., (1982). Spatial distribution of the households of different income classes in the metropolitan area: A case study of the Tokyo region. *Studies in Regional Science, 13*, 107–118.

Okiyama, M., & Tokunaga, S., (2009). Analysis of disparity reduction among people in different regions and different income classes by exporting products of biofuel industry: Utilizing a SAM/I-O linked model for Thailand. *Studies in Regional Science, 39*(4), 893–909.

Ozawa, M. N., & Yon, H. S., (2002). The economic benefit of remarriage: Gender and income class. *Journal of Divorce and Remarriage, 36*(43528), 21–39.

Ozawa, M. N., Kim, J., & Joo, M., (2006). Income class and the accumulation of net worth in the United States. *Social Work Research, 30*(4), 211–222.

Pandey, J., Gupta, M., & Naqvi, F., (2016). Developing decision making measure a mixed method approach to operationalize Sankhya philosophy. *European Journal of Science and Theology*, *12*(2), 177–189.

Park, J. C., Lee, K. G., Hong, W. H., & Joo, C. H., (2014). Energy consumption of rental houses for the low-income class in South Korea. *Advanced Materials Research*, *935*, 122–125.

Pechman, J. A., (1972). Distribution of federal and state income taxes by income classes. *The Journal of Finance*, *27*(2), 179–191.

Piketty, T., & Saez, E., (2003). Income inequality in the United States, 1913–1998. *The Quarterly Journal of Economics*, *118*(1), 1–41.

Piketty, T., (2015). About capital in the twenty-first century. *American Economic Review*, *105*(5), 48–53.

Pucher, J., (1981). Equity in transit finance: Distribution of transit subsidy benefits and costs among income classes. *Journal of the American Planning Association*, *47*(4), 387–407.

Rao, T. D., (1984). Are the union excise duties regressive: An analysis of their burden by income class (India). *Artha Vijnana*, *26*(3), 232–242.

Ren, Y., Zhang, Y., Loy, J. P., & Glauben, T., (2018). Food consumption among income classes and its response to changes in income distribution in rural China. *China Agricultural Economic Review*, *10*(3), 406–424.

Rich, B. L., Lepine, J. A., & Crawford, E. R., (2010). Job engagement: Antecedents and effects on job performance. *Academy of Management Journal*, *53*(3), 617–635.

Richman, A. L., Civian, J. T., Shannon, L. L., Jeffrey, H. E., & Brennan, R. T., (2008). The relationship of perceived flexibility, supportive work-life policies, and use of formal flexible arrangements and occasional flexibility to employee engagement and expected retention. *Community, Work and Family*, *11*(2), 183–197.

Robinson, D., Perryman, S., & Hayday, S., (2004). *The Drivers of Employee Engagement*. Report-Institute for Employment Studies.

Rothbard, N. P., (2001). Enriching or depleting? The dynamics of engagement in work and family roles. *Administrative Science Quarterly*, *46*(4), 655–684.

Rynes, S. L., Gerhart, B., & Minette, K. A., (2004). The importance of pay in employee motivation: Discrepancies between what people say and what they do. *Human Resource Management*, *43*(4), 381–394.

Saks, A. M., (2006). Antecedents and consequences of employee engagement. *Journal of Managerial Psychology*, *21*(7), 600–619.

Saks, A. M., (2008). The meaning and bleeding of employee engagement: How muddy is the water. *Industrial and Organizational Psychology*, *1*(1), 40–43.

Sasaki, K., (1990). Income class, modal choice, and urban spatial structure. *Journal of Urban Economics*, *27*(3), 322–343.

Schaufeli, W. B., Leiter, M. P., & Maslach, C., (2009). Burnout: 35 years of research and practice. *Career Development International*, *14*(3), 204–220.

Schaufeli, W. B., Salanova, M., González-Romá, V., & Bakker, A. B., (2002). The measurement of engagement and burnout: A two sample confirmatory factor analytic approach. *Journal of Happiness Studies*, *3*(1), 71–92.

Shaw, K., & Bastock, A., (2005). *Employee Engagement: How to Build a High-Performance Workforce*. Chicago: Melcrum Publishing Limited.

Shuck, B., (2011). Integrative literature review: Four emerging perspectives of employee engagement: An integrative literature review. *Human Resource Development Review*, *10*(3), 304–328.

Shuck, B., Reio, Jr. T. G., & Rocco, T., S., (2011). Employee engagement: An examination of antecedent and outcome variables. *Human Resource Development International, 14*(4), 427–445.

Silva, A. C., & Yakovenko, V. M., (2005). Temporal evolution of the "thermal" and "superthermal" income classes in the USA during 1983-2001. *Europhysics Letters, 69*(2), 304–310.

Smoluk, H. J., & Voyer, J., (2014). The spirit of capitalism among the income classes. *Review of Financial Economics, 23*(1), 1–9.

Son, J., & Park, J., (2019). Effects of financial education on sound personal finance in Korea: Conceptualization of mediation effects of financial literacy across income classes. *International Journal of Consumer Studies, 43*(1), 77–86.

Sonnentag, S., (2003). Recovery, work engagement, and proactive behavior: A new look at the interface between nonwork and work. *Journal of Applied Psychology, 88*(3), 518–528.

Tachibanaki, T., & Shimono, K., (1991). Wealth accumulation process by income class. *Journal of the Japanese and International Economies, 5*(3), 239–260.

Tims, M., Bakker, A. B., & Xanthopoulou, D., (2011). Do transformational leaders enhance their followers' daily work engagement? *The Leadership Quarterly, 22*(1), 121–131.

Torlak, O., & Koc, U., (2007). Materialistic attitude as an antecedent of organizational citizenship behavior. *Management Research News, 30*(8), 581–596.

Tucker, R. S., (1943). The composition of income and ownership of capital by income classes in the United States in 1936. *Journal of the American Statistical Association, 38*(222), 187–200.

Turpeinen, O., (1952). Nutrition among lower income classes in Finland. *Annales Medicinae Internae Fenniae, 41*(3/4), 216–233, 305–315.

Turpeinen, O., (1953). Nutrition among lower income classes in Finland. III. General features of the diets. *Annales Medicinae Internae Fenniae, 42*(1), 75–88.

Wang, J., Chen, Y., Zheng, Z., & Si, W., (2014). Determinants of pork demand by income class in urban western China. *China Agricultural Economic Review, 6*(3), 452–469.

Watarai, F., & Romanelli, G., (2009). Stepfamilies among Brazilian low-income classes: Relationships and roles. *International Journal of Interdisciplinary Social Sciences, 4*(7), 235–245.

Weicher, J. C., (1971). The allocation of police protection by income class. *Urban Studies, 8*(3), 207–220.

Whelan, C. T., Nolan, B., & Maître, B., (2017). The great recession and the changing intergenerational distribution of economic stress across income classes in Ireland: A comparative perspective. *Irish Journal of Sociology, 25*(2), 105–127.

White, H. C., (1981). Where do markets come from? *American Journal of Sociology, 87*(3), 517–547.

Wollard, K. K., & Shuck, B., (2011). Antecedents to employee engagement: A structured review of the literature. *Advances in Developing Human Resources, 13*(4), 429–446.

Yotopoulos, P. A., (1983). *Middle-Income Classes and Food Crises: The 'New' Food-Feed Competition* (p. 31). The University of Warwick, Department of Economics, Development Economics Research Center, Discussion Paper.

Yotopoulos, P. A., (1985). Middle-income classes and food crises: The 'new' food-feed competition. *Economic Development and Cultural Change, 33*(3), 463–483.

Yotopoulos, P. A., (1989). Distributions of real income: Within countries and by world income classes. *Review of Income and Wealth, 35*(4), 357–376.

CHAPTER 11

Is Work Engagement Gender Oriented? A Man/Woman Perspective

SINDHU RAVINDRANATH

Assistant Professor, ICFAI Business School (IBS), Hyderabad, The ICFAI Foundation for Higher Education (IFHE), Hyderabad, Telangana, India, Tel.: (+91) 9032865740, E-mail: sindhur.inc@gmail.com

ABSTRACT

This chapter is about understanding the different perspectives of man and woman at work. It has primarily had to do with discussing whether engagement at work gender-specific. The chapter elaborates on the women in the workplace and issues faced by them. The stereotypes of the society and the way both genders can be engaged in the work.

11.1 INTRODUCTION

Discussing "gender," we will need to first understand the word "gender." While "sex" is a biological feature attributed to a living being, the "Gender" seems to be something that has been assigned socially or by society than by nature. "Gender" is the concept that shows the differences in biological sex (Ogunbodede, 2015). "Gender" the way it has been expressed is more of an economic and social layering which could lead to inclusion or exclusion. Irrespective of the socio-economic class differences, the gender differences are very much evident in the materialist lifestyle and this is different in different parts of the world as well as changes with time.

There are many studies which explain why women have taken the brunt of these classifications but what matters to this chapter is that they have and have been reduced to a lifetime of being underutilized irrespective of their qualifications and talents as well as lower earnings. Feminist economists have pointed out time to time about this aspect recording that as public sector, where women are found to hold more jobs are being reduced and private sectors increases women become had hit by the losses of jobs. Moreover, the public sector jobs also came with a posy of benefits while in the private sector the jobs for women are in the low paid low skilled areas where benefits do not include. Hence, women tend to be hard hit in this aspect as well.

11.1.1 GENDER IS A MAJOR POINT OR AREA OF SOCIAL AND ECONOMIC STRATIFICATION AND, AS A RESULT OF EXCLUSION

Inequality is a situation that shows disparity or in effective an absence of equality. For this discussion, we will consider "gender equality" as the act where men and women are treated equally and are considered equally important for every aspect of everyday life on earth and technically for the development and progress of nations (Hussain and Kirmani, 2010); this given we have to face the real-life scenario where we find women lag behind in comparison to men in terms of literacy, and such other factors which denote human development. And given that, women maybe half of the population of the nation or a little less than that but if such significant number of the population is underdeveloped where can a nation claim "development"? (Madhok, 2014). Research by scientists have proved that when there is discrimination of workforce there will be a dearth of talent in the economy which will show the way to consequences that prove to be negative to the economy (Esteve-Volart (2004). Some of the countries like India have its social practices which stem from either religious or cultural backgrounds wherein women are kept away from the mainstream economic contributions thus depriving the nation form the talent women could bring in for its economic development.

When we explain "gender inequality," it is the act of treating men and women differently due to aspects taken in to account that has no input into the characteristics or the requirements of the job they are expected to do (Kaushik, Sharma, and Kaushik, 2014). When the gender equality in the workforce is bought for discussion, it is not about the women choosing between work and her household activities like taking care of the children or family but it is about men and women having equal choice in deciding to work and what or where to work in (Klugman, Kolb, and Morton, 2014). The gender inequality exists in almost all the countries in the world when women are compensated 10–30% less for the same jobs that men do and are paid more (Comyn et al., 2014)

Attempts have been made since 1976 to document and bring out how the women who have advanced in career and their advancement been "handled" (Bennet, 1976). It was said, by Bennet that it would be very difficult to solve the "attitude-based problems" that they will face than the "sex-based differences in pay." In a recent gender diversity conducted by Mckinsey, where they have surveyed 1,421 executives globally, it emerges that the cultural factors still constitute the most important player in achieving the goals in gender diversity (Devillard et al., 2014). It shows the ancientness of the problem in women's development, gender diversity, and acceptability of women in the area of corporate management or working class as such.

A McKinsey report mentions that "women are as ambitious as men but they have doubts about the support from the corporate culture for their success." This has some truth in it (Devillard et al., 2014).

11.2 NEED FOR DISCUSSING THIS THEME

Recent researches on the engagement differences between men and women show that women in academics engage less with the industry (Tartari and Salter, 2015). This is being attributed to quite a few aspects like the fact that there are fewer women to interact with and discuss the facts on hand.

The question arises "why does that happen?" Is that because women are less engaged than men or is there any other reason? Researching of literature that has been published for the recent few years on the topic brings forth some very interesting information.

Female achievement or success is not at all question here. Men and women are at par when it comes to educational qualifications and success in achievements relating to education. In many countries like the US and a few Asian countries, studies have found women to have surpassed men in terms of educational achievements (Devillard et al., 2017). The disparity and differences in views arise as the younger generations who are either at the beginning of the career or a step ahead of the beginners try to balance their work and home life, start a family. This stage the working women find themselves having to carry on with the major bulk of responsibilities in caregiving. The studies done by McKenzie as well as few other researchers have found that once women begin to devote more hours to the home or family-care, their workplace performance is at a disadvantage. This is because there is an unwritten unrealistic expectation from women to be the first caretaker to the family in their emotional and wellbeing. This, in turn, brings forth the fact that when the society and the culture expect women to be the primary caregiver in the domestic arena, the rules and expectations from women at the organizational level become unrealistic. The dual roles demand that women cannot be available to the workplace all the time. They will have to compartmentalize and delegate often.

In one of the studies (Brinton, 1993) we come across the effects of this in Japan. The cultural demand for the ideal mother clubbed with the long working hours and commitment which the organizations ask makes motherhood a very difficult reality for Japanese working women. This has contributed to a very low birth rate in Japan. The study also notes that in the countries where the female workforce has the acceptance and freedom as equal as their male counterparts, the birth rates are almost replacement level. This happens in most post-industrial countries. These countries, the study emphasis "have made it possible for women (and men) to balance work and family." For negating the effects of the reducing birth rates and bring it back to normal, it is expected that the culture and laws of the countries should bring on more gender equality at home as well as at the workplace.

For a country to maintain a healthy population naturally, a birth rate of 2.1 is required minimum. Since the advent of the eighties, the birth rate has been declining drastically. Currently, in the United States, the fertility rate is 1.9 and Europe is down more so in the south of Europe at the birth rate of 1.3.

In Japan and in other countries like India, women are looking to be more educated and have rising careers. Here the "second shift" at home as the caregiver, when the home life does not have the gender quality aspect weaved in; affects their work and commitment to work. This results in the fact that women wait longer to commit to having families and maybe marry but have fewer or no children. This impacts the birth rates of the countries with less gender equality (Brinton, 1993).

Recent studies like the one in Harvard show how much the stereotyping and the biases that are ingrained in people have placed women in a very unfair positions and the effect of those biases in the very unfair practices that are prevalent in the current societies and working life. The most unfair of all this is that people have been aware of these biases and their effects on women and their life. It is known in the annals of every industry that women earn 77% of what men take home as salaries. In other words, to repeat what the UN has already said-women take home 23% less than what men take home. The unseen here is the loss of potential wherein you lose a perfectly trained, capable employee when women do not make it to the top. "A highly educated woman who steps out of the labor force when she has kids, would ensure that the labor force loses a worker with very high human capital," said Mary Brinton, the Reischauer Institute Professor of Sociology and the department chair at FAS. Let's imagine many such workers who are lost. The effective calculations could be staggering. According to Brinton, this cost gets more staggering because the longer a woman is out of the workforce, her wages, when she decides to go back to work, stagnate much more than it would have in a six-month gap. It is and would be difficult to change the mindset of the industry especially those like Wall Street or information technology but the call is out to consider the expectation on the family care to be modified.

The proof that shared family care works successfully can be seen from the practices in the Scandinavian countries where men consider it their right to take child care leave. This practice of granting a couple more months of leave if the father, as well as the mother, takes leave time for childcare was started in the 1970s.

11.3   REVIEW OF THIS THEME

When Taylor (1911) introduced the concept of scientific management, it gave a picture of humans who came with a set of ready capacities to perform the required work like machines. It was expected that the workers have physical labor and other than that nothing else to contribute to the job in hand which has set parameters with expected outcomes (Kahn, 1992). The further research conducted within the organizational behavior has been concentrating on these assumptions through the constructs of work motivation which embed these assumptions. The exploration of the factors that bring forth the workers to deliver their service and thereby contributing to the growth of the organization has been less (Kahn, 1992). What has happened on the other hand is that progress has been there in research in the concepts of central life interest (Dubin, 1956), organizational commitment (Mowday et al., 1982), job involvement (Lawler and Hall, 1970) and which are areas that focus on about how people are encouraged externally as well as internally to be productive in the work they are assigned in the organization.

In 1990, Kahn explored the concept of the roles by the instance of how the roles are being owned by people. He suggested that people "occupy" roles at their jobs in varying degrees. Kahn elaborates on Goffman's (1961a) idea of "people's attachment to and detachment" from and the variation of the same towards the job roles. According to Kahn, though he was motivated by Goffman's (1961) study on people and their temporary attachments to the roles they play, the examples Goffman has presented show that he was focused on nonverbal language-the body's natural reaction to the situations and happenings and people around it. Given that, Kahn concedes that organizational study will require a different method of analysis because organizational experiences are more complex-psychologically as well as emotionally (Diamond and Allcorn, 1985).

Kahn went ahead and defined personal engagement as connecting people to their roles at work. People express themselves emotionally, cognitively, and physically while they are personally engaged while performing their work or role at work. Kahn's personally disengagement definition states that when disengaged, a person uncouples himself/herself from the role they are handling thus withdrawing themselves emotionally, physically, and cognitively during the performance of the role. Following Alderfer (1972) and Maslow (1954), Kahn exposits that the ideas of personnel engagement and disengagement bring forth the fact that people require self-employment and self-expression are both important in the work lives of the employees. In Kahn's study (1990) which was two-fold in the technique, he was a participant and an observer in the respective part of the experiment. He concluded that personal engagement and disengagement are two ends of a continuum. Given Kahn's premise that there are different dimensions of self which people turn to depending on the environment they are in, the variable of personal engagement is defined as the effort, involvement, flow, mindfulness, and intrinsic motivation that a person has related to their work role (Hackman and Oldham, 1980; Lawler and Hall, 1970; Csikszentmihalyi, 1982; Langer, 1989; Deci, 1975). Personal engagement is about the person expressing and employing the preferred self at the job or while performing the job role. When preferred self is employed and expressed during the same act the relationship of self to the role performed is created (Kahn, 1990). Personal disengagement, on the other hand, is where the person had withdrawn the preferred dimensions and hence the personal, leading to burned out, effortless, apathetic or detached and automatic behaviors as researchers have called them being expressed while performing the work role (Kahn, 1990; Hochschild, 1983; Hackman and Oldham, 1980; Maslach, 1982; Goffman, 1961a). During detachment, the withdrawal of personal self would exhibit role performance lacking physical and cognitive connections, absence of emotions and incomplete or passive performance at work role. Detachment also creates defending behavior where true identity, feelings, and thoughts are hidden.

Kahn (1990) arrived at three psychological conditions that are the influencing factors for personal engagement and disengagement for people, (a) meaningfulness, (b) safety, and (c) availability. Accordingly, there are three questions each member of the organization asks to themselves. The questions are to give the person an understanding on:

- The meaningfulness of the role/action to the person;
- The safety the person is accorded to do the role/act;
- The availability of the person to complete the role/act.

Kahn reiterates that the given questions actually is a reflection of what the actual logic of workplace contracts which are expected to be clear and contain desired guarantees of benefits and protection. These contracts are agreed upon by the employees/persons entering the contract with a deep belief that they have the necessary competencies to fulfill the obligations required for the contract. This belief propels the person towards the roles. The engagement level of each person is depended on the perception they have as to the benefits or the guarantees and meaningfulness and safety according to the person as well as their perceived availability. An experience is about acts like resources, benefits, and guarantees that influence the thoughts and decisions.

Psychological-meaningfulness (Kahn, 1990) is associated to those work aspects that give incentives or disincentive which will be a cause for engaging or disengaging personally. The meaningfulness made people feel more emotionally or cognitively energized. This was felt when they viewed the role they were involved in as worthwhile, valuable or useful. The meaningfulness meant that the people were able to believe they were giving back to the people who they took from. The main aspects that had a direct influence on meaningfulness are work-interactions, role characteristics, and task characteristics.

Psychological-safety (Kahn, 1990) was related to those factors of social systems that are non-threatening consistent and predictable for the person to decide to engage. Safety in this context would mean that the employee assumes that there will be no negative consequences to their career, status, or self-image if they employ their "self" to the job role at hand. This would induce a person to involve more of self to the work role with less reservations. This creates more trust with the relationships in this case-organization and co-workers. Psychological safety is seen to be affected by organizational norm, interpersonal-relationships, management style, and process as well as group and intergroup dynamics.

Psychological availability, on the other hand, was related to distractions that preoccupy individuals-the need of having all the faculties-emotional, psychological, and physical at hand to be engaged completely into the role, given the distractions experienced by them in the daily life. This includes work as well as non-work activities and how they cope with the demands placed on them (Kahn, 1990; Pearlin, 1983). Outside lives, individual insecurity, depletion of emotional energy, depletion of physical energy are the four types of distractions that influence psychological availability (Kahn, 1990).

This would explain Brinton's (1993) study where it was found that Japanese women concentrating on careers have less plans of family and children. When constantly needed to prove themselves more than their counterparts, humans tend to adjust parts of life which they perceive to be not important at that point of time. In this case, it is the women who put off having family or children on the back burner so that they can keep up with the demands of a career and the perceptions of being focused on it.

When Kahn (1992) bought forth the concept of "psychological presence," he explained it as "being fully present as a person" while conducting the work assigned which meant that a person's thoughts, beliefs, and feelings are involved with the job for performing it. In this aspect, Kahn (1992) gives a picture of the variable as an amalgamation of the feelings of a person in being open to oneself and others, while connected to work, complete as a person within the job role boundaries. This description (Kahn, 1992) of psychological presence describes the experience of the organization members when they bring the deeper personal selves into the job role performances thereby increasing the depth of the experience. Kahn (1992) quotes Argyris (1982) when he explains the dimension of being fully aware and non-defensive as an aspect of psychological presence Argyris (1982) had given that this dimension will be possible only if people do not have to be consciously or unconsciously afraid of the results or implications of being oneself within the job role.

The other aspect that researchers have brought forth is where psychological presence is empathy (Kahn, 1992). Empathy happens when a person is able to identify with another creating a flow of experience (Kahn, 1992; Csikszentmihalyi, 1982). The flow according to Csikszentmihalyi would mean that people identify and feels one with the activity in the role and oneself. Hence, when fully present, the person will feel in tandem with the role activity and other people who are a part of the activity. Nothing or none is outside, different or separate to them and the role activity. This connection to the activity and surroundings results in the person "loosing themselves" in the role activity (Kahn, 1992). This connection leads people to feel that they are accessible to the colleagues, teammates, and the activities involved in their work. They feel "authentic" and "connected" (Kahn, 1992; cf. Goffman, 1959) with a sense of "giving and receiving" (Rogers, 1958) as related to the people involved in the job role or activity.

Another aspect that seems to be involved in a personal engagement to work is the integration of different dimensions of oneself to the activity one is involved in. Dimensions like emotional, physical, and intellect combine together to give an appropriate action and involvement to the act being completed by the person. Integration thus is an experience involving one's sense of being whole at a given situation. When this presence is felt people, feel complete rather than a broken part or left out part of a whole (Kahn, 1992; Ollmann, 1971). This also merges the boundaries (Miller and Rice, 1967) people would have erected between themselves and the external aspects of their work life. When faced with uncomfortable situations or feeling vulnerable during the roles they have to play in the work-life, for self-defense, people tend to wall off their real self from the external aspects thus creating a compartmentalized situation whereby the self is protected from being vulnerable in perceived potential threats, whereby causing the self to never be fully present in the given situations (Menzies, 1975; Miller and Rice, 1967; Klein, 1959).

Havens, (1989) indicated that when a person is completely involved, it's indicated by the verbal behaviors of the person involved. They would be more searching out the ways to complete the activity than intellectualizing about it. They are more of themselves to the table, indicating a more profound psychological presence. The authenticity of the people's response to the task performance context can measure the psychological presence (Kahn, 1992). This total involvement Kahn (1992) says, is the manifestation of humanity in its most "deepest core." This kind of involvement in the long term would bring forth a system that is dynamic and authentic as well as creates a shared understanding of the systems. According to Minuchin (1974) and Smith and Berg (1987) these authentic systems help (Schein, 1987; Shapiro and Carr, 1991) organizations as well as families to become more productive systems of work and life.

The concept of work engagement has been around since 1990 when Gallup had used it for its research. Though it's been around since then, the origin nor the definition of the term was highly debated. Definitions have evolved over time with Kahn being the foremost to bring forth the antecedents and the effects of employee engagement forth, the origin still is not very clear. Employee engagement though considered to be one of the five most important challenges a CEO has to face (Wah, 1990). The antecedents of employee engagement are-organizational commitment and extra-role behavior (Schaufeli, 2012). Engagement is also viewed as an autonomous, clear concept which is the opposite of burnout. A 2002 paper by Schaufeli et al. defines it as a "positive, fulfilling, work-related state of mind that is characterized by vigor, dedication, and absorption." The fully involved employee is expected to show elevated levels of mental resilience and energy which is what Vigor is in this concept. Being deeply involved in work being done attaching a lot of significance pride, enthusiasm, and challenge to one's job is one of the most identifying characteristics of an engaged person. This indicates dedication. Absorption is when one forgetting oneself in the job and find it difficult to have a difference between self and the job. This is in tandem with what Kahn (1990) defined as "harnessing of organization members' selves to their work roles: in engagement, people employ and express themselves."

Research has indeed time and again shown it clearly that employees who are engaged outperform even the employees who are satisfied because in comparison to satisfaction, because engagement relates to-enthusiasm, elation, excitement, and alertness while satisfaction relates to contentment, relaxation, serenity, and calmness which its more of satiation and not activation which is what engagement points to (Rich, Lepine, and Crawford, 2010). There is also a very clear demarcation between engaged employees and workaholics (Van Beek et al., 2012). Research also has shown that leaders have a key role in ensuring that engagement is happening because of the infectious nature it has been proven to have (Bakker, Van Emmerik, and Euwema, 2006). As long as employees are able to re-invent themselves through self-development processes, employees are expected to be engaged highly (Schaufeli, 2012). Engaged employees are able to modify the job design or content by negotiating job contents, selecting specific tasks to work with or by giving more meaning to work tasks, whereby they are able to keep a dynamic work environment, otherwise known as job crafting (Wrzeniewski and Dutton, 2001). While Bakker (2011) did say that these strategies of job crafting are used by engaged employees indeed, a peek into the reality workspace also gives the idea that it can happen only if certain factors are in place in the work environment.

There are daily aspects of work which can affect and employees' levels of engagement. Vigor, absorption, and dedication within a person are affected over a short period of time by these daily aspects. Variance in engagement up to 40% can be attributed to these dynamics in the work environment (Sonnetag, Dormann, and Demerouti, 2010). Daily interactions or events such as colleagues being supportive or autonomy in one's work and positive feedbacks received can create a variance in the work engagement of a person. Targeted HRM strategies can help increase and sustain employee engagement.

Employee engagement, research says can be built by utilizing the job design factors and motivations aspects of the job through the design and resources (Bakker, Oerlemans, and Ten Bummelhuis, 2012; Schaufeli and Salanova, 2008, 2010) by challenging employees whereby stimulating learning and motivation factors.

11.3.1 GENDER AND WORK ENGAGEMENT

A few studies that have been done in the direction of gender versus work engagement, it has been seen that work engagement is assumed to be neutral of gender implicitly (Pitt-Catsouphes and Matz-Costa, 2009). This aspect of neutrality of genders would then mean that effects on work engagement are actually individual and that irrespective of gender, a person can indeed be engaged in the work or job role depending on the personal experiences and result thereof. Gender is assumed to be the social differences between men and women in the studies (Tshilongamulenzhe and Takawira, 2015). The social system mentioned is more of an institutionalized system. Some research on the other hand also argues that work engagement is indeed gender-sensitive (Banihani et al., 2013). Men, the researchers say are regarded as more useful for having the characteristics of men and women and the womanly characteristics other hands are not regarded as useful for organizations. Hence it seems organizations are designed in such a way as to utilize the manly characteristics more and whereby delivering psychological meaningfulness to men whose characteristics are considered valuable than women. This would mean that men would be more engaged at work than women (Banihani et al., 2013, p. 412).

Williams (1993) has gone one more step forward to mention that women are penalized for being women while men are rewarded for being men. Crompton et al. (2007) also mention that women and men have very distinctive experiences throughout life and these experiences show that they may impact a woman's' capacity to be fully engaged at work and a man's capacity to be engaged at home. This then would indicate that being the main caregiver at home women might find themselves less available for work and this might be seen as reduced work engagement. Researchers (Banihani et al., 2013, p. 415) also argue that the fact that work engagement in the current corporate scenario is more male-oriented and thus gives women very less chance of experiencing psychological meaningfulness, availability, and safety which men enjoy given that the most cure for organizational problems which lie in work engagement are gendered. Schaufeli et al. (2006) study across the countries bought through a set of varying results in the relationship between gender and work engagement.

11.3.2   WOMEN AND WORK SPACE

Mckinsey's study on women in the workplace 2018 (Krivkovich et al., 2018) gives a birds eye view on the women who are exploring and living the career dream in the corporate world. The 2018 data involved 279 organizations and more than 13 million employees. Adding another 7 million from different avenues, 20 million people were said to have participated in the survey which includes people from different races, LGBTQ, and other ethnicities.

The BCG research series titles *women@bcg* have also found that organizations are failing to keep its senior women executives engaged. They argue that its good for the overall organization if the women employees can be engaged.

Since few years, organizations have been committing to have greater gender diversity among their employee pool. But the Mckinsey study shows the progress in this direction has been stunted. The study shows it's not the women. It's about the action the organizations have to take to support the women who have done their part by achieving the required credentials to work with them.

Research shows that women are under-represented in corporates especially in the top management. The organizations should also be aware of their decisions during the recruitment-training-promotion stages of the employees work-life for the actual progress to happen in the diversity area.

While the attrition levels between men and women are the same at 15% in 2017, it is evident that the numbers in the final employee count between men and women do not match because due to the fact the hiring of women at entry levels are very less and further promoting or hiring women to the managerial levels are also less likely. Hence, ultimately, we find that women are represented at 38% in the managerial levels while men are represented at 62%. Now, the less represented women cannot definitely match up in the ranks at the c-suite because they have few women in the managerial levels to begin with and much lesser that are promoted to reach the higher levels. The performance bias mentioned earlier from the researches also catches up here with the reality (Banihani et al., 2013; William, 1993).

Given that the organizations' framework is favoring men, (Banihani et al., 2013, p. 412) women also find that they have less support from the managers than their men counterparts, especially in the matter of organizational politics.

Women also are seen to have much less access to senior's organizational leaders or management than their male counterparts. And most opportunities to the higher levels of management are recommended by these senior leaders, women end up having no access to such considerations.

Another barrier women come across is the microaggression(s); they can be subtle, sexual, racist, and sometimes unintentional. Unfortunately, this shines a light on the inequality-women having to prove more for their competencies to be accepted than men, mistaken a junior employee much lower than their ranks, hear demeaning remarks about a promotion or personal choices such that of lesbian women are a few examples. Above all this, there was only 62% of employees who participated in the Mckinsey survey said that their organization put a strong step forward in declaring sexual harassment as no tolerated under any circumstances, even though 98% of organizations have policies in place to counter the harassment of sexual nature. Moreover if, women are less represented in a particular organization or position, then end being the "only" ones around and hence the representative for the entire race of women. This puts a lot of stress on the employee to ensure that she does not get women blamed or branded, making a good pathway to burnt out. Where would engagement be here?

11.4 IMPLICATIONS

The research has shown that quite a few countries are currently having or predicted to have more percentage of younger workforces as this century goes forward. India is one such country touted to have a billion people aged between 16 and 64 and is termed as "highest number of the young workforce" in by 2027. This would mean that countries like India would have very young people working with older age employees with vast experience.

The requirement for any organization is that all the employees work together to achieve the ultimate success. Apart from this, the female labor force participation rate is low in the world, and in countries like India, it's at 27%, according to the International Labor Organization (ILO). This would be caused by the still present patriarchy which is highly seen in most Asian and some Mediterranean cultures (Ghai, 2018). Patriarchy and social norms are seen in some parts of the world that have been seen to decrease female participation. It's necessary that the huge amount of loss occurred by nations across due to the suppression of women caused by gender has to be dealt with and measures bought out to reduce the losses. There is also the glass ceiling which women face. These issues are also faced by the minorities such as LGBT and colored people. Unless the issues are bought to fore and worked towards reducing the bipartisan outlook organizations, countries, and women as well as the minorities. Studies like this are necessary for this move towards a more inclusive workplace.

11.5 CONCLUSION

The director of the Women and Public Policy Program at Harvard Kennedy School and professor of public policy, Iris Bohnet encourage "gender equality nudges" – which are little changes that can be made to have a larger impact on bringing the organizations out of the gender bias in their policies and procedures. There are also strategies such as Job sharing, child care options, ensuring that sexiest language usage is noted, dealt with appropriately and most likely the most important would be job flexibility.

It would take immense learning and cultural changes to make women match the numbers or in this case the engagement levels and numbers of working employees in the organizational environment. What could be done is to take one sure strong foot at a time and progress surely? Each has its own strengths and competencies. Irrespective of the fact that few women reach the leadership level, there are enough data to show that women leaders are preferred more than their counterparts. Gender diversity will become a reality.

11.6 CASE: AMEYA AND/OR GENDER BIAS

Ameya was working with a very famous organization which was into the field of information technology. Ameya was the Marketing strategist. She reported to the GM Marketing. Ameya felt the job had possibilities for her to grow in learning as well as experience. She had a brilliant experience and record with the earlier firm. The GM marketing, let's call him Bob, was highly experienced and been in the organization for 12 years now. He started here as a junior marketing associate and has gone on to become the GM. Everyone was in awe of him. Ameya was the first lady in his team. The organization has a policy of having 10% of female members in every team.

Excited as Ameya was, every meeting the GM would look at her kindly and tell her to have a seat near him, anything she would say he will hear, then without any comment or reaction, he would move on to the next person or topic as if he never heard her say anything at all. She was never given any job or project and all were asked to do was to assist the other team members. She was feeling like she was s glorified assistant to the team members.

One day she was called by the GM to his cabin and given a project which was till then being handled by Mark. Mark was being let go due to some financial discrepancies found in the handling of the project's funds. Bob expressed instructed Ameya that the financial aspects he will personally handle (because "girls! Finances are beyond them. I have three girls you see and they cannot handle funds on their own anytime" to quote him) while she just needed to make sure the reports are typed in time and the client is kept happy "these are something a woman can do well?" in his words.

Ameya was given a team of two men who were experienced and much older to her. One of them told her – "you just sit there Ameya. We will handle the whole thing. Don't worry your pretty brain about it. Shocked she went to Bob and mentioned this. Bob told her-they are right! Why would you worry? I had given you the position because you are an MBA which they are not and this position required an MBA there on the paper. They are experienced in these kinds of projects. They will make sure it is done. You do not need to worry."

Ameya decided that this is not something she can fight openly. She slowly and surely started learning the way work was handled, and took over the reins of the project. Finally, one day the team member who earlier came to her and asked her not to worry said – "you get your way around here and with the client because you are a woman. Lucky you!"

Ameya had enough. She resigned. Bob called her to ask why. She said harassment. He was shocked. "How can you say that? We have protected you and supported you all the while!"

APPENDIX 11.1: CASE DISCUSSION

SYNOPSIS

This chapter is about gender diversity and its effect on the workspace as well as and women. The most frequently identified reasons for women being out of work or moving out of work even though highly educated are the social norms and prevalent patriarchy is most societies. Women have come forward in the last few years and have broken the said glass ceiling it has still been a much-less number of women coming up to the c-suite and more dropping off in between their career. There have been studies which found that women are competitive in the managerial profiles and in work assigned to them. Still due to the social norms prevailing in the societies, e.g., women have the work at home too while the man "tired" from office sits and sees the TV or the woman is expected to take a leave to take care of a sick child or parent. The world has to move on from these stereotyped roles assigned for women and allow them to move forward and achieve the full potential they could.

TEACHING OBJECTIVES

The chapter has the following four teaching objectives:
- This chapter gives an introduction about gender and its classification.
- The chapter lets the readers understand the challenges women face at the workspace.
- It gives analyzed reasons brought to light why women are found to stop their careers in the middle or never start a career ever.
- To make readers wonder about what could be done to have more women working and stay working and have fulfilling successful career.

KEYWORDS

- **diversity**
- **gender**
- **gender bias**
- **gender equality**
- **man**
- **perspectives**
- **woman**
- **women vs. men**
- **work engagement**

REFERENCES

Alderfer, C. P., (1972). *Existence, Relatedness, and Growth: Human Needs in Organizational Settings.*

Argyris, C., & Schon, D., (1982). *Theory in Practice*. San Francisco: Jossey-Bass.

Argyris, C., (1982). *Reasoning, Learning, and Action*. San Francisco: Jossey-Bass.

Bakker, A. B., (2011). An evidence-based model of work engagement. *Current Directions in Psychological Science*, 20, 265–269.

Bakker, A. B., Oerlemans, W. G., & Ten, B. L. L., (2013). Becoming fully engaged in the workplace: What individuals and organizations can do to foster work engagement. *The Fulfilling Workplace: The Organization's Role in Achieving Individual and Organizational Health*, pp. 55–69.

Bakker, A. B., Van, E. H., & Euwema, M. C., (2006). Crossover of burnout and engagement in work teams. *Work and Occupations*, 33, 464–489.

Banihani, M., Lewis, P., & Syed, J., (2013). Is work engagement gendered? *Gender in Management: An International Journal*, 28(7), 400–423.

Batra, R., & Reio, Jr. T. G., (2016). Gender inequality issues in India. *Advances in Developing Human Resources*, 18(1), 88–101.

Brinton, M. C., (1993). *Women and the Economic Miracle: Gender and Work in Postwar Japan* (Vol. 21). University of California Press.

Burke, R. J., (2016). *The Fulfilling Workplace: The Organization's Role in Achieving Individual and Organizational Health*. Routledge.

Crawford, E. R., Jeffery, A. L., & Bruce, L. R., (2010). "Linking job demands and resources to employee engagement and burnout: A theoretical extension and meta-analytic test." *Journal of Applied Psychology*, 95(5), 834.

Crompton, R., Lewis, S., & Lyonette, C., (2007). *Women, Men, Work, and Family in Europe* (pp. 159–182). Basingstoke: Palgrave Macmillan,

Csikszentmihalyi, M., (1982). *Beyond Boredom and Anxiety*. San Francisco: Jossey-Bass.

Deci, E. L., (1975). *Intrinsic Motivation*. New York: Plenum Press.

Devillard, S., Sancier-Sultan, S., & Werner, C., (2014). *Why Gender Diversity at the Top Remains a Challenge* (Vol. 2, pp. 1–3). McKinsey Quarterly.

Devillard, S., Sancier-Sultan, S., De Zelicourt, A., & Kossoff, C., (2016). *Women Matter 2016: Reinventing the Workplace to Unlock the Potential of Gender Diversity*. Retrieved from: Women-Matter-2016-Reinventing-the-Workplace-to-Unlock-Thepotential-of-Gender-Diversity.pdf, McKinsey & Co.

Devillard, S., Sancier-Sultan, S., De Zelicourt, A., & Kossoff, C., (2017). *Women Matter 2017: Reinventing the Workplace for Greater Gender Diversity*. McKinsey & Co.

Diamond, M. A., & Allcorn, S., (1985). Psychological dimensions of role use in bureaucratic organizations. *Organizational Dynamics*, 14(1), 35–59.

Dubin, R., (1956). Industrial workers' worlds: A study of the "central life interests" of industrial workers. *Social Problems*, 3, 131–142.

Ghai, S., (2018). *"The Anomaly of Women's Work and Education in India."*

Goffman, E., (1959). *The Presentation of Self in Everyday Life*. New York: Doubleday Anchor.

Goffman, E., (1961a). *Encounters: Two Studies in the Sociology of Interaction*. Indianapolis: Bobbs-Merrill Co.

Hackman, J. R., & Oldham, G. R., (1980). *Work Redesign*. Reading, Mass.: Addison-Wesley.

Hall, D. T., & Richter, J., (1989). Balancing work life and home life: What can organizations do to help? *Academy of Management Executive*, 2, 212–223.

Hochschild, A. R., (1983). *The Managed Heart*. Berkeley and Los Angeles.

http://ecowgen.ecreee.org/index.php/gender-inequality-a-case-study-of-women-under-represen-tation-in-the-energy-sector-an-analytical-approach-standtall/ (accessed on 25 February 2020).

https://news.harvard.edu/gazette/story/2016/03/the-costs-of-inequality-for-women-progress-until-they-get-near-power/ (accessed on 25 February 2020).
https://www.bcg.com/en-in/careers/women-at-bcg/improving-the-engagement-of-women-in-the-workforce.aspx (accessed on 25 February 2020).
https://www.mckinsey.com/about-us/new-at-mckinsey-blog/ten-years-of-women-matter-research (accessed on 25 February 2020).
https://www.mckinsey.com/business-functions/organization/our-insights/why-gender-diversity-at-the-top-remains-a-challenge (accessed on 25 February 2020).
https://www.summer.harvard.edu/inside-summer/gender-inequality-women-workplace (accessed on 25 February 2020).
https://www.theatlantic.com/business/archive/2015/12/gender-equality-workplace-2015/422328/ (accessed on 25 February 2020).
Kahn, W. A., (1992). To be fully there: Psychological presence at work. *Human Relations*, *45*(4), 321–349.
Klein, M., (1959). Our adult world and its roots in infancy. *Human Relations*, *12*, 291–303.
Klugman, J., Rodríguez, F., & Choi, H. J., (2011). The HDI 2010: New controversies, old critiques. *The Journal of Economic Inequality*, *9*(2), 249–288.
Krivkovich, et al., (2018). *Women in Work Place*. Mckinsey & Company Report.
Langer, E. J., (1989). *Mindfulness*. Reading, Mass.: Addison-Wesley.
Lawler, E. E., & Hall, D. T., (1970). Relationships of job characteristics to job involvement, satisfaction, and intrinsic motivation. *Journal of Applied Psychology, 54*, 305–312.
Maslach, C., (1982). *Burnout: The Cost of Caring*. Englewood.
Maslow, A., (1954). *Motivation and Personality*. New York: Harper & Row.
Menzies, I., (1975). A case study in the functioning of social systems as a defense against anxiety. In: Colman, A., & Bexton, W., (eds.), *Group Relations Reader*. Sausalito, CA: GREX.
Miller, E. J., & Rice, A. K., (1967). *Systems of Organization*. London: Tavistock.
Minuchin, S., (1974). *Families and Family Therapy*. Cambridge, MA: Harvard University Press.
Mowday, R., Porter, L., & Steers, R., (1982). *Organizational Linkages: The Psychology of Commitment, Absenteeism, and Turnover*. New York: Academic Press.
Ogunbodede, G., (2015). *Gender Inequality, A Case Study of Women Under-Representation in the Energy Sector (an Analytical Approach) Stand-Tall*.
Ollmann, B., (1971). *Alienation*. Cambridge, MA: Cambridge University Press.
Pitt-Catsouphes, M., & Matz-Costa, C., (2009). *Engaging the 21st Century Multigenerational Workforce: Findings from the Age and Generations Study*. Metlife Mature Market Institute Report.
Rogers, C. R., (1958). The characteristics of a helping relationship. *Personnel and Guidance Journal, 37*, 6–16.
Schaufeli, W. B., & Salanova, M., (2008). Enhancing work engagement through the management of human resources. In: Näswall, K., Sverke, M., & Hellgren, J., (eds.), *The Individual in the Changing Working Life* (pp. 380–404). Cambridge: Cambridge University Press.
Schaufeli, W. B., & Salanova, M., (2010). How to improve work engagement? In: Albrecht, S., (ed.), *The Handbook of Employee Engagement: Perspectives, Issues, Research and Practice* (pp. 399–415). Northampton, MA: Edwin Elgar.
Schaufeli, W. B., & Salanova, M., (2011). Work engagement. On how to better catch a slippery concept. *European Journal for Work and Organizational Psychology, 20*, 39–46.
Schaufeli, W. B., Bakker, A. B., & Salanova, M., (2006). The measurement of work engagement with a short questionnaire: A cross-national study. *Educational and Psychological Measurement, 66*(4), 701–716.

Schaufeli, W., (2012). Work engagement: What do we know and where do we go? *Romanian Journal of Applied Psychology, 14*(1), 3–10.
Schein, E. H., (1987). *Process Consultation* (Vol. I). Reading, MA: Addison-Wesley.
Shapiro, E. R., & Carr, A. W., (1991). *Lost in Familiar Places*. New Haven, CT: Yale University Press.
Smith, K. K., & berg, D. N., (1987). *Paradoxes of Group Life*. San Francisco: Jossey-Bass.
Sonnentag, S., Dormann, C., & Demerouti, E., (2010). Not all days are created equal: The concept of state work engagement. In: Bakker, A. B., & Leiter, M. P., (eds.), *Work Engagement: A Handbook of Essential Theory and Research* (pp. 25–38). New York: Psychology Press.
Tartari, V., & Salter, A., (2015). The engagement gap: Exploring gender differences in University-Industry collaboration activities. *Research Policy, 44*(6), 1176–1191.
Taylor, F. W., (1911). *Principles of Scientific Management*. New York: Harper & Row.
Tshilongamulenzhe, M. C., & Takawira, N., (2015). Examining gender influence on employees' work engagement within a South African university. *Risk Governance and Control, 5*(2).
Van Beek, I., Hu, Q., Schaufeli, W. B., Taris, T., & Schreurs, B. H., (2012). For fun, love or money. What drives workaholic, engaged and burned-out employees at work? *Applied Psychology: An International Review, 61*, 30–55.
Williams, C., (1993). *Doing Women's Work: Men in Non-Traditional Occupations*. Sage, London.

FURTHER READING

Abbott, R. D., Donahue, R. P., Kannel, W. B., & Wilson, P. W. F., (1988). The impact of diabetes on survival following myocardial infarction in men vs. women: The Framingham study. *JAMA: The Journal of the American Medical Association, 260*(23), 3456–3460.
Acharya, A., & Gupta, M., (2016a). An application of brand personality to green consumers: A thematic analysis. *Qualitative Report, 21*(8), 1531–1545.
Acharya, A., & Gupta, M., (2016b). Self-image enhancement through branded accessories among youths: A phenomenological study in India. *Qualitative Report, 21*(7), 1203–1215.
Barnes, T. D., & Beaulieu, E., (2017). Engaging women: Addressing the gender gap in women's networking and productivity. *PS - Political Science and Politics, 50*(2), 461–466.
Bell, F., & Bell, A., (2017). Legal and medical aspects of diverse gender identity in childhood. *Journal of Law and Medicine, 25*(1), 229–247.
Brinia, V., (2012). Men vs. women, educational leadership in primary schools in Greece: An empirical study. *International Journal of Educational Management, 26*(2), 175–191.
Burke, W., Beeker, C., Kraft, J. M., & Pinsky, L., (2000). Engaging women's interest in colorectal cancer screening: A public health strategy. *Journal of Women's Health and Gender-Based Medicine, 9*(4), 363–371.
Byron, P., & Hunt, J., (2017). 'That happened to me too': Young people's informal knowledge of diverse genders and sexualities. *Sex Education, 17*(3), 319–332.
Cheng, C. I., Yeh, K. H., Chang, H. W., Yu, T. H., Chen, Y. H., Chai, H. T., & Yip, H. K., (2004). Comparison of baseline characteristics, clinical features, angiographic results, and early outcomes in men vs. women with acute myocardial infarction undergoing primary coronary intervention. *Chest, 126*(1), 47–53.
Clark, N., (2008). Nutrition: The battle of the bulge: Men vs. women. *National Strength and Conditioning Association Journal, 15*(6), 55–58.

Coakley, M., Fadiran, E. O., Parrish, L. J., Griffith, R. A., Weiss, E., & Carter, C., (2012). Dialogues on diversifying clinical trials: Successful strategies for engaging women and minorities in clinical trials. *Journal of Women's Health*, *21*(7), 713–716.

Crawford-Browne, S., & Kaminer, D., (2012). The use of concept mapping in engaging women to identify the factors that influence violence. *Journal of Psychology in Africa*, *22*(4), 527–535.

Davidson, S., Morrison, A., Skagerberg, E., Russell, I., & Hames, A., (2018). A therapeutic group for young people with diverse gender identifications. *Clinical Child Psychology and Psychiatry*, 1359104518800165.

Dischinger, P. C., Kufera, J. A., Ho, S. M., Ryb, G. E., & Wang, S., (2016). On equal footing: Trends in ankle/foot injuries for men vs. women. *Traffic Injury Prevention*, *17*, 150–155.

Elnakat, A., & Gomez, J. D., (2015). Energy engenderment: An industrialized perspective assessing the importance of engaging women in residential energy consumption management. *Energy Policy*, *82*(1), 166–177.

Evans, J. M. M., Ryde, G., Jepson, R., Gray, C., Shepherd, A., Mackison, D., et al., (2016). Accessing and engaging women from socio-economically disadvantaged areas: A participatory approach to the design of a public health intervention for delivery in a Bingo club. *BMC Public Health*, *16*(1).

Gardiner, P. M., McCue, K. D., Negash, L. M., Cheng, T., White, L. F., Yinusa-Nyahkoon, L., Jack, B. W., & Bickmore, T. W., (2017). Engaging women with an embodied conversational agent to deliver mindfulness and lifestyle recommendations: A feasibility randomized control trial. *Patient Education and Counseling*, *100*(9), 1720–1729.

Giscard, D. S., (2017). Engaging women in countering violent extremism: Avoiding instrumentalization and furthering agency. *Gender and Development*, *25*(1), 103–118.

Glick, D. A., Krishnan, M. C., Fisher, S. K., Lieberman, R. E., & Sisson, K., (2016). Redefining residential: Ensuring competent residential interventions for youth with diverse gender and sexual identities and expressions. *Residential Treatment for Children and Youth*, *33*(2), 107–117.

Grote, N. K., Zuckoff, A., Swartz, H., Bledsoe, S. E., & Geibel, S., (2007). Engaging women who are depressed and economically disadvantaged in mental health treatment. *Social Work*, *52*(4), 295–308.

Gupta, M., & Kumar, Y., (2015). Justice and employee engagement: Examining the mediating role of trust in Indian B-schools. *Asia-Pacific Journal of Business Administration*, *7*(1), 89–103.

Gupta, M., & Pandey, J., (2018). Impact of student engagement on affective learning: Evidence from a large Indian University. *Current Psychology*, *37*(1), 414–421.

Gupta, M., & Ravindranath, S., (2018). Managing physically challenged workers at micro sign. *South Asian Journal of Business and Management Cases*, *7*(1), 34–40.

Gupta, M., & Sayeed, O., (2016). Social responsibility and commitment in management institutes: Mediation by engagement. *Business: Theory and Practice*, *17*(3), 280–287.

Gupta, M., & Shaheen, M., (2017a). Impact of work engagement on turnover intention: Moderation by psychological capital in India. *Business: Theory and Practice*, *18*, 136–143.

Gupta, M., & Shaheen, M., (2017b). The relationship between psychological capital and turnover intention: Work engagement as mediator and work experience as moderator. *Journal Pengurusan*, *49*, 117–126.

Gupta, M., & Shaheen, M., (2018). Does work engagement enhance general well-being and control at work? Mediating role of psychological capital. *Evidence-Based HRM*, *6*(3), 272–286.

Gupta, M., & Shukla, K., (2018). An empirical clarification on the assessment of engagement at work. *Advances in Developing Human Resources, 20*(1), 44–57.

Gupta, M., (2017). Corporate social responsibility, employee-company identification, and organizational commitment: Mediation by employee engagement. *Current Psychology, 36*(1), 101–109.

Gupta, M., (2018). Engaging employees at work: Insights from India. *Advances in Developing Human Resources, 20*(1), 3–10.

Gupta, M., Acharya, A., & Gupta, R., (2015). Impact of work engagement on performance in the Indian higher education system. *Review of European Studies, 7*(3), 192–201.

Gupta, M., Ganguli, S., & Ponnam, A., (2015). Factors affecting employee engagement in India: A study on offshoring of financial services. *Qualitative Report, 20*(4), 498–515.

Gupta, M., Ravindranath, S., & Kumar, Y. L. N., (2018). Voicing concerns for greater engagement: Does a supervisor's job insecurity and organizational culture matter? *Evidence-Based HRM, 6*(1), 54–65.

Gupta, M., Shaheen, M., & Das, M. (2019). Engaging employees for quality of life: mediation by psychological capital. *The Service Industries Journal, 39*(5–6), 403–419.

Gupta, M., Shaheen, M., & Reddy, P. K., (2017). Impact of psychological capital on organizational citizenship behavior: Mediation by work engagement. *Journal of Management Development, 36*(7), 973–983.

Hawkins, N., & Ronchi, E., (2008). Seeking a seat at the policy table: Engaging women in biotechnology research and in decision making. *Women in Biotechnology: Creating Interfaces,* 71–91.

Kim, K. A., Fann, A. J., & Misa-Escalante, K. O., (2011). Engaging women in computer science and engineering: Promising practices for promoting gender equity in undergraduate research experiences. *ACM Transactions on Computing Education, 11*(2).

Kusumadevi, M. S., Dayananda, G., Veeraiah, S., Elizabeth, J., & Kumudavathi, M. S., (2011). The perception of intramuscular injection pain in men vs. women. *Biomedical Research, 22*(1), 107–110.

MacInnis, C. C., & Hodson, G., (2015). Why are heterosexual men (vs. women) particularly prejudiced toward gay men? A social dominance theory explanation. *Psychology and Sexuality, 6*(3), 275–294.

Malins, P., (2016). How inclusive is "inclusive education" in the Ontario elementary classroom?: Teachers talk about addressing diverse gender and sexual identities. *Teaching and Teacher Education, 54,* 128–138.

Martin, L. A., Neighbors, H. W., & Griffith, D. M., (2013). The experience of symptoms of depression in men vs. women: Analysis of the national comorbidity survey replication. *JAMA Psychiatry, 70*(10), 1100–1106.

McDougall, C. L., Leeuwis, C., Bhattarai, T., Maharjan, M. R., & Jiggins, J., (2013). Engaging women and the poor: Adaptive collaborative governance of community forests in Nepal. *Agriculture and Human Values, 30*(4), 569–585.

McEwan, J. R., (2010). Engaging women in academic medicine in the UK: Report of a workshop at the association of physician's annual meeting. *QJM, 103*(9), 635–639.

Myors, K. A., Johnson, M., Cleary, M., & Schmied, V., (2015). Engaging women at risk for poor perinatal mental health outcomes: A mixed-methods study. *International Journal of Mental Health Nursing, 24*(3), 241–252.

Ngamvithayapong-Yanai, J., Luangjina, S., Nedsuwan, S., Kantipong, P., Wongyai, J., & Ishikawa, N., (2013). Engaging women volunteers of high socioeconomic status in supporting socioeconomically disadvantaged tuberculosis patients in Chiang Rai, Thailand. *Western Pacific Surveillance and Response Journal: WPSAR, 4*(1), 34–38.

Opare, S., (2005). Engaging women in community decision-making processes in rural Ghana: Problems and prospects. *Development in Practice, 15*(1), 90–99.

Osnes, B., (2013). Engaging women's voices through theatre for energy development. *Renewable Energy, 49,* 185–187.

Pandey, J., Gupta, M., & Naqvi, F., (2016). Developing decision-making measures a mixed-method approach to operationalize Sankhya philosophy. *European Journal of Science and Theology, 12*(2), 177–189.

Riggs, D. W., (2012). Talking about 'diverse genders and sexualities' means talking about more than White middle-class queers. *Educational Diversity: The Subject of Difference and Different Subjects,* 219–235.

Riggs, D. W., Taylor, N., Signal, T., Fraser, H., & Donovan, C., (2018). People of diverse genders and/or sexualities and their animal companions: Experiences of family violence in a binational sample. *Journal of Family Issues, 39*(18), 4226–4247.

Rodriguez, M. T., & Roberts, T. G., (2013). Engaging women through common initiative groups in Cameroon. *Journal of International Agricultural and Extension Education, 20*(2), 135–137.

Searle-Chatterjee, M., (2011). Men vs. women. *New Scientist, 211*(2824), 33.

Sheehan, R., & Vadjunec, J. M., (2016). Roller derby's publicness: Toward greater recognition of diverse genders and sexualities in the Bible belt. *Gender, Place and Culture, 23*(4), 537–555.

Sindik, J., Mikić, Z. K., Dodigović, L., & Čorak, S., (2016). Analysis of the relevant factors for the engaging women in various sports in Croatia. *Montenegrin Journal of Sports Science and Medicine, 5*(1), 17–28.

Sydó, N., Abdelmoneim, S. S., Mulvagh, S. L., Merkely, B., Gulati, M., & Allison, T. G., (2014). Relationship between exercise heart rate and age in men vs. women. *Mayo Clinic Proceedings, 89*(12), 1664–1672.

Takano, S., (2000). The myth of a homogeneous speech community: A sociolinguistic study of the speech of Japanese women in diverse gender roles. *International Journal of the Sociology of Language, 146,* 43–85.

Taylor, N., Riggs, D. W., Donovan, C., Signal, T., & Fraser, H., (2018). People of diverse genders and/or sexualities caring for and protecting animal companions in the context of domestic violence. *Violence Against Women.*

Vishram, S., Crosland, A., Unsworth, J., & Long, S., (2007). Engaging women from South Asian communities in cardiac rehabilitation. *British Journal of Community Nursing, 12*(1), 13–18.

Visram, S., Crosland, A., Unsworth, J., & Long, S., (2008). Engaging women from South Asian communities in cardiac rehabilitation. *International Journal of Therapy and Rehabilitation, 15*(7), 298–304.

Walsh, C. A., Rutherford, G., & Kuzmak, N., (2010). Engaging women who are homeless in community-based research using emerging qualitative data collection techniques. *International Journal of Multiple Research Approaches, 4*(3), 192–205.

Woolhouse, S., Brown, J. B., & Thind, A., (2011). 'Meeting people where they're at': Experiences of family physicians engaging women who use illicit drugs. *Annals of Family Medicine, 9*(3), 244–249.

Xiaoning, S., (2007). Why gender matters in CMC? Supporting remote trust and performance in diverse gender composition groups via IM. *Lecture Notes in Computer Science (Including Subseries Lecture Notes in Artificial Intelligence and Lecture Notes in Bioinformatics)*, *4663* LNCS(Part 2), pp. 626–627.

Zolait, A. H., Isa, S. M., Ali, H. M., & Sundram, V. P. K., (2018). Men vs. women: Study of online shopping habits and factors influencing buying decisions in Bahrain. *International Journal of E-Services and Mobile Applications*, *10*(4), 61–73.

CHAPTER 12

Volunteering for Community: Learning and Challenges in Diversity

TRILOK KUMAR JAIN

Professor, School of Business and Commerce, Manipal University Jaipur, Rajasthan, India, Tel.: (+91) 9414430763, E-mail: jain.tk@gmail.com

ABSTRACT

The present chapter is an attempt to understand the diversity and principles of diversity management. The author contends that every society and every organization must try to foster diversity. Diversity requires training of people, which can be done through any means, and volunteering is one such means to impart required training. The author tries to present volunteering as a tool for diversity management. The author presents a few small cases to let the reader think and analyze common situations relating to diversity.

12.1 INTRODUCTION

There is no one perfect answer to the following cases discussed in this chapter; however, the purpose is just an analysis of similar situations. The chapter also discusses the possible strategies of the chapter instructors. The author shares some perspectives, which may or may not be appropriate in every situation.

The diversity here refers to relative differences among people due to age, gender, race, caste, religion, skills, background, values, and ethnicity. Diversity at the place of work refers to having employees from different backgrounds working together. Diversity is an opportunity as well as a challenge. While diversity enables people to learn from each other and to develop specialized skills, it also creates possibilities of increased conflicts and confrontations.

Diversity at the place of work has an important role to play in the increasingly globalized world. The diversity in population increases due to workforce-migration and due to the increasing polarization of skill centers. Due to increasing workforce diversity, there is also a need of initiatives towards diversity management. Diversity management refers to all the tools that directly or indirectly increase capabilities of people to work in diversified environment.

12.1.1 WHAT IS A COMMUNITY?

A group of people who have close interaction among themselves can be called a community. This community can have physical proximity or may have some other form of proximity-which permits frequent interaction of ideas, opinions, and judgments. In the present day world of modern urban societies, we have more communities online than in the physical world due to reducing physical interaction among people. However, there are still closely knit communities that we can find in tribal societies and rural areas. These communities still follow age-old systems of development. They are slow to change, and take years to accept or reject a practice. However, online communities are very fast to adapt, evolve, and develop. Whatever may be format of community, the development of community is a very important issue.

Community development is a continuous process and requires investment of time and energy on the part of community. Community development is an outcome of sustained efforts towards the development of a belief system, ecosystem, and a method of living together. Community development can take place through the management of culture, fostering diversity, and enabling cross-cultural learning. Thus, community management or organizational development are similar processes in some ways. Both these processes focus on building the required culture, ecosystem, and resilience.

12.2 NEED FOR DISCUSSING THIS THEME

Diversity is very important in the growth of organizations. Diversity can help in fostering an environment of creativity, innovation, and tolerance in an organization. Diversity creates an opportunity for individuals to experiment, innovate, and be different. When employees work in a diversified work environment, their tolerance for differences increases and they start appreciating differences in ideas, opinions, and presentations. Diversity helps an organization in fostering the environment of creativity, which is the need of the hour for survival and growth of organizations in modern competitive industrial environments. Diversity introduces the values that can help the organization in its growth and development. There are many advantages of diversity, which include the following:

- Diversity introduces the possibility for innovation and creativity.
- Diversity enables people to accept divergence.
- Diversity enables people to develop a wider perspective.
- Diversity enables people to develop a global vision and prepares their competitive strengths.
- Diversity management is an essential tool for the development of organizations and society.
- Volunteering is an important tool for diversity management. Through volunteering, people can develop a better understanding and can create a positive environment around them.
- Every employee wants to contribute something to the society, so volunteering can be used to induce employees to help the society. Similarly, volunteering can be used in educational institutions and other forums.

While there are many approaches and models towards diversity management, this chapter highlights the role of volunteering as a tool of diversity management. Volunteering refers to participation by people in social development initiatives without any financial or other rewards or compulsions. Thus, it is something that a person undertakes for sheer passion, joy, and purpose. Thus, volunteers are driven by inner motivation to serve, a desire to participate in developmental work, and a broader perspective. Educational institutions have various forums to foster volunteering, for example, Scouts, Guides, National Service Scheme (NSS), National Credit Corps, and Clubs, etc. They carry a noble idea with them-which may directly or indirectly help the society in its development. The present chapter introduces the role of volunteering in diversity management.

12.3 REVIEW OF THIS THEME

Diversity is defined as a characteristic of a group, where people come from different demographic backgrounds. These members of the group exhibit different backgrounds in terms of religion, gender, ethnicity, or other such marked differences (Ely and Roberts, 2008; Ely and Thomas, 2001).

Consistent with labor predictions, the workforce of the 21st century may be characterized by increased numbers of women, minorities, ethnic backgrounds, intergenerational workers, and different lifestyles (Langdon, McMenamin, and Krolik, 2002). The ability of organizations to handle diversity well results in their overall efficiency and effectiveness (Harvey, 1999; Kuczynski, 1999). 75% of the Fortune 1000 companies have institutionalized diversity management initiatives (Daniels, 2001). Thus, diversity management has become a corporate necessity.

Diversity includes both differences and similarities and therefore diversity is a broader concept than just the word difference (Thomas, 1995). Wentling and Palma-Rivas (1997) propose a broad and one narrow perspective of diversity (Mor Barak, 2005). Broader understand may include personality, education, language, etc., while the narrow perspective may focus on the issues covered in the equal employment opportunity (EEO) law, and focuses on issues like race, gender, ethnicity, age, national origin, religion, and disability. Diversity can take up a variety of issues in its broader meaning including health, income, age, and other such categories for grouping people. Diversity is simply "all the ways in which we differ."

Morrison (1992) categorizes diversity in four levels:

- Diversity as racial/ethnic/sexual balance;
- Diversity as understanding other cultures;
- Diversity as culturally divergent values; and
- Diversity as broadly inclusive (cultural, subcultural, and individual).

Griggs (1995) offers another classification of diversity in terms of primary and secondary dimensions. The issues which are hard to change and may continue with us throughout our lives have been categorized as primary diversity and includes issues like race, gender, etc. While those issues, which can change over a period of time may be covered in the secondary dimensions and include issues like education, place of work, etc. Norton and Fox (1997) argue that employee diversity and organizational change are inextricably linked, and that these two elements have rarely been integrated sufficiently to meet the demands of today's fast-paced economy.

Grobler (2002, p. 46) asserts that each individual is unique. Diversity can be on primary or secondary dimensions. Primary dimensions can be primary factors including age, gender, etc. Secondary dimensions can be religion, education, income, etc. Globalization has enabled people to interact with different cultural and ethnic diversity. This has helped people in their pursuit of growth and development. Diversity can be a challenge also. When people are not able to accept, appreciate, and support diversity, there can be a challenge. Every society or organization has to create structures and processes to foster acceptance of diversity.

Diversity can have both positive and negative impacts. It may nurture the creativity and innovative capabilities of the group. However, it may also create a lack of desired bonding and thereby may cause the group members to experience dissatisfaction and isolation (Milliken and Martins, 1996). It can also be a source of new ideas. Diversity can also be a source of development of people. Various research studies show that various types of team and organizational diversity sometimes increase conflict, reduce social cohesion, and increase employee turnover (Jackson, Joshi, and Erhardt, 2003; Webber and Donahue, 2001). With the increase in globalization, the diversity among work-force is increasing (Johnson, 2002; Yaprak, 2002). We are witnessing greater number of younger employees and greater number of women in workforce these days (Mor-Barak, 2005; Gorski, 2002). Every organization is today preparing itself for the increasing diversity of tomorrow and the resultant challenges.

Organizations need to become innovative in order to compete and survive. There are only one way-make people innovative. There is increase reliance on workforce diversity for gaining organizational innovative capabilities (Mumford and Licuanan, 2004; West and Anderson, 1996). Workforce diversity also enables organizations to create fair practices, greater gender participation and better appreciation of workforce related laws, thus it ultimately helps in improving organizational performance. Workforce diversity has a positive impact on overall organizational performance (Williams and O'Reilly, 1998). Richard (2000), Richard (1999) and Richard and Johnson (2000) have suggested that HR practices can further stimulate organizational performance through promotion of workforce diversity.

Watson, Kumar, and Michaelsen, (1993) studies relation between diversity and performance and finds positive effects on organizational effectiveness, but, Ancona and Caldwell, (1992b) finds opposite impact on organizational performance (Chatman, Polzer, Barsade, and Neale, 1998; Richard and Johnson, 2001) assert that there are contextual factors that influence performance. Diversity has an impact on organizational performance and organizational dynamism (Jackson, Joshi, and Eisenhardt, 2003; Webber and Donahue, 2001). Thus, workforce diversity management has become an important issue for the management of the companies (Mor-Barak, 2005).

Workplace diversity is a matter of study for organizational researchers (Janssens and Steyaert, 2003). It has been conceptualized by researchers from several viewpoints. Diversity is an issue which has attracted both narrow and broad perspectives (Nkomo, 1995). Those scholars who are focusing on issues like race, gender, etc. are carrying a narrow perspective (e.g., Cross,

Katz, Miller, and Seashore, 1994). Some scholars focus on diversity based on the functional skills of workers (Nkomo, 1995). Jackson, May, and Whitney (1995) state that diversity includes differences in age, gender, income, race, values, attitude, and various other parameters, which can influence the overall personality of people. Diversity includes differences that enable differences in expressions (Thomas, 1991). McGath, Berdahl, and Arrow (1995) classify diversity on the basis of following criteria: age, ethnicity, gender, sexual orientation, physical status, religion, and education; task-related knowledge, skills, and capacities; values, attitudes, positions, and hierarchical status.

Williams and O'Reilly (1998) say that there are three ways to look at workplace diversity including social factors (for example ethic groups), similarity or attraction related factors and decision making, and information processing factors. Turner (1987) uses gender, ethnicity in classifying people. Some scholars use race, similarity, attraction, etc. as criteria for classification (Berscheid and Walster, 1978). Some scholars use workplace expertise, decision-making criteria in classifying workers (Wittenbaum and Stasser, 1996). Thus, the classification of people on different criteria can have different impacts on a study relating to diversity.

Some scholars believe that identifying a group with some distinct characteristics and classifying that group as "out-group" may disrupt the working of the overall working of the organization and the overall group dynamics (Pelled, Cummings, and Kizilos, 1999; Tsui, Egan, and O' Reilly, 1992). Consistent with this, research on self-categorization theory has shown that out-group members evoke more disliking, distrust, and competition than in-group members (Hogg, Cooper-Shaw, and Holz worth, 1993). Moreover, biases against out-group members seem to unfold automatically: the perception of a salient quality (e.g., race, sex) more or less inevitably triggers a corresponding categorization (Fiske and Neuberg, 1990).

In addition, if out-group members come from cultures or subcultures with which in-group members are unfamiliar, linguistic or paralinguistic differences may foster miscommunication and misunderstanding (Hambrick, Davison, Snell, and Snow, 1998; Palich and Gomez-Mejia, 1999). Issues like attitude, perceptions, etc. may cause subtle differences among groups, whereby these differences may reduce the performance of the organization and may cause communication problems in the organization (Palich and Gomez-Mejia, 1999).

Some researchers find that diversified workforce has less commitment to their organizations (Harrison, Price, and Bell, 1998), they communicate less among themselves and may perform less efficiently and slowly and may experience disruptions due to conflicts and discords (Watson et al., 1993; Tsui, Egan, and O' Reilly, 1992; Pelled, Eisenhardt, and Xin, 1999; Hambrick, Cho, and Chen, 1996).

Diversity management is the use of tools to promote equal opportunities, to promote participatory approach and to foster greater opportunities for expression and it has a positive impact on the overall performance of the organization (for example, Christensen, 1993; Elmuti, 1993; Kandola, 1995; Liff, 1997, 1999; Kramer, 1998; Hughes, 1999; Devine et al., 2007). Kossek and Pichler (2006) argued that the best practices for diversity focus on selecting for diversity, reducing workplace discrimination, and generating financial effectiveness. Diversity Management refers to the use of organized practices to promote greater understanding among members of the organization. Milliken and Martins (1996), diversity management refers to the ability to manage heterogeneous groups.

Woods and Sciarini (1995) suggest that those organizations, which promote diversity and institutionalize diversity management, emerge as successful and innovative organizations. NASSCOM-Mercer, have carried out extensive research in the IT Industry in India. They have presented reports including "Gender Inclusivity and Diversity in the Indian IT-BPO Industry" (2008), "Gender inclusivity in India: building empowered organizations" (2009), and "Workforce inclusiveness in Indian IT industry" (2009) has highlighted that India has more working women than any other country in the world in the IT industry. The Indian IT and BPO industry has given plenty of opportunities to women, so out of their total employment, over 35% are women.

NASSCOM-PWC (2010) reports on "Diversity in action," identified that gender is an important diversity issue in the Indian context, while cultural diversity may be a bigger issue in developed countries in the west. The report presents findings from the study, which reveals that the workers feel a need of better treatment to women.

The research work by Kulik et al. (2011) supports the idea that there is a positive relation between workforce diversity, proper treatment to gender and organizational success and impact. Diversity management requires commitment from the top management, and an active participation by the HR manager (McDonald, 1999). Cox (1993). Hanaoka (1999) have added three diversity promoting aspects in human resource (HR) management: (1) the diversity climate-consisting of individual, inter-group, and organizational-level factors, (2) Individual career outcomes-containing the two measures of affective outcomes and achievement outcomes, and (3) organizational effectiveness-which is concerned with improving workforce quality and achievement of organizational goals.

12.3.1 VOLUNTEERING

John Wilson describes volunteering in the Annual Review of Sociology (2000) as "Volunteering is any activity in which time is given freely to benefit another person, group, or cause. Volunteering is part of a cluster of helping behaviors, entailing more commitment than spontaneous assistance but narrower in scope than the care provided to family and friends. Although developed somewhat independently, the study of volunteerism and of social activism has much in common." Volunteering is the involvement of youth in helping others-therefore it is the manifestation of philanthropic spirit on the part of people. It is primarily for supporting others in need. The volunteers are those persons, who spare their time and resources for helping others. They don't derive any financial or other benefits from this help. They extend their help with the intention of helping others and deriving the pleasure of being a part of philanthropic pursuits.

12.3.2 EVOLUTION OF THE THEME

During the 1960s and 1970s, diversity management was introduced in the USA. President John F. Kennedy in 1961 introduced a President's Committee on Equal Employment Practices with the goal of ending discrimination in employment by the government. The Civil Rights Act (of the USA) of 1964 was another step in this direction, which tried to curb any kind of discrimination. The initiatives to promote diversity have been increasing since then. Many countries have enacted laws for equal opportunities, fair recruitment practices and for equal pay for equal work. Many organizations are today promoting diversity based on gender, caste, background, etc.

Diversity has always been valued in India. India is a country of diversity, and this diversity has contributed to the rich culture and heritage of the Indian subcontinent. People from different castes, religions, backgrounds are appreciated and recognized for their specialized contributions. The willingness of people to appreciate diversity and promote it has been the cardinal value of Indian culture and civilization. This spirit of furthering diversity has continued today in modern-day organizations also and is increasing due to growing worldwide movement towards diversity. Companies like Infosys,

ICICI, etc. have started ensuring that at least one-third of their employees including top management are women. Similar initiatives are voluntarily taken by almost every important company in India. The government is also planning to introduce reservation of seats for women in parliament-but the proposal is still lying pending. When this will be implemented, it will give a boost to gender participation. Reservation for women at the Panchayat level election is already there, which has certainly improved participation of women in political processes. Educational institutions have been using various means of volunteering to foster creativity, innovation, and learning appetite among the students.

12.3.3 DESCRIPTION OF THE TOPIC

Diversity can be developed and nurtured through continuous efforts. Diversity can enable people to prepare their own competitive capabilities. Volunteering can become one method for developing the capabilities of people for their own society. If volunteering is based on diversity, it would foster an overall global vision and will help in the development of a global character among the students.

Various companies have started an employee volunteering program to engage employees in community development. These initiatives are based on the voluntary initiatives of the companies. These initiatives enable companies to transform their employees into social development initiators. These employees spare some time to help the society in its development. Companies are able to engage a large number of their employees in such positive initiatives. These initiatives have two advantages-the employees engage themselves in a work which they want to do, and the company is able to contribute towards social development through its employees. Both are able to get benefits in terms of positive vibrations, positive thinking, social connect and better public image. Such initiatives have given indirect rewards to the company.

The initiatives towards employee volunteering are also helpful tools in creating a better bondage among employees. Employees work in informal groups in proving their services to the society. These informal groups are based on interest; there is a possibility of creating a good bondage among the employees. These voluntary groups enable employees to create bondage across diversity. Thus, employees from different regions (but with common interests) come together as volunteers and end up as great teams pursuing similar goals. These initiatives can contribute substantially towards diversity management.

12.4 IMPLICATIONS

Managers have to take up leadership roles to stretch the organizations beyond their present domains. The roles that managers have to take up will depend on their ability to understand and adapt to new situations. The managers must continuously strive to acquire new skills and capabilities. Fostering a culture of innovation and tolerance is crucial for desired growth. In order to bring about desired changes in the mindset of people, managers can take up many paths including volunteering. If the workers are keen to take up volunteering for some social cause, the managers should create a supportive environment and try to help the employees in their initiatives. If some relaxation and administrative support can help people in pursuing their volunteering initiatives, the managers must come forward to support such employees. There will be a long term positive impact on all the stakeholders through such initiatives. The organization will be considered a proactive and positive organization, while the volunteers will earn respect, self-pride, and a sense of purpose and better appreciations of society at large. The organizations that foster volunteerism will be able to develop true leaders for the society, who will take up important roles in the growth of the organization itself in the future. Thus, there is a need for concerted efforts to promote volunteerism among employees.

12.5 CONCLUSION

Diversity is a blessing for organizational and societal growth-provided it is properly nurtured. There is a need for well-planned actions to foster diversity. Diversity management revolves around various tools and approaches to create an environment for learning, tolerance, and mutual acceptance. This requires carefully planned actions on the part of leaders. Volunteering is an important tool to foster diversity. Volunteering is based on initiatives on the part of the volunteers to immerse themselves in society and learn and give back to society through some planned initiatives. Volunteers learn from society or groups and they try to codify their learning. They also try to help in the routine processes of society. Society and organizations pass through many problems. The most common problems are rising narrow-mindedness and rising clashes among people from different backgrounds. This is an opportunity for volunteers to spread good thoughts and solidarity among people. They can also contribute to the growth of society by creating shared success stories. Volunteering revolves around planned initiatives for developing the resilience, values, and learning appetite in the cultures.

12.6 CASE: SIX CASES RELATING TO WORKFORCE DIVERSITY

- **Case Situation Number 1:** Assume that you are in a city where some form of communal violence has taken place. This communal violence could be on the basis of caste or religion or other such factors. The overall sentiments of people are that of hatred, fear, and distrust. It is difficult to work in this environment. What will you do? How will you solve this problem? How will you contribute to the development of this city? How can you help the society in regaining its normal status? How can you participate as a volunteer in such a situation? How will you take up initiatives to foster a culture of sharedness and a culture of inclusivity among such persons, who are torn apart by an environment of hatred and fear.
- **Case Situation Number 2:** Assume that you are a part of an online community, where people have been divided into two groups, and both the groups are using vulgar language for each other and they are constantly involved in passing negative comments on each other. How can you make people collaborative and reduce their bitterness and animosity? How can volunteers help in defusing an environment of distrust in such situations?
- **Case Situation Number 3:** Assume that you are a part of a backward city where most of the people are living in small suburban colonies. People are grouped on the basis of caste and there is very little interaction among people from different castes or groups. You wish to develop this community, but you don't know the path ahead. How will you proceed?
- **Case Situation Number 4:** You are a resident of a village, which is very backward in comparison to modern cities or developed regions. Age-old practices, rituals, and taboos still dominate. You wish to take up developmental issues as priorities among the villagers, but you get surprised to see that developmental issues are not discussed by people. They still continue to discuss only traditional rituals and old practices, which are no longer relevant.
- **Case Situation Number 5:** Identify the reasons for the development of a diversified society and find the factors that could have helped in its growth and development. Identify those aspects of social development, which have enabled people to create an egalitarian and open society, where people value divergence, tolerate differences in ideas and opinions, and welcome creativity and innovations. How can such practices be continued and promoted in the future? You may also take up a comparison of a well-diversified group with a contrasting group

where people are intolerant for differences and do not welcome or accommodate different people. Such a comparison can help in understanding the role of diversity in broadening the attitudes of the group members.

- **Case Situation Number 6:** Community forum is an opportunity for students of Suresh Gyan Vihar University Jaipur to participate in volunteering initiatives. There are 11 community groups, which offer volunteering opportunities. The students, who join these groups, participate in different social development activities. Each group tries to induct students from different groups fostering diversity and wider involvement. These groups try to induct students from each state of India and also try to take up a few students from out of India in their group to prepare a heterogeneous group. Most of these groups have representation from each state of India. These initiatives have enabled these groups to achieve their purpose of fostering social development through student creativity. Each of these groups competes to create impact and therefore their ability to create a creative and innovative workplace is crucial. They have to create an ecosystem, which can foster a creative and innovative environment. Through constant efforts to welcome diversity, these groups are able to have an impact across the university. They compete among themselves to attract the best volunteers, who can bring innovations, zeal, enthusiasm, and dedicated efforts. They are able to broadcast their contribution through their members, who are also a part of wide groups of volunteers.

S. No.	Community (Voluntary Experience)	Area of Work
1.	AXIOM	Food, cultures, and fun and promotion of varieties of foods
2.	CAC	Entrepreneurial pursuits and innovative business ideas
3.	GREEN	Tree plantation, greenery, etc.
4.	Hyphen	Creative writing, poetry, etc.
5.	I-TECH	Technology-based advances
6.	MAP	Skill enhancement of people
7.	SEEDS	Promotion of rural development (a village has been adopted by this community)
8.	SMILES	To bring smiles to people
9.	SPORTY	Sports and related activities
10.	TCL (The creative league)	Creativity
11.	Technovations	Innovations

Each community manages its financial resources on its own. Each community organizes voluntary activities on the basis of participation of the students and their collective decision making. Thus, the students learn the art of democracy through these communities and volunteering experiences. These communities are opportunities for students to mingle with others from different backgrounds and work collectively for some good cause. Such initiatives have enabled students to have a better understanding about each other through an organized system.

The management of communities is undertaken by the students, who prepare their own president, secretary, and other office bearers and recruit new members and plan and undertake social development activities as per their agenda.

12.6.1 CASE ASSIGNMENT

You should try to find out other such sectors where volunteering can be used to induct diverse people together for social development?

APPENDIX 12.1: CASE DISCUSSION

- **General Guidelines:** There is no perfect answer to these case situations. These case situations are common situations that we encounter in our everyday life. The ultimate purpose is to enable the participants to discuss and evolve new solutions based on discussions in a diversified team. The solution that the participants develop may look simple, but the essential element is that it should be based on their discussions and experience sharing. It is not appropriate to impose any solution upon the participants and it is also not appropriate to completely reject any solution proposed by the participating teams. The case writer has presented discussion points and possible solutions, which may be presented just as another alternative to the participants.
- **Case Situation Number 1:** A similar situation took place with an NGO in Ahmadabad during a communal riot. They had volunteers and employees from different religions working together. Their employees and volunteers joined together and created an amicable atmosphere through their initiatives. They were able to bring people from different religions together and create an open and trusting environment. The

primary purpose of this case is to let the participants think and find solutions for the most common problems relating to diversity.
- **Case Situation Number 2:** In a similar situation, volunteers can be encouraged to share positive messages, which can foster a positive environment among different groups and can create an environment for mutual collaboration.
- **Case Situation Number 3:** Volunteers can create a solution here. Volunteers from different areas can be invited to live with other groups so that slowly there is better interaction between different groups. Many volunteers have successfully undertaken this initiative in South Africa and other such regions where people are divided into different groups.
- **Case Situation Number 4:** Study Indian society and its characteristics. How has Indian society evolved into a highly diversified and tolerant society, where diversity, differences in ideas, and differences in beliefs are approved? How has Indian society absorbed diverse ideas into its culture and has evolved out of the confluence? Let your students discuss and share their observations. Prepare some volunteering initiatives for the students, whereby the students are able to foster tolerance among society through the sharing of stories of integrity and inclusive culture.
- **Case Situation Number 5:** Identify those situations, where volunteering has been successfully used to create a better work culture, better understanding, and better teamwork among employees. You can enable the students to identify the various advantages of volunteering and how to use it to create passion and purpose among volunteers. There are some areas where people can join together for volunteering, for example: creating cleanliness and raising awareness; tree plantation and environment promotion; hobby; helping the poor; and counseling services.

Volunteering is possible on the basis of choice, passion, and interest of the volunteer. If a person is very keen to help, others-he can take up volunteering as another opportunity to immerse in social development. Volunteering should not be confused with assignments or internships or other similar projects, which are mandatory in nature. Volunteering is purely based on the intrinsic desire to serve.

KEYWORDS

- academic training
- community development
- diversity
- engagement
- learning
- social development
- team-building
- volunteering

REFERENCES

Barney, J., Wright, M., & Ketchen, D. J., (2001). "The resource-based view of the firm: Ten years after 1991." *Journal of Management, 27*, 625–641.

Bhatnagar, D., (1987). "*A Study of Attitudes Towards Women Managers in Banks," Working Paper, No. 1987/668.* Indian Institute of Management, Ahmadabad.

Cho, & Rainey, (2010). "Managing diversity in U.S. federal agencies: Effects of diversity and diversity management on employee perceptions of organizational performance." *Public Administration Review, 70*(1), 109–121.

Cox, T., (1991). "Managing cultural diversity: Implications for organizational competitiveness." *Academy of Management, 5.*

Dobbs, M. F., (1996). "Managing diversity: Lessons from the private sector." *Public Personnel Management, 25*(3), 351–367.

Dora and Kieth, (1998). "Demographic diversity and faultiness: The compositional dynamics of organizational groups." *Academy of Management Review, 23*(2), 325–340.

Ely, R., & Thomas, D., (2001). "Cultural diversity at work: The effects of diversity perspectives on work group processes and outcomes." *Administrative Science Quarterly, 46*(2), 229–273.

Esty, K., Richard, G., & Marcie, S. H., (1995). *Workplace Diversity.* A manager's guide to solving problems and turning diversity into a competitive advantage. Avon Publishers.

Khandelwal, P., (2002). "Gender stereotypes at work: Implications for organizations." *Indian Journal of Training and Development, 32*(2), 72–83.

Kulik, A. M., C. T., & Metz, I., (2011). "The gender diversity performance relationship in services and manufacturing organizations." *The International Journal of Human Resource Management, 22*(7), 1464–1485.

Kundu, S. C., & Turan, M. S., (1999). "Managing cultural diversity in future organizations." *The Journal of Indian Management and Strategy, 4*(1), 61–72.

Milliken, F. J., & Martins, L. L., (1996). "Searching for common threads: Understanding the multiple effects of diversity in organizational groups." *Academy of Management Review, 21*, 402–403.

NASSCOM & Mercer Report, (2008), *"Gender Inclusivity and Diversity in the Indian IT-BPO Industry, and Gender Inclusivity in India."* Building Empowered Organizations (2009).

NASSCOM Strategic Review, (2012). Retrieved from: www.nasscom.in (accessed on 25 February 2020).

NASSCOM-PWC Survey Report which Constitutes, (2001). *"The Corporate Awards for Excellence in Diversity Among IT/ITES Companies of India"* and *NASSCOM-PWC Report.* "Changing landscape and emerging trends, Indian IT/ITES Industry."

O'leary, B. J., & Weathington, (2006). "Beyond the business case for diversity." *Employee Response Rights Journal, 18*, 283–292.

Quinetta, M. R., (2004). Cornell University, *"Disentangling the Meanings of Diversity and Inclusion in Organizations"* (CAHRS Working Paper #04-05).

Ramarajan, L., & Thomas, D., (2010). *"A Positive Approach to Studying Diversity in Organizations."* Harvard Business School.

Schwind, H., Das, H., & Wagar, T., (2007). "Diversity management." In: Schwind, H., Das, H., & Wagar, T., (eds.), *"Canadian Human Resource Management: A Strategic Approach"* (8th edn., pp. 486–524). Toronto: McGraw-Hill Ryerson.

Thomas, K., Katerina, B., Robin, E., Susan, J., Aparna, J., Karen, J., Jonathan, L., David, L., & David, T., (2003). "The effects of diversity on business performance: Report of the diversity research network." *Human Resource Management, Spring, 42*(1), 3–21.

Thomas, R. R., (1991). *"Beyond Race and Gender."* AMACOM, New York.

Wentling, R. M., & Palma-Rivas, N. (1998). Current status and future trends of diversity initiatives in the workplace: Diversity experts' perspective. *Human Resource Development Quarterly, 9*(3), 235–253.

Williams, & O''Reilly (1998). "Demography and diversity in organizations: A review of 40 years of research." *Research in Organization Behavior, 20*, 77–140.

Woodard, N., & Debi, S. S., (2005). "Diversity management issues in USA and India: Some emerging prospective." In: Pritham, S., Jyotsna, B., & Asha, B., (eds.), *Future of Work, Mastering Change.* Excel Publishers.

Woods, R. H., & Sciarini, M. P., (1995). "Diversity programs in chain restaurants." *Cornell Hotel and Restaurant Administration Quarterly, 36*(3), 18–23.

FURTHER READING

Acharya, A., & Gupta, M., (2016a). An application of brand personality to green consumers: A thematic analysis. *Qualitative Report, 21*(8), 1531–1545.

Acharya, A., & Gupta, M., (2016b). Self-image enhancement through branded accessories among youths: A phenomenological study in India. *Qualitative Report, 21*(7), 1203–1215.

Alcázar, F. M., Fernández, P. M. R., & Gardey, G. S., (2013). Workforce diversity in strategic human resource management models: A critical review of the literature and implications for future research. *Cross Cultural Management, 20*(1), 39–49.

Andrews, R., & Ashworth, R., (2017). Feeling the heat? Management reform and workforce diversity in the English fire service. *Fire and Rescue Services: Leadership and Management Perspectives*, 145–158.

Benavides-Vaello, S., Katz, J. R., Peterson, J. C., Allen, C. B., Paul, R., Charette-Bluff, A. L., & Morris, P., (2014). Nursing and health sciences workforce diversity research using Photo Voice: A college and high school student participatory project. *Journal of Nursing Education*, *53*(4), 217–222.

Brunow, S., & Miersch, V., (2015). Innovation capacity, workforce diversity, and intra-industrial externalities: A study of German establishments. *The Rise of the City: Spatial Dynamics in the Urban Century*, 288–219.

Byrd, M. Y., & Lloyd-Jones, B., (2017). Developing a social justice-oriented workforce diversity concentration in human relations academic programs. *Organizational Culture and Behavior: Concepts, Methodologies, Tools, and Applications*, *43500*, 474–491.

Carter, S. D., (2018). Increased workforce diversity by race, gender, and age and equal employment opportunity laws: Implications for human resource development. *Gender and Diversity: Concepts, Methodologies, Tools, and Applications*, *1*, 380–405.

Cho, S., Kim, A., & Mor Barak, M. E., (2017). Does diversity matter? Exploring workforce diversity, diversity management, and organizational performance in social enterprises. *Asian Social Work and Policy Review*, *11*(3), 193–204.

Choi, J. N., Sung, S. Y., & Zhang, Z., (2017). Workforce diversity in manufacturing companies and organizational performance: The role of status-relatedness and internal processes. *International Journal of Human Resource Management*, *28*(19), 2738–2761.

Choi, S., (2017). Workforce diversity and job satisfaction of the majority and the minority: Analyzing the asymmetrical effects of relational demography on whites and racial/ethnic minorities. *Review of Public Personnel Administration*, *37*(1), 84–107.

Colville, J., Cottom, S., Robinette, T., Wald, H., & Waters, T., (2015). A community college model to support nursing workforce diversity. *Journal of Nursing Education*, *54*(2), 65–71.

Craft-Blacksheare, M., (2018). New careers in nursing: An effective model for increasing nursing workforce diversity. *Journal of Nursing Education*, *57*(3), 178–183.

Cui, J., Jo, H., Na, H., & Velasquez, M. G., (2015). Workforce diversity and religiosity. *Journal of Business Ethics*, *128*(4), 743–767.

Demie, F., (2019). Raising achievement of black Caribbean pupils: Good practice for developing leadership capacity and workforce diversity in schools. *School Leadership and Management*, *39*(1), 5–25.

DeWitty, V., (2014). Keeping a sharp focus on advancing workforce diversity, nurses' education. *The American Nurse*, *46*(5), 12.

Ellis, K. M., & Keys, P. Y., (2013). Workforce diversity and shareholder value: A multi-level perspective. *Review of Quantitative Finance and Accounting*, *44*(2), 191–212.

Essary, A. C., & Wade, N. L., (2016). An innovative program in the science of health care delivery: Workforce diversity in the business of health. *Journal of Allied Health*, *45*(2), e21–e25.

Garcia, A. N., Kuo, T., Arangua, L., & Pérez-Stable, E. J., (2018). Factors associated with medical school graduates' intention to work with underserved populations: Policy implications for advancing workforce diversity. *Academic Medicine*, *93*(1), 82–89.

Garcia, R. I., Blue, S. G., Sinkford, J. C., Lopez, M. J., & Sullivan, L. W., (2017). Workforce diversity in dentistry-current status and future challenges. *Journal of Public Health Dentistry*, *77*(2), 99–104.

Garnero, A., (2017). Workforce diversity, productivity, and wages: The role of managers and shareholders in France [Diversité de la main-d'oeuvre, productivity et salaries: Le role des managers et des propriétaires en France]. *Travail et Emploi.*, *152*, 59–87.

Garnero, A., Kampelmann, S., & Rycx, F., (2014). The heterogeneous effects of workforce diversity on productivity, wages, and profits. *Industrial Relations, 53*(3), 430–477.

Garnero, A., Kampelmann, S., & Rycx, F., (2017). Is workforce diversity always performance-enhancing?: A literature review. *Reflets et Perspectives de la Vie Economique, 55*(4), 81–91.

Ghaffarzadegan, N., Hawley, J., & Desai, A., (2014). Research workforce diversity: The case of balancing national versus international post docs in US biomedical research. *Systems Research and Behavioral Science, 31*(2), 301–315.

Giang, V., (2013). Canada's mainstream media workforce diversity as an indicator of Canadian society's acceptance and inclusion of ethno-cultural communities: A case study of promising practices developed by the United Nations Association in Canada. *International Journal of Organizational Diversity, 12*(1), 73–85.

Glazer, G., Tobias, B., & Mentzel, T., (2018). Increasing healthcare workforce diversity: Urban universities as catalysts for change. *Journal of Professional Nursing, 34*(4), 239–244.

Greer, B. M., & Hill, J. A., (2012). Leveraging workforce diversity in practice: Building successful global relationships with minority-owned suppliers. *Handbook of Research on Workforce Diversity in a Global Society: Technologies and Concepts* (pp. 323–340).

Grissom, J. A., Kern, E. C., & Rodriguez, L. A., (2015). The "representative bureaucracy" in education: Educator workforce diversity, policy outputs, and outcomes for disadvantaged students. *Educational Researcher, 44*(3), 185–192.

Guajardo, S. A., (2013). Workforce diversity: An application of diversity and integration indices to small agencies. *Public Personnel Management, 42*(1), 27–40.

Guajardo, S. A., (2014). Workforce diversity: Ethnicity and gender diversity and disparity in the New York City police department. *Journal of Ethnicity in Criminal Justice, 12*(2), 93–115.

Guajardo, S. A., (2015). Assessing organizational efficiency and workforce diversity: An application of data envelopment analysis to New York City agencies. *Public Personnel Management, 44*(2), 239–265.

Guerrero, E. G., (2013). Workforce diversity in outpatient substance abuse treatment: The role of leaders' characteristics. *Journal of Substance Abuse Treatment, 44*(2), 208–215.

Gupta, M., & Kumar, Y., (2015). Justice and employee engagement: Examining the mediating role of trust in Indian B-schools. *Asia-Pacific Journal of Business Administration, 7*(1), 89–103.

Gupta, M., & Pandey, J., (2018). Impact of student engagement on affective learning: Evidence from a large Indian university. *Current Psychology, 37*(1), 414–421.

Gupta, M., & Ravindranath, S., (2018). Managing physically challenged workers at micro sign. *South Asian Journal of Business and Management Cases, 7*(1), 34–40.

Gupta, M., & Sayeed, O., (2016). Social responsibility and commitment in management institutes: Mediation by engagement. *Business: Theory and Practice, 17*(3), 280–287.

Gupta, M., & Shaheen, M., (2017a). Impact of work engagement on turnover intention: Moderation by psychological capital in India. *Business: Theory and Practice, 18,* 136–143.

Gupta, M., & Shaheen, M., (2017b). The relationship between psychological capital and turnover intention: Work engagement as mediator and work experience as moderator. *Journal Pengurusan, 49,* 117–126.

Gupta, M., & Shaheen, M., (2018). Does work engagement enhance general well-being and control at work? Mediating role of psychological capital. *Evidence-Based HRM, 6*(3), 272–286.

Gupta, M., & Shukla, K., (2018). An empirical clarification on the assessment of engagement at work. *Advances in Developing Human Resources, 20*(1), 44–57.

Gupta, M., (2017). Corporate social responsibility, employee-company identification, and organizational commitment: Mediation by employee engagement. *Current Psychology*, *36*(1), 101–109.

Gupta, M., (2018). Engaging employees at work: Insights from India. *Advances in Developing Human Resources*, *20*(1), 3–10.

Gupta, M., Acharya, A., & Gupta, R., (2015). Impact of work engagement on performance in Indian higher education system. *Review of European Studies*, *7*(3), 192–201.

Gupta, M., Ganguli, S., & Ponnam, A., (2015). Factors affecting employee engagement in India: A study on off shoring of financial services. *Qualitative Report*, *20*(4), 498–515.

Gupta, M., Ravindranath, S., & Kumar, Y. L. N., (2018). Voicing concerns for greater engagement: Does a supervisor's job insecurity and organizational culture matter? *Evidence-Based HRM*, *6*(1), 54–65.

Gupta, M., Shaheen, M., & Das, M. (2019). Engaging employees for quality of life: mediation by psychological capital. *The Service Industries Journal*, *39*(5–6), 403–419.

Gupta, M., Shaheen, M., & Reddy, P. K., (2017). Impact of psychological capital on organizational citizenship behavior: Mediation by work engagement. *Journal of Management Development*, *36*(7), 973–983.

Haggins, A., Sandhu, G., & Ross, P. T., (2018). Value of near-peer mentorship from protégé and mentor perspectives: A strategy to increase physician workforce diversity. *Journal of the National Medical Association*, *110*(4), 399–406.

Ibidunni, A. S., Falola, H. O., Ibidunni, O. M., Salau, O. P., Olokundun, M. A., Borishade, T. T., Amaihian, A. B., & Peter, F., (2018). Workforce diversity among public healthcare workers in Nigeria: Implications on job satisfaction and organizational commitment. *Data in Brief*, *18*, 1047–1053.

Idrees, R. N., Abbasi, A. S., & Waqas, M., (2013). Systematic review of literature on workforce diversity in Pakistan. *Middle East Journal of Scientific Research*, *17*(6), 780–790.

Jonsen, K., Tatli, A., Özbilgin, M. F., & Bell, M. P., (2013). The tragedy of the uncommon: Reframing workforce diversity. *Human Relations*, *66*(2), 271–294.

Joshi, A. M., Inouye, T. M., & Robinson, J. A., (2018). How does agency workforce diversity influence Federal R&D funding of minority and women technology entrepreneurs? An analysis of the SBIR and STTR programs, 2001–2011. *Small Business Economics*, *50*(3), 499–519.

Joshua-Gojer, A. E., Allen, J. M., & Huang, T. Y., (2018). Workforce diversity in volunteerism and the peace corps. *Gender and Diversity: Concepts, Methodologies, Tools, and Applications*, *1*, 141–159.

Kang, J., (2015). Effectiveness of the KLD social ratings as a measure of workforce diversity and corporate governance. *Business and Society*, *54*(5), 599–631.

Katz, J. R., Barbosa-Leiker, C., & Benavides-Vaello, S., (2016). Measuring the success of a pipeline program to increase nursing workforce diversity. *Journal of Professional Nursing*, *32*(1), 6–14.

Kermanshachi, S., & Sadatsafavi, H., (2018). Predictive modeling of U.S. transportation workforce diversity Trends: A study of human capital recruitment and retention in complex environments. *International Conference on Transportation and Development 2018: Planning, Sustainability, and Infrastructure Systems-Selected Papers from the International Conference on Transportation and Development*, pp. 105–114.

Kisaka, L. G., Jansen, E. P. W. A., & Hofman, A. W. H., (2019). Workforce diversity in Kenyan public universities: An analysis of workforce representativeness and heterogeneity by employee gender and ethnic group. *Journal of Higher Education Policy and Management*, *41*(1), 35–51.

Köllen, T., (2016). Sexual orientation and transgender issues in organizations: Global perspectives on LGBT workforce diversity. *Sexual Orientation and Transgender Issues in Organizations: Global Perspectives on LGBT Workforce Diversity*, pp. 1–560.

Kundu, S. C., & Mor, A., (2017). Workforce diversity and organizational performance: A study of IT industry in India. *Employee Relations*, *39*(2), 160–183.

Labedz, Jr. C. S., & Berry, G. R., (2013). Emerging systemic-structural threats to workforce diversity: Beyond inadequate agency. *Journal of Organizational Transformation and Social Change*, *10*(3), 218–237.

Laveist, T. A., & Pierre, G., (2014). Integrating the 3Ds-social determinants, health disparities, and health-care workforce diversity. *Public Health Reports*, *129*(2), 9–14.

Lopes, M. A., Moraes, F., Carvalho, F. M., Bruhn, F. R. P., Lima, A. L. R., & Reis, E. M. B., (2019). Effect of workforce diversity on the cost-effectiveness of milk production systems participating in the "full bucket" program. *Semina: Ciencias Agrarias*, *40*(1), 323–338.

Malik, P., Lenka, U., & Sahoo, D. K., (2018). Proposing micro-macro HRM strategies to overcome challenges of workforce diversity and deviance in ASEAN. *Journal of Management Development*, *37*(1), 6-26.

Marfelt, M. M., & Muhr, S. L., (2016). Managing protean diversity: An empirical analysis of how organizational contextual dynamics derailed and dissolved global workforce diversity. *International Journal of Cross Cultural Management*, *16*(2), 231–251.

Maturo, F., Migliori, S., & Paolone, F., (2018). Measuring and monitoring diversity in organizations through functional instruments with an application to ethnic workforce diversity of the U.S. Federal Agencies. *Computational and Mathematical Organization Theory* (pp. 1–32).

McDougle, L., Way, D. P., Lee, W. K., Morfin, J. A., Mavis, B. E., Matthews, D., Latham-Sadler, B. A., & Clinchot, D. M., (2015). A national long- term outcomes evaluation of US. premedical post-baccalaureate programs designed to promote health care access and workforce diversity. *Journal of Health Care for the Poor and Underserved*, *26*(3), 631–647.

McGee, R., (2016). "Biomedical workforce diversity: The context for mentoring to develop talents and foster success within the 'pipeline.'" *AIDS and Behavior*, *20*, 231–237.

Memduhoglu, H. B., (2016). Perceptions of workforce diversity in high schools and diversity management: A qualitative analysis. *Egitim ve Bilim.*, *41*(185), 199–217.

Mensah, M. O., & Sommers, B. D., (2016). The policy argument for healthcare workforce diversity. *Journal of General Internal Medicine*, *31*(11), 1369–1372.

Mertz, E., Wides, C., Cooke, A., & Gates, P. E., (2016). Tracking workforce diversity in dentistry: Importance, methods, and challenges. *Journal of Public Health Dentistry*, *76*(1), 38–46.

Munyeka, W., (2014). Employees' discernment of workforce diversity and its effect on job satisfaction in a public service department. *Mediterranean Journal of Social Sciences*, *5*(15), 37–48.

Murray, T. A., Pole, D. C., Ciarlo, E. M., & Holmes, S., (2016). A nursing workforce diversity project: strategies for recruitment, retention, graduation, and NCLEX-RN success. *Nursing Education Perspectives*, *37*(3), 138–143.

Narayanan, S., & Raina, K., (2018). Effect of primary dimensions of workforce diversity on employee engagement: A literature review. *International Journal of Mechanical Engineering and Technology*, *9*(3), 655–670.

Nivet, M. A., & Berlin, A., (2014). Workforce diversity and community-responsive healthcare institutions. *Public Health Reports, 129*(2), 15–18.
O'Brien, K. R., Scheffer, M., Van Nes, E. H., & Van Der Lee, R., (2015). How to break the cycle of low workforce diversity: A model for change. *PLoS One, 10*(7).
Ogbo, A. I., Anthony, K. A., & Ukpere, W. I., (2014). The effect of workforce diversity on organizational performance of selected firms in Nigeria. *Mediterranean Journal of Social Sciences, 5*(10), 231–236.
Pachter, L. M., & Kodjo, C., (2015). New century scholars: A mentorship program to increase workforce diversity in academic pediatrics. *Academic Medicine, 90*(7), 881–887.
Pandey, J., Gupta, M., & Naqvi, F., (2016). Developing decision making measure a mixed method approach to operationalize Sankhya philosophy. *European Journal of Science and Theology, 12*(2), 177–189.
Park, J., & Kim, S., (2015). The differentiating effects of workforce aging on exploitative and exploratory innovation: The moderating role of workforce diversity. *Asia Pacific Journal of Management, 32*(2), 481–503.
Pletcher, B. A., Rimsza, M. E., Basco, W. T., Hotaling, A. J., Sigrest, T. D., Simon, F. A., et al. (2013). Enhancing pediatric workforce diversity and providing culturally effective pediatric care: Implications for practice, education, and policymaking. *Pediatrics, 132*(4), e1105–e1116.
Podsiadlowski, A., Gröschke, D., Kogler, M., Springer, C., & Van Der Zee, K., (2013). Managing a culturally diverse workforce: Diversity perspectives in organizations. *International Journal of Intercultural Relations, 37*(2), 159–175.
Printz, C., (2016). Cancer, biomedical science leaders strive to improve workforce diversity. *Cancer, 122*(6), 823–824.
Saeed, R., Lodhi, R. N., Ashraf, H., Riaz, S., Dustgeer, F., Sami, A., Mahmood, Z., & Ahmad, M., (2013). Effect of workforce diversity on the performance of the students. *World Applied Sciences Journal, 26*(10), 1380–1384.
Scott, C. L., & Sims, J. D., (2014). Workforce diversity career development: A missing piece of the curriculum in academia. *Impact of Diversity on Organization and Career Development, 129*–150.
Sims, J. D., Shuff, J., Lai, H. L., Lim, O. F., Neese, A., Neese, S., & Sims, A., (2018). Diverse student scholars: How a faculty member's undergraduate research program can advance workforce diversity learning. *Teacher Training and Professional Development: Concepts, Methodologies, Tools, and Applications, 4*, 1804–1826.
Siotos, C., Payne, R. M., Stone, J. P., Cui, D., Siotou, K., Broderick, K. P., Rosson, G. D., & Cooney, C. M., (2019). Evolution of workforce diversity in surgery. *Journal of Surgical Education*.
Sourouklis, C., & Tsagdis, D., (2013). Workforce diversity and hotel performance: A systematic review and synthesis of the international empirical evidence. *International Journal of Hospitality Management, 34*(1), 394–403.
Srinivasan, M. S., (2015). Integrating workforce diversity in global business: A psycho-spiritual perspective. *Journal of Human Values, 21*(1), 1–10.
Starr-Glass, D., (2017). Workforce diversity in small and medium-sized enterprises: Is social identification stronger than the business case argument? *Managing Organizational Diversity: Trends and Challenges in Management and Engineering*, pp. 95–117.
Sung, S. Y., & Choi, J. N., (2019). Contingent effects of workforce diversity on firm innovation: High-tech industry and market turbulence as critical environmental contingencies. *International Journal of Human Resource Management*.

Tab, M., (2016). Helping minority students from rural and disadvantaged backgrounds succeed in nursing: A nursing workforce diversity project. *Online Journal of Rural Nursing and Health Care*, *16*(1), 59–75.

Tobias, B., Glazer, G., & Mentzel, T., (2018). An academic-community partnership to improve health care workforce diversity in greater Cincinnati: Lessons learned. *Progress in Community Health Partnerships: Research, Education, and Action*, *12*(4), 409–418.

Tran, B., (2014). Assistive technology: Human capital for mobility (Dis)abled workforce diversity development. *International Journal of Ambient Computing and Intelligence*, *6*(2), 15–28.

Tran, B., (2017). Organizational diversity: From workforce diversity to workplace inclusion for persons with disabilities. *Handbook of Research on Organizational Culture and Diversity in the Modern Workforce*, pp. 100–131.

Tran, B., (2018). Organizational diversity: From workforce diversity to workplace inclusion for persons with disabilities. *Gender and Diversity: Concepts, Methodologies, Tools, and Applications*, *1*, 160–191.

Travers, J., Smaldone, A., & Gross, C. E., (2015). Does state legislation improve nursing workforce diversity? *Policy, Politics, and Nursing Practice*, *16*(43528), 109–116.

Valantine, H. A., (2017). 50 years to gender parity: Can stem afford to wait?: A cardiologist and NIH chief officer of scientific workforce diversity reflect on what it will take to keep women in biomedicine. *IEEE Pulse*, *8*(6), 46–48.

Valantine, H., (2016). NIH's essential 21st-century research challenge: Enhancing scientific workforce diversity. *Journal of Investigative Dermatology*, *136*(12), 2327–2329.

Vandenberghe, V., (2016). Is workforce diversity good for efficiency? An approach based on the degree of concavity of the technology. *International Journal of Manpower*, *37*(2), 253–267.

Vapiwala, N., & Winkfield, K. M., (2018). The hidden costs of medical education and the impact on oncology workforce diversity. *JAMA Oncology*, *4*(3), 289–290.

Vardeman-Winter, J., & Place, K. R., (2017). Still a lily-white field of women: The state of workforce diversity in public relations practice and research. *Public Relations Review*, *43*(2), 326–336.

Vasconcelos, A. F., (2016). Mapping Brazilian workforce diversity: A historical analysis. *Management Research Review*, *39*(10), 1352–1372.

Vishwanatha, J. K., Basha, R., Nair, M., & Jones, H. P., (2019). An institutional coordinated plan for effective partnerships to achieve health equity and biomedical workforce diversity. *Ethnicity and Disease*, *29*, 129–134.

Walsh, R. M., Jeyarajah, D. R., Matthews, J. B., Telem, D., Hawn, M. T., Michelassi, F., & Reid-Lomardo, K. M., (2016). White Paper: SSAT commitment to workforce diversity and healthcare disparities. *Journal of Gastrointestinal Surgery*, *20*(5), 879–884.

Walters, T., & Irby-Butler, T., (2017). The kaleidoscope view of workforce diversity and inclusion. *Journal - American Water Works Association*, *109*(8), 66–69.

Williams, S. D., Hansen, K., Smithey, M., Burnley, J., Koplitz, M., Koyama, K., Young, J., & Bakos, A., (2014). Using social determinants of health to link health workforce diversity, care quality and access, and health disparities to achieve health equity in nursing. *Public Health Reports*, *129*(2), 32–36.

Winkfield, K. M., & Gabeau, D., (2013). Why workforce diversity in oncology matters. *International Journal of Radiation Oncology Biology Physics*, *85*(4), 900–901.

Winkfield, K. M., Flowers, C. R., & Mitchell, E. P., (2017). Making the case for improving oncology workforce diversity. American society of clinical oncology educational book. *American Society of Clinical Oncology, Annual Meeting*, *37*, 18–22.

Xierali, I. M., Castillo-Page, L., Zhang, K., Gampfer, K. R., & Nivet, M. A., (2014). AM last page: The urgency of physician workforce diversity. *Academic Medicine, 89*(8), 1192.

Yamada, Y., Iwaasa, T., Ebara, T., Shimizu, T., & Mizuno, M., (2019). Relationship between acceptance of workforce diversity and mental health condition among Japanese nurses. *Advances in Intelligent Systems and Computing, 818*, 563–567.

Yamada, Y., Mizuno, M., Shimizu, T., Asano, Y., Iwaasa, T., & Ebara, T., (2017). Elements of workforce diversity in the Japanese nursing workplace. *Advances in Intelligent Systems and Computing, 487*, 167–176.

Zaballero, A. G., & Kim, Y., (2014). Theoretical frameworks and models supporting the practice of leveraging workforce diversity. *Cross-Cultural Interaction: Concepts, Methodologies, Tools and Applications, 1*, 266–283.

Zaballero, A. G., Tsai, H. L., & Acheampong, P., (2014). Leveraging workforce diversity and team development. *Cross-Cultural Interaction: Concepts, Methodologies, Tools and Applications, 3*, 1138–1150.

Zaitouni, M., & Gaber, A., (2017). Managing workforce diversity from the perspective of two higher education institutions. *International Journal of Business Performance Management, 18*(1), 82–100.

Index

A

Ableism/disablism, 110
Absenteeism, 140, 183, 190, 195, 225
Academic
　circles, 36
　literature, 270
　routine, 143
　training, 333
Accenture India, 251, 256
Acceptability, 3, 297
Accessibility, 243, 250, 252
Accountability, 249
Acculturation, 216, 217, 229
　stress, 216, 217
Achieving competitive edge, 223
Acquisitions, 213–215, 218, 268
Active aging, 34–36, 38–40, 43–49, 52, 53, 66
　index (AAI), 43–46
Adidas-Reebok merger, 218
Adjustments, 221, 229, 247, 253–255
Administrators, 132, 140
Advantages, 10, 77, 192, 194, 238, 256, 321, 327, 332
Age, 48, 58, 61, 110, 113, 185, 329
　discrimination, 48, 54, 110
　　Employment Act (ADEA), 48, 113
Akin fragmenting task, 141
American
　Disabilities Act (ADA), 113
　Express, 16
　multinational
　　beverage corporation, 281
　　technology, 19, 280
Annual Social and Economic Supplement, 189
Antidiscrimination laws, 7, 112
Anti-nepotism policies, 191
Appraisal, 21, 121, 195, 243
Article 2, 243, 244
Article 6, 243, 244
Assessment day, 122
Assistive technology, 242
Attitude-based problems, 297
Aulwin bank, 230
Autism
　Society of Indian, 251
　spectrum disorder (ASD), 251
Automobile companies, 222
Autonomy, 137, 141, 150, 151, 168, 176, 177, 221, 227, 228, 232, 280, 283, 304
Awareness
　raising activities, 248
　training, 242

B

Baby boomers, 43, 56, 57, 64, 65, 133, 134, 160–162, 166, 169, 175, 178
Banking, 19, 56
Behavioral engagement, 48
Biases, 14, 23, 63, 84, 86, 87, 89, 173, 185, 248, 249, 298, 324
Bisexual, 1, 3, 25
Bourgeoning stage, 11
BPO industry, 325
Brainstorming meeting, 91
Brand image, 239
BRICS nations, 74
Bridge employment, 40, 51
Broad learning plan, 148
Budding role (team member engagement), 278
　commitment, 278
　creativity, 279
　environment, 279
　modifications, 279
　programs, 279
　suggestions, 278
　technology, 279
　training, 278
Bundle of expectations, 143
Bureaucratic organizations, 82
Business
　environment, 4, 220

innovation, 89
localities, 90
organizations, 74, 80, 84, 88, 132
strategies, 282

C

Capacity augmentation, 222
Capitalist system, 275
Cardio-vascular diseases, 191
Career
  advancement, 24
  development opportunities, 89
  experiences, 159
  jobs, 40
  opportunities, 227
  progression, 11, 13
  security, 163
Case
  analysis, 286
  assignment, 331
  discussion, 18, 54, 90, 92, 116, 142, 177, 196, 198, 230, 253, 284
  synopsis, 22, 147, 309
Caste, 110
Categorization elaboration model (CEM), 76
Cause-focused passion, 135
Central point, 243
Cloud
  based contact management solutions, 284
  computing, 19
Coca Cola Company, 281, 282
Cognitive
  ability, 35
  biasing, 87
  connections, 300
  emotional involvement, 131
  impact, 232
  solutions, 19
  style, 158, 213
Cohort, 35, 36, 54, 57, 58, 60, 133, 157–162, 165–167, 169, 171–175, 177
Collaboration, 220, 248
Colleagues, 6, 9, 10, 18, 20, 22, 55, 115, 132, 163, 185, 187–189, 219, 220, 254, 278, 279, 281, 302, 304
Comfy zone generation, 136

Commission of the European Communities (CEC), 36
Commitment, 2, 8, 17, 65, 85, 133, 136, 137, 147, 161, 170, 176, 183, 185, 186, 188, 190, 192, 193, 219, 224, 225, 230, 249, 270–272, 278, 283, 287, 298, 324–326
Communal approval, 11
Communicate corporate culture, 10
Communication, 10, 43, 57, 63, 87, 137, 148, 162, 172, 188, 214–217, 221, 224, 226, 228–230, 232, 243, 245, 250, 276, 277, 282, 324
  biases, 87
  problems, 57, 87, 324
Communities online, 320
Community, 1–3, 5–18, 20–24, 35, 39, 42, 53, 63, 65, 75, 81, 83, 88, 92, 111, 162–164, 170, 185, 187, 246, 280, 281, 320, 327, 329–331
  advocacy, 10
  business, 3, 10, 14, 15
  development, 320, 333
  engagement campaign, 89
  participation, 53
Compensation, 252, 253
Conflict, 79, 84, 108, 111, 115, 141, 185, 188, 190, 191, 197, 217, 229, 230, 323
Consultancy services pioneer, 75
Contemporary
  organizations, 134
  workers, 174
Contextual references, 188
Continuity theory, 50
Corporate
  culture, 191, 218, 219, 226, 280, 297
  social responsibility, 219
Cost cutting, 223
Counseling, 191, 332
Critical developmental stages, 160
Cross-cultural, 45, 66, 78, 222, 320
  aspect, 45
  diversity, 232
  mergers, 221
  trainings, 221
Cultural
  ambassadors, 228
  attributes, 214

backgrounds, 296
changes, 227, 308
conflicts, 215
differences, 188, 217, 218
diversity, 77, 213, 214, 325
factors, 218, 297
geographical boundaries, 244
leadership lessen casualties, 215
related stress, 218
social practices, 74
values, 108, 217, 221
   systems, 158
Culturally divergent values, 322
Curricular diversity, 82
Customer
   card, 283
   lead generation, 176
   service, 283
Custom-made engagement strategies, 171
Cutting-edge competition, 138
Cynicism, 270

D

Data-driven, 176
Decision
   makers, 133, 221
   making, 54, 81, 86, 108, 147, 150, 158, 166, 184, 324, 331
      ability, 188
      criteria, 324
      process, 89, 221
Demographic
   attributes, 108
   characteristics, 77
   homogeneity, 87
   shifts, 151
   trends, 45
Demotivation, 24
Department of Labor (DOL), 248
Depression, 3, 161, 175, 191, 217
Designation, 120–123, 141, 286
Developmental opportunities, 282
Digital natives, 135, 141
Dilemma, 22, 52, 139, 146, 151, 175, 184
Disabilities, 48, 77, 110, 112–114, 136, 146, 237–239, 242–255, 322
Disabled-friendly workplace, 242
Discomfort, 24, 199, 238, 248

Discontentment, 24, 192
Discrimination, 6, 48, 110–114, 241, 243, 275
   at work, 111, 126
   discriminatory questions, 111
   types, 109, 116
Divergence, 214, 321, 329
Diversification, 223
Diversity, 1, 2, 4–6, 8–11, 13–19, 22, 24, 43, 46, 48, 61, 64, 65, 73–78, 80–90, 92, 94, 107–109, 111, 113, 115, 123, 125, 126, 131–133, 136, 138, 139, 141, 146, 157–159, 163, 165, 166, 168, 169, 174, 175, 177, 196, 200, 213–215, 221, 222, 224, 230, 231, 240, 251, 280, 297, 306, 308–310, 319–328, 330, 332, 333
   concept, 123, 125
   engagement, 94
   inclusion aspects, 131
   management, 77, 85, 88, 107, 126, 200, 319–323, 325–328
   oriented vision, 89
   policies, 88, 230
   rules, 2, 4
   structure, 11, 14
   training, 9, 11, 15, 88
Drug use, 111

E

Economic
   activities, 36
   business, 248
   contributions, 35, 36, 296
   crisis, 49
   political/social environments, 159
   security, 53, 243
   social environment, 138
Ecosystem, 17, 165, 320, 330
Education, 23, 42, 44, 53, 57, 77, 89, 109–116, 122, 123, 126, 135, 139, 158, 165, 213, 244–246, 267, 272–274, 281, 297, 322, 324
   discrimination, 112
   level, 158
   practitioners, 23
Educational
   achievement, 273
   discrimination, 112, 116, 126
   diversification, 107

diversity, 108, 115, 126
  occupational wage, 274
  institute's reputation, 116
  institutional background, 107, 116
  qualification, 107, 285, 297
    discrimination, 107
Efficiency, 185, 271, 322
Egalitarian structure, 85
Elder's well-being, 66
Electrical producers, 119
Emotional
  attachment, 220
  stability, 39
  untethered, 186
Empirical evidences, 177
Employee
  Alliance for Gay, 14
  business network, 89
  engagement, 3, 4–6, 11, 12, 25, 46, 56, 59, 60, 148, 213–215, 219–228, 231, 232, 268–272, 274–276, 279–283, 287, 303–305
    communication, 276
    culture, 277
    during change, 220
    growth opportunities, 276
    identify achievements, 276
    interventions, 232
    merged entities, 221
    organizational culture, 221
    process, 12
  feedback, 56
  motivation, 17
  participation, 242
  productivity, 132
  rated work-family conflict, 190
  relations, 228
  resource group (ERG), 14
  retention, 219, 225, 271
  segment, 168
  turnover, 46, 183, 276, 323
  value proposition, 227
  well-being, 132
Employee's
  education qualification, 111
  educational institutional value, 107
  skills, 115, 116
  state insurance act, 246

Employer's
  law, 255
  perspective, 51
  point-of-view, 220
Employment, 14, 34, 40, 45–47, 49, 54, 76, 80, 81, 87, 107, 110, 112, 113, 161, 170, 239, 243–249, 252, 254, 255, 274, 325, 326
  discrimination, 110, 113
  opportunities, 54, 80, 247, 249, 252
  practices, 87
  process with disability, 247
  related discrimination, 112
  training, 252
  Tribunal
    claim, 255
    disability discrimination, 255
Empowered employees, 277
Empowering
  employees, 227
  trade unions, 49
Empowerment, 4, 13, 44, 77, 162, 278
Engagement, 1, 2, 4, 5, 8, 11–13, 17, 35, 36, 39, 40, 46–48, 54, 58–60, 63, 64, 66, 74, 75, 88, 90, 131–134, 140–142, 147–150, 160, 166–176, 213, 215, 217, 219–222, 224, 225, 227, 228, 230, 242, 256, 267–270, 272, 274–280, 282, 283, 295, 297, 300, 301, 303–308, 333
  aging workforce, 46, 66
  disability, 242
  drivers, 224
  framework, 173
  practices, 60, 174, 283
  process, 174
  strategic goals, 172
  strategies, 171–175
  suffers, 220
  tool, 174
Engaging racial diversity, 73, 81
Entrepreneurial, 134, 163, 165, 167, 214
  approach, 165
Environment, 248, 279
  accessibility, 53
  challenges, 159
Executive post-graduate diploma program (EPGDP), 122, 125
Equal
  employment opportunity (EEO), 81, 322

Commission (EEOC), 80, 81
  opportunity, 245
  Remuneration Act, 287
Equality, 7, 9, 10, 14, 15, 20, 21, 24, 65, 77, 243, 244, 253–255, 274, 275, 296, 298, 308
  Act, 253–255
Ergonomic workspaces, 281
Ethical
  standards, 220
  work practices, 174
Ethics
  curriculum, 256
  group, 213
Ethnicity, 2, 17, 18, 73, 74, 76–79, 108, 110, 319, 321, 322, 324
Ethnocentrism, 217
Evolution of active aging, 36
Extramarital affairs, 191
Extrinsic motivation, 150

F

Facebook, 20
Factor productivity, 113
Fad management program, 162
Family
  care, 297
  response, 192
Fare wages, 256
Federal laws, 112
Female
  board members, 188, 189
  colleagues, 186
  employees, 49, 185, 187, 189, 190
  participation, 307
Feminist economists, 295
Financial
  effectiveness, 325
  performance, 46
  resources, 39, 331
  strain, 193
Financially untroubled managers, 186
Flexibility, 86, 186, 196, 277
Flexible schedules, 168, 169
Fortune 500 company, 91
Foster volunteerism, 328
Friendly work environment, 53, 199
FullContact, 284
Functional specialization, 108

Fundamental rights, 5, 245

G

Gallup organization, 268
Gays, 1, 9, 21, 25
Gender, 2–4, 7, 9, 11, 13, 14, 18, 20, 49, 73, 74, 76–78, 108, 110–112, 115, 132, 137, 138, 158, 162, 188–190, 200, 213, 230, 281, 295–298, 305–310, 319, 321–327
  bias, 308, 310
  biased recruitment, 188
  disparity, 138
  diversity, 297
  equality, 132, 310
  identity, 2, 3, 11, 18, 110, 111
  issues, 49
  sensitive, 305
  wage gap, 189
  work engagement, 305, 306
General guidelines, 331
Generation
  cohorts, 157, 159, 166, 178
  Gen X, 43, 62, 64, 65, 134, 137, 146, 160–163, 167, 178
    cohort, 56
  Gen Y, 56, 62, 64, 137, 159, 160, 164, 168, 170, 174, 175, 178
  Gen Z, 62, 131, 133–139, 149–151, 165, 178
    cohort, 56
    engagement, 150
Generational
  demographics, 62
  diversities, 171
Global
  business scenario, 73
  businesses, 7
  canvas, 162
  character, 327
  community, 75
  competition, 147
  diversity, 73, 81, 86, 88
    management, 94
  economic fronts, 138
  economy, 223
  exposure, 133
  non-discrimination policy, 15
  phenomenon, 215
  techno-oriented world, 134

thinking process, 165
vision, 321, 327
workforce, 159
Globalization, 74, 94, 138, 250, 274, 323
Glocal, 94
GM marketing, 308
Goalposts, 220
Godrej, 8, 9, 13, 16
Goldman Sachs, 14
Google, 9, 15, 16, 280
Government
groups, 36
institutions, 274
Grievance
handling, 11
meetings, 221
Grooming policy, 199, 200
Group
discussion, 59, 256, 286
performance, 164, 170
Growth, 74, 227
Guardianship certificate, 247
Guidance, 136, 139, 143, 145, 148

H

Harassment
discrimination, 256
free work environment, 8
issues, 186
Harvard Business School, 92
Health, 16, 33–35, 39, 40, 44, 45, 49, 51–54, 79, 171, 184, 185, 190, 192, 193, 242–244, 249, 254, 279, 280, 322
health and marriage, 200
care, 35, 40, 51, 249
Healthy discussion, 191
Herfindahl index, 113
Heterogeneity, 77, 85, 132, 137, 138, 142
Heterogeneous group, 325, 330
Heteronormative knowledge, 3
Hetero-normativity, 12
Heterosexual, 3, 6, 7, 12, 14
consortium, 3
Hierarchical
positions, 109
status, 324
High-performing employers, 282
Hippocratic treatment, 23

Hiring escape, 118
Homogeneous, 173
Homosexual acts, 5–7
Homosexuality, 5, 6
Honest and confident, 136
Hormonal changes, 185
Human
cultural aspects, 215
resource (HR), 4, 10, 20, 23, 24, 48, 54, 56, 58, 61, 74–76, 78, 87, 92, 111, 112, 115–117, 119, 120, 123, 124, 132, 175, 186, 197, 200, 213, 214, 224, 227, 229, 231, 241, 248, 256, 268, 269, 275, 281, 323, 325
management (HRM), 24, 112, 142, 174, 304
rights, 15, 244, 245
campaign, 15
Hygiene factors, 219
Hypocritical
character, 24
product, 144

I

Idealistic and forward thinking, 136
Identity, 249
I-Generation, 165
Impediments strategies, 248
change, 248
education, 248
incentives, 248
Implications, 49, 73, 92, 116, 248, 302
Importance of educational degree, 126
Impulse Private Limited, 19
Income, 5, 19, 34, 39, 163, 190, 193, 247, 255, 267–269, 272–275, 284–287, 322, 324
employee engagement, 272
tax concession, 247
Indian
Business Studies Institution, 142
constitution, 245
corporations, 1, 2, 5, 18
culture and civilization, 326
laws, 245, 246
multinational conglomerate, 251
organizations, 2, 4, 5, 17
penal code (IPC), 5, 6

railways, 252
society, 13, 332
traditions, 146
Indiscrimination act, 10
Individual assignment, 150
Industrialized revolution period, 175
Industry requirement activity, 147
Inequality, 9, 17, 18, 24, 74, 78, 109, 275, 269, 272–275, 285, 287, 296, 307
Information gathering, 281
Infosys, 8, 9, 13, 14, 16, 326
Infrastructure, 49, 81, 221, 227, 239, 249
Innovation, 16, 65, 73, 76, 132, 158, 164, 220, 223, 249, 250, 276, 280, 320, 321, 327–330
Innovative
  capabilities, 323
  intra-team engagement processes, 279
Institutional
  discrimination, 126
  memory, 47
Institutionalized system, 305
Intelligent systems, 151
Intended audience, 124, 125
Interaction method, 82
Intergenerational
  transmission, 54
  value differences, 137
Intergroup conflict, 216
International
  Covenant on
    Civil and Political Rights (ICCPR), 244
    Economic, Social, and Cultural Rights (ICESCR), 243
  Labor Organization (ILO), 36, 237, 243, 244, 307
  Law, 243
  organizations, 36
  perspective, 81
  practices, 15
Internet
  related products, 280
  solutions, 19
Interpersonal
  communication, 78, 84
  dialogue, 83
  relationships, 301
  socio-cultural categorizations, 43

Interview
  day, 118
  panel, 176
Intrinsic motivation, 148, 150, 300
IT
  industry, 196, 325
  infrastructure, 19

J

Jarzabkowski, 84, 88
Job
  changes, 217
  characteristics, 219
  control, 219
  design, 177, 229, 280, 304, 305
  dissatisfaction, 24
  fit, 219
  hopping, 140
  market, 114, 275
  opportunities, 275
  positions, 167
  profile, 93, 119, 167, 171, 188
  requirement, 216, 239
  resources, 132
  responsibility, 167, 196
  retention, 186
  role, 83, 220, 300–302, 305
  rotation, 279
  salary packages, 188
  satisfaction, 4, 11, 188, 190, 192, 271
  security, 122, 146, 161, 163, 170
  sharing, 308
  specification, 117

K

Kahn's personally disengagement, 300
Karan Thapar, 20
Key
  determinant, 134
  issue, 198–200
  learning points, 24
Knowledge divide, 126

L

Labor
  force participation, 40, 307
  market

organizations, 83
situations, 132
Lack of patience, 134
Language, 110
LaSalle Network, 281
Law curriculum, 286
Leader's support, 221
Leadership style, 279
Learning, 8, 11, 24, 55, 59, 65, 76, 83, 92, 114, 135, 143, 144, 147, 148, 150, 161, 164, 165, 168, 217, 220, 282, 305, 308, 309, 320, 327, 328, 333
Legal
  consequences, 111
  guardian, 247
  guardianship certificate, 247
Legislative arrangement, 81
Lemon tree hotels, 251
Lengthening retirement, 36
Lesbians, gays, bisexual, and transgender (LGBT), 1–18, 20–25, 146, 308, 25
  adult population, 5
  agenda, 21
  bisexual, and transgender empowerment, 14
  community, 1–3, 11, 15, 18
  consumers, 16
  employees, 2–5, 8–10, 12–16
    network, 8
    recruitment, 11
  friendly policies, 9, 10
  inclusive workplace culture, 16
  people, 1- 3, 5, 11, 13, 24
  person, 8
  problems, 14
  relationship, 20
  role model, 2
  sexual identity, 1, 2, 4
  talents, 2, 4
  workforce, 9
Less payment of paid-leave, 241
Leveraging intergenerational talent, 56
Life
  expectancy, 35, 44, 45, 48
  satisfaction, 184, 192, 193
Linguistic, 324
Lobo, 240
Loss of identity, 216

Lower
  organizational ranks, 109
  turnover, 240
Loyalty, 47, 57, 65, 80, 86, 93, 140, 161, 163, 166, 170, 175, 183, 186, 188, 225–227, 240, 268, 283
Lucrative
  bonuses, 168
  promotions, 167

M

Maltese government, 46
Man, 3, 118, 190, 295, 305, 310
Management
  cooperation and participation, 230
  graduate, 125
  practices and leadership, 63
  style, 301
Manager
  expectations, 219
  self-efficacy, 219
Managerial
  designations, 121
  implication, 75
  issues, 183
  supervisory ranks, 109
Managing
  diversity, 158, 177, 214
  married
    men, 185
    women, 184
  racial diversity, 73, 75, 85
Mandatory retirement age, 48, 54
Manufacturing, 222, 281
  operations, 222
Marital status, 10, 77, 187, 188, 192
  commitment, 187
Market
  innovation, 16
  positioning, 11
  scenario, 76, 166
Marketing
  communications, 143
  domain, 119
  strategies, 117, 123
Married
  counterparts, 187
  employee, 184, 186

Index 351

employee management, 196–198
men, 185, 186, 189, 190, 196, 200
women, 185, 189, 195–197, 200
Master's group, 279
Materialist lifestyle, 295
McDonald, 88, 89, 325
Mckinsey survey, 307
Medical
  coverage, 10
  expenses, 39
  facilities, 51
  leave, 256
  treatment, 246
Mediterranean cultures, 307
Mental
  capacity, 39
  health, 40, 192
    act, 246
  physical maladaptation, 217
  resources, 39
  social well-being, 34
Mentoring program, 279
Merger process, 215, 217
Mergers and acquisition, 215, 232
Micro-individual level, 79
Microsoft whitepaper, 193
Millennials, 17, 141, 147, 151, 165
Millennium
  babies, 138
  development goals, 250
Miniscule span of attention, 134
Mobile technology, 19
Monitoring, 11, 141
Motivation, 4, 11, 13, 60, 150, 160, 164, 177, 194, 217, 248, 250, 299, 305, 321
  beware, 217
  commitment, 13
Motivational resources, 39
Multi-channel retailer, 283
Multi-color bangles, 197
Multicultural environment, 85, 86
Multiculturalism, 81
Multi-generational
  organizations, 132
  workforce, 63, 64
  workplace context, 131
Multi
  generations, 169

multitasking, 165
nation business operations, 74
national
  companies (MNCs), 4
  insurance company, 175
Multiple phases, 184, 185, 195
Musculoskeletal problems, 190
Mutual commitment building, 278

N

Nabet, 252
National
  Association for the Blind, 252
  Credit Corps, 321
  dementia strategy, 46
  economy, 275
  entrance exam, 122
  level, 123, 125
  Management Institute, 122, 125
  origin, 79, 112–114, 322
  pension legislation, 53
  policy on education, 246
  service scheme (NSS), 321
  social context, 134
  stock exchange, 188
  strategic policy, 46
Nationality, 4, 76, 77, 110
Natural resources, 135
Nexters, 160, 164
NG pharmaceuticals, 116–119, 123, 124
NIFTY 500, 188, 189
Non-discrimination, 15, 20, 81, 243
Non-governmental organizations, 248
Non-human species, 135
Non-monetary rewards, 167
Normate's
  process, 238
  recurrent postulation, 238
No-spouse policies, 191

O

Occupation and employment, 243
Online communities, 320
Open
  door policy, 221
  participative discussion, 149
Operationalization, 80, 81, 83, 89

Operationalize diversity, 83, 89
Opportunity policies, 11
Optimal distinctive theory (ODT), 94, 80
Organization
   Economic Co-operation and Development (OECD), 36, 52, 274
   members, 219, 270, 302, 304
Organizational
   commitment, 132, 133, 188, 267, 270, 299, 303
   culture, 10, 61, 74, 83–85, 89, 168, 171, 186, 214, 215, 220, 221
   culture strategies, 277
      consistency, 277
      empowerment, 277
      set expectations, 277
   demography, 7, 108, 115
   discrimination, 114, 124
   diversity, 82, 85, 323
   drift, 217
   dynamism, 323
   effectiveness, 323
   experiences, 300
   factors, 187
   function, 213
   groups, 84
   growth, 47, 116
   level, 76, 78, 219, 298
      factors, 325
   performance, 10, 63, 92, 183, 323
   policies, 49, 53, 78, 109
   practices, 24, 107, 138
   problems, 305
   processes, 84, 88
   productivity, 183
   prosperity, 132
   settings, 12, 85
   solutions, 85
   structure, 79, 82, 84, 85, 90
   variables, 109
   workforce requirement, 114
Orientation processes, 278
Over burdening married employees, 185

P

Parental attention, 164
Participatory approach, 325

Paternalistic employment relationship, 161, 170
Payment of gratuity act, 246
Pedagogical objectives, 147
Pension, 39, 49, 52, 53, 247, 253
   crisis, 49
Perceived
   control, 39
   unfairness, 216
Perceptual biases, 24
Performance, 1, 2, 4, 6, 8, 10, 11, 17, 21–23, 47, 48, 81, 83, 86, 87, 91, 93, 108, 119–121, 138, 143, 148, 149, 167, 176, 177, 183, 185, 186, 188, 193, 219, 224, 230, 243, 267, 269, 271, 276–278, 280, 281, 297, 300, 303, 306, 323–325
   appraisals, 6, 11, 48
   management
      process, 276
      system, 276
Personal
   commitments, 198
   disengagement, 270, 300
   engagement, 300
   expression, 199
   Interview, 122
   safety, 53
   situations, 200
Personality, 9, 77, 108, 158, 160, 184, 188, 213, 223, 283, 322, 324
Perspectives, 43, 77, 83, 85, 86, 145, 148, 158, 224, 295, 310, 319, 323
Pew research center analysis, 159
Pharmaceutical
   company, 176
   industry, 117
Pharmaceuticals, 119, 124
Philanthropic spirit, 326
Photographic reminiscence, 251
Physical abilities and disabilities, 158
Physical
   aging, 184
   disability, 237, 238, 244, 256
   fitness, 192
   health, 34, 53, 192, 196
   physiological and psychological consequences, 48
   psychological health, 184, 192

Pink money, 5, 25
Polarization, 217
Policies, 248
Policy
  amendments, 52
  makers, 52, 134, 137, 244
Political affiliation, 110
Positive
  affect, 39
  aging, 35
  environment, 148
Postgraduate, 125, 140, 147, 231
Post-industrial countries, 298
Post-lunch interviews, 120
Post-merger culture
  adaptation, 215
  clash, 215
  employees, 215
  integrations, 223
  politics, 218
  psychological issues, 231
Post-millennials, 136, 142, 147, 165
Power
  centralization, 87
  conflict, 87
  relations, 77, 86
Practical learning, 149
Premier
  Education Institutes, 116
  Institutes, 113, 114, 116–118, 123
  Management Institute, 117
Prime database, 188
Private sector, 189, 296
Problem-solving, 81, 184, 185
Productivity, 13, 35, 46, 62, 82, 93, 113, 141, 183, 185, 220, 225, 227, 228, 269, 278, 280, 281
  speed, 241
Professional
  development, 167
  wardrobes, 194
Profitability, 271
Promotion, 6, 11, 21–23, 35, 48, 52, 109, 114, 115, 119, 121–123, 125, 141, 162, 168, 185, 194, 281, 285, 286, 306, 307, 323, 330, 332
  condition, 123, 125
  processes, 109
  swag, 119
Psychological
  availability, 12, 301
  effect, 41
  issues, 191, 215, 216, 219, 231
  meaningfulness, 12, 305
  presence, 270, 302, 303
  resources, 12
  safety, 12, 301
Purchasing power, 11
PwD
  champions network, 251
  employees, 251

Q

Qualification and experience, 117
Queen Victoria's rule, 5

R

Race, 4, 7, 11, 13, 73, 75–80, 83, 86, 87, 93, 94, 110, 112, 114, 115, 132, 136, 158, 213, 307, 319, 322–324
  races at work, 75, 94
Racial
  diversity, 73–75, 81–88, 90, 92, 94
    congruence, 81
    engagement, 75, 90
  ethnic
    differences, 110
    discrimination, 110
    sexual balance, 322
  inclusion, 90
  representation approach, 83
  representativeness, 82
Rajat's hiring race, 117
Recruiting disabled workers, 239
Recruitment, 10, 18, 20, 86, 87, 115, 124, 176, 188, 189, 306, 326
  process, 124, 195
Redlining, 111
Rehabilitation Council of India Act, 246
Religious
  activities, 53
  beliefs, 111, 199, 200
Relocating employees, 195
Relocation decisions, 188
Remuneration, 164, 170, 188, 282, 286, 287

Repercussions, 12, 24
Resource
  allocation, 227
  based dynamic model, 39
Retail, 19, 251, 281
Retirement, 10, 33, 39–42, 46–52, 54, 133, 167, 184
  adjustment, 39, 49–51
    quality, 39, 51
  aspirations, 42
  planning, 66
  resources inventory (RRI), 39
  schemes, 10, 52
Reverse discrimination, 111
Review meeting, 176
Revolutionary innovations, 15
Rewards, 56, 220, 282
  productivity, 163, 170
Rights, 13, 14, 18, 65, 79, 110, 199, 216, 243–246, 251, 255, 256
Robust practices, 174
Role theory, 50, 192

S

Salary
  package, 119
  tax reduction, 239
Sales manager, 121, 123
SAP, 250, 251
Savings, 39, 41, 49
Savvy strategies, 175
Scholarly literature, 131
Scouts, 321
Screwfix, 283, 284
SDT theory, 148
Security, 34, 50, 137
  solutions, 19
Selection discrimination, 107, 126
Self determination theory, 148
Self-defense, 303
Self-dependent, 135
Self-employment, 40, 300
Self-expression, 300
Self-identification, 248
Self-identify, 248
Self-scrutiny, 112
Sen's capability, 44
Senior employment, 53

Sense-making management, 90
Sex
  based differences, 297
  characteristics, 111
  discrimination, 113
Sexism, 43, 111
Sexual
  abuse, 138
  discrimination, 111
  diversity, 136
  minority, 1, 3, 5–8, 12, 17, 18, 24, 25
  orientation, 2–4, 6–9, 11–13, 16–18, 21, 77, 110, 132, 136, 324
Sexually harass, 191
Short span assignments, 149
Sick leaves, 53, 190
Silver skills, 47, 66
Single
  employee vs married employee, 186
  married employees, 186, 187
Situational leadership, 148, 149
Skewness, 87, 189
Skill
  affirmative action, 214
  levels, 249
Social
  acceptance, 17
  aging, 35
    phenomenon, 35, 66
  capital, 109
  cause, 328
  cognition, 76
  community, 250
  control imperatives, 275
  cultural identities, 76
  demographic shifts, 138
  development, 321, 327, 329–333
  differences, 74, 76, 305
  discriminatory structure, 78
  distance, 84
  economic stratification, 296
  instrumental exchanges, 108
  events, 9
  interaction, 39, 40, 53, 165
  networking, 134, 135
  networks, 115
  norms, 161, 170, 307, 309, 310
  participation, 44

platforms, 135
practices, 296
resources, 39, 40
responsibilities, 135
roles, 192, 193
   accomplishment, 192
security, 41, 243
stigma, 2, 9
support, 39–41, 240, 242
trends, 43
Societal
   changes, 175
   rejection, 12
Society
   organizations, 3
   political participation, 53
Socio-contextual factors, 43
Sociocultural
   aspects, 83
   constraints, 4
Socio-economic
   class differences, 295
   development, 33
   status, 40, 84, 158, 192
Southwest Airlines, 282, 283
Speech-recognition program, 241
Speed Lovers, 135, 139
Spouse
   appraisals, 193
   policy, 191
Stage theory, 50
Stereotypes, 7, 43, 63, 79, 110, 173, 185, 186, 187, 191, 217, 248, 295, 298
   roles, 310
Stigmatization, 43
Stock exchanges, 19
Stockholm targets, 46
Stressful scenarios, 195
Structural diversity, 82
Student's internal assessment, 147
Stumpy work ethics, 136
Supervisor's perception, 190
Supervisory relationship, 191
Surface-level diversity, 77, 79, 94

T

Talent management, 175, 220
Talented workforce, 239

Tangible resources, 39
Tanishq, 251
Target
   audience, 147, 256, 286
   group, 231
Tata
   group, 251
   Motors Titan, 251
Tax benefits, 239
Teaching
   experience, 125
   objectives, 286, 310
   scheme, 286
   strategy, 59
Team
   building, 227, 333
   dynamics, 142, 148
   members, 16, 55, 91, 93, 110, 115, 176, 177, 197, 230, 276, 278, 309
Tech-based solutions, 176
Technological work environment, 163
Technology, 64, 65, 134, 135, 138, 141, 161, 163, 165, 177, 215, 222, 223, 230, 241, 242, 244, 249, 251, 273–275, 279, 284, 299, 308
   infrastructures, 241
Tech-savvy generation, 57, 163
Telecommunications, 19
Telecommuting, 168
Tenure, 76, 77, 108, 213
Theoretical
   base, 22
   model, 39
Ticket concession, 247
Top
   management involvement, 248
   performing employee, 277
Trade union perspective, 49
Traditional
   business organization, 268
   education system, 165, 169
   generation, 161
   management policies, 175
   routines, 142
   values, 23, 146
Training
   policies and programs, 246
   process, 195

programs, 55, 173, 227
Trait engagement, 47
Transformational leadership, 85, 148, 279
Transgender employments, 14
Transition speed, 217
Travel concession, 247
Tree plantation, 332
Trust, 63, 84, 93, 121, 134, 183, 220, 228, 230, 240, 274, 275, 277, 278, 284, 287, 301

U

Ultra-modern, 136
Undergraduate, 24, 125, 140, 231
Unhealthy relationship, 184
Uninformed absenteeism, 190
Urban
 dictionaries, 137
 language, 137
US Labor Department, 81

V

Value-in-diversity hypothesis, 81
Vantage point, 47
Vasudhaiv Kutumbakamb, 250
Verbal
 behaviors, 87, 303
 communication, 251
 non-verbal patterns, 77
Verdict, 252
Veterans, 160, 161, 166, 167, 170, 174, 175, 178
 status, 112, 114
Vigilant generation, 138
Volatile
 environment, 158
 uncertain, complex, and ambiguity (VUCA world), 165
Voluntary
 basis, 246
 groups, 327
Volunteering, 41, 43, 53, 135, 319, 321, 326–328, 330–333
 experiences, 331
 opportunities, 330
Volunteers, 321, 326–332

W

Wage structure, 274
Walk-in-interview, 119
Watch-making division, 251
Wax, 184
Well-being, 33, 35, 38, 39, 44–47, 52, 132, 184, 192, 193, 220, 243, 271, 297
Whistleblower rights, 256
Williams Institute, 15
Woman, 3, 20, 186, 188, 190, 196, 295, 298, 299, 305, 309, 310
 concentrating, 302
 development, 297
 vs. men, 310
 work space, 306
Woodruffe-Burton, 2, 4
Work
 aspirations, 175–177
 autonomy, 132, 136
 awards, 193
 centric attitude, 166
 culture, 9, 15, 21, 185, 191, 243, 332
 engagement, 171, 175, 178, 304, 305, 310
  strategies, 178
 experience, 11, 132, 267, 273, 285
 family
  balance., 185
  conflict, 190, 191
 home and married life, 193
 performance, 21, 82, 177, 184, 187, 190, 192, 193
 personal values, 166
 schedules, 168, 170
 without walls, 193
Workforce, 4, 8, 17, 18, 33, 35, 36, 41, 42, 46–49, 52, 53, 56, 58, 60, 61, 63, 73–76, 79–83, 88–90, 111, 113, 114, 116, 131–134, 137, 138, 158–177, 183, 184, 186, 188, 193, 214, 216, 221–223, 230, 240, 242, 251, 256, 271, 273, 277, 296, 298, 299, 307, 320–326
 composition, 183, 200
 diversity, 158, 183, 200, 320, 323, 329
Working
 conditions, 109, 161, 168, 170, 280
 men, 200
 women, 185, 192, 200, 297, 298, 325
Work-life balance, 167, 171, 188, 280

Workmen's compensation act, 246
Workplace
  culture, 1, 2, 4–6, 8–10, 13, 14, 16–18, 20, 22, 25
  discrimination, 109, 110, 112, 126, 241, 325
  diversity, 1, 2, 4, 5, 7, 18, 25, 78, 158, 196, 213, 323, 324
  dynamics, 166
  eccentricities, 12
  engagement efforts, 172
  environment, 4, 219
  expertise, 324
  intergenerational climate (WIC), 42, 43
  policies, 15
  relationships, 172, 173
World
  class merit, 92

Health Organization (WHO), 34, 36, 38, 133, 242

X

Xers, 134, 137, 163, 164, 167, 168, 170

Y

Younger
  employees, 142
  generations, 40, 133, 146, 297
Youngest generation, 133, 164

Z

Zeal, 142, 330
Zefo culture, 20
Zers, 134–142, 146, 149, 151, 169